671.56
MACLF

D1414257

Solders and Soldering

Solders and Soldering

Materials, Design, Production, and Analysis
for Reliable Bonding

HOWARD H. MANKO
Consultant, Solder Consulting Services, Teaneck, N.J.

Second Edition

McGRAW-HILL BOOK COMPANY
New York   St. Louis   San Francisco   Auckland   Bogotá
Düsseldorf   Johannesburg   London   Madrid   Mexico
Montreal   New Delhi   Panama   Paris   São Paulo
Singapore   Sydney   Tokyo   Toronto

Library of Congress Cataloging in Publication Data

Manko, Howard H
    Solders and soldering.

    Includes index.
    1. Solder and soldering. I. Title.
TT267.M26 1979   671.5'6   79-9714
ISBN 0-07-039897-6

34567890 HDHD   86543210

*The editors for this book were Tyler G. Hicks and Elizabeth P. Richardson,
the designer was Naomi Auerbach, and the production supervisor
was Sally Fliess. It was set in Baskerville
by University Graphics, Inc.*

Printed and bound by Halliday Lithograph.

Again to Mira

Contents

3. The Metallurgy of Solder   41

6. Special Applications   234

7. Hand Soldering and the Soldering Iron   262

Preface

The purpose of this book is to present the new science of soldering in its proper dimension. Every attempt has been made to make this book as technically accurate and up to date as possible and to present the information in a form understandable to the wide range of solder specifiers, users, and inspectors. These include the product or systems designer, the production engineer, quality-control personnel, the purchasing agent, and the operator as well as the student. The book covers theory as well as practice, as it builds a strong base of solder understanding in the hope that this will help to achieve even higher degrees of reliability.

There are three compelling reasons why solder is not going to be replaced in the foreseeable future. In order of importance they are the stress-coupling ability of solder, the economy and reliability of the process, and the ease of manufacturing, repair, and inspection. Let us look at these one by one.

Stress Coupling. Pure solder at room temperature[1] is a ductile self-annealing alloy capable of absorbing stresses without work hardening and/or fatigue failure upon cycling. This unique property enables the industry

---

[1]Stress-coupling ability is reduced at very low temperatures.

to join together materials of varying coefficients of expansion, levels of rigidity, and strength without the special design requirements typical of mechanical assemblies. The average printed-circuit board, for example, violates all structural design rules for stress concentration, strength of materials, load distribution, behavior under vibration, etc. If it were not for the solder joint and its stress-coupling ability, the printed-circuit board as we know it today could not exist. With any stresses unyielding joints, like those in brazing or welding, would cause crumbling in the laminate, peeling in the circuitry, and damage in the glass-to-metal seals of components.

Economy and Reliability. Because solder joints can be made at relatively low temperatures, where the normal plastic materials and assemblies we handle do not undergo heat deterioration, the process is very economical. Relatively simple tools, low-cost materials, and a controllable process make soldering especially economical.

Economies, however, can be achieved only if they are associated with high reliability. This is an area that has suffered greatly in the past from a combination of poor material selection, lack of design, inadequate processes, and a general lack of understanding of soldering as a technology. Most large organizations who have taken this premise seriously have been able to obtain reliability levels in soldering that cannot be matched by any other known interconnection method. With some planning, touchup levels of 1 in 100,000 joints have been achieved by such companies as IBM in their automatic soldering, while Autonetics, a Division of North American Aviation, has reported 15 billion solder-joint-hours without failure on the Minuteman program. But this does not just happen; it must be designed and planned from the beginning and carefully monitored throughout the production process.

The ease of manufacture, repair, and inspection. Relative to other metallurgical joining methods, soldering is an easy process, requiring relatively simple tools and techniques. Wave soldering, for instance, enables one to solder a large number of coplanar joints with a high degree of uniformity. Even hand soldering, which is slower, will yield high reliability with the appropriate motivation, training, and materials.

Ease of repair is another advantage of soldering that makes it unique among the metallurgical joining methods. Only soldering is truly reversible, thus making repair simple and easy. Replacement of components, correction of manufacturing faults, and field installation require little skill. With correct procedures and materials, a repaired solder joint will be as reliable as the original connection.

Finally, we cannot overemphasize the importance of the inspectability of solder, which yields itself to 100 percent visual inspection. This is a strength rather than a disadvantage. Final visual inspection, under the correct guidelines and with in-process control, is extremely meaningful, especially because it can be carried out on 100 percent of the joints without affecting them physically.

This ease of manufacture, repair, and inspection is often also the downfall of soldering technology. Unfortunately, management often pays little attention to this part of the assembly process, committing basic errors which cause a lack of reliability later on. The author sometimes wonders if a more difficult process, requiring much more sophistication, would lead to more "respect" in soldering. If *you* cannot carry out the soldering process reliably and economically, it is *you* who have committed a basic error.

Solder is an ancient joining method, mentioned in the Bible (*Isaiah* 41:7). There is also evidence of its use in Mesopotamia some 5000 years ago, and later in Egypt, Greece, and Rome.[1] For the near future it appears that as long as we use a combination of conductors, semiconductors, and insulators to build our circuitry, based on electrical and magnetic impulses, solder is indispensible. Newer technologies, like fiber optics, fluidics, and the like, have not developed up to their expectations. The author feels that the next truly revolutionary step will relate to a biological nerve-brain type of system. Once we understand the mechanism behind these marvelous body functions, we can abandon our clumsy metal-electrical technology. Until then, however, solder is here to stay.

To acknowledge adequately those who have contributed to the completion of this book would require endless pages for the listing of technical articles I have read, lectures and symposia attended, correspondence and discussions with men in the field, and the continuous interchange of ideas on soldering and soldering methods to which I have been exposed. Specifically, however, I wish to acknowledge the help and cooperation of my associates at IBM, Poughkeepsie, N.Y., and Alpha Metals, Inc., Jersey City, N.J., who helped in the preparation of the first edition. Furthermore, I wish to acknowledge the contributions of the technical staff of Hollis Engineering, Inc., Nashua, N.H., Hexacon Electric Co., Roselle Park, N.J., and the members of the International Electrotechnical Commission (WG3–Soldering) for their help with the second edition.

Howard H. Manko

---

[1]Jochem Wolters, "Zur Geschichte der Lottechnik," Degussa, Hanau, West Germany, 1977.

Solders and Soldering

ONE

Solder-Bond Formation

1-1 Metallurgical Bonds Although soldering is such a well-known joining method and has been in general use for so many years, the average engineer and user of this process finds difficulty in defining it properly. However, on the basis of the information presented in this book it is possible to define soldering as follows:

Soldering. A metallurgical joining method using a filler metal (the solder) with a melting point below 600°F (315°C),[1] soldering relies on wetting for the bond formation and requires neither diffusion nor intermetallic compound growth with the base metal to achieve bonding.

Let us analyze this definition a bit further. Metallurgical bonds, in general, are those bonds in which metallic continuity from one metal to the other is established. The other metallurgical joining methods are brazing, welding, and some of the more recent developments which are a combination of these, e.g., diffusion bonding and thermal-compression bonding. The limit of 600°F was set arbitrarily, and many people consider 800°F (427°C) the cutoff point for soldering. In comparison with soldering, brazing is defined as follows:

---

[1] Equivalent temperatures are approximations.

1

Brazing. A metallurgical joining method using a filler metal which melts over 800°F and which relies on wetting as well as diffusion for the bond strength.

Under the same conditions, welding could be defined as follows:

Welding. A metallurgical joining process which relies on the diffusion of the base metals with or without the filler metal for the joint formation.

Let us compare these three major joining methods (see Table 1-1). The heat distortion during bond formation is small for soldering, large for brazing, and very large for structural welding. For electronic-component welding, provided the welding time is short, the heat distortion is also very small.

Next let us consider the equipment cost. It is smallest for soldering, larger for brazing, and highest for welding. This difference is even more pronounced when ease of automation for simultaneous bonding is considered. This is easiest in soldering, much more difficult in brazing, and not practical in welding.

Finally, let us consider the finished product. Whereas soldering can be inspected visually for wetting, which reveals the soundness of the joint, brazing cannot be checked for diffusion without destroying the joint, and welding cannot be analyzed visually at all. Furthermore, when we find a faulty solder joint through nondestructive visual inspection, we can easily rework it. A brazed joint is more difficult to reopen, and welded joints must be physically destroyed in the base metal, thus making repair impractical.

1-2 Classification of Bonding Methods   When we discuss soldering in relation to other bonding methods, it is difficult to list the advantages and disadvantages without a brief survey of other methods. For this purpose, the three major categories of bonding methods are listed, with short descriptions for each.

Metallurgical Bonds. These have been described in detail in Sec. 1-1. However, only soldering and brazing are truly reversible processes. By the mere application of heat, it is possible to open the joints and separate the components. This is not true in welding, thermal-compression bonding, or any similar process. These processes require the actual destruction of the components going into the joint itself. The ease of opening a soldered connection is one of the major advantages of soldering because it gives soldering flexibility in replacing components, repair, and rework. The mechanism of diffusion, which is more prominent in brazing, as described in Sec. 1-1, makes the opening of brazed joints more difficult. Design of repeat make-and-break connections in brazing is not recommended and not too practical.

Mechanical Bonds. The most common mechanical bonds are the screw and the rivet for mechanical connections and the crimp, the wire wrap,

TABLE 1-1 Soldering vs. Other Bonding Methods*

| Factor | Type of bond | | | | | | |
| | Metallurgical | | | Mechanical | | | Chemical |
	Solder	Braze	Weld	Crimp	Screw	Wrap	Conductive cement
Temperature limit of joint (melting or breakdown), °F	100–800	800–1600	Conductor melting temp.	No limit except that of wire			160–300
Heating effect on assembly	Small	Large	Small (quick)	None			Cures at ambient to 250°F
Ease of rework and rebonding of permanent joint	Simple		Not practical	Not practical	Simple		Not practical
Size of joint relative to conductor	Small			Medium	Large	Small	Small
Process economy; Equipment cost	Low	Medium	High	Low			Low
Ease of automation	Easiest	More difficult	More difficult	More difficult			More difficult
Extra hardware?	No			Yes		No	No
Joint stable? Vibration	Yes			Yes	No	Yes	Yes
Oxidation	Yes			No			Yes

*From H. H. Manko, How to Design the Soldered Electrical Connection, *Prod. Eng.*, June 12, 1961.

3

and the screw for electrical connections. Of all these methods, only the screw connections are truly reversible bonds. The other methods require the destruction of the hardware involved in the bonding and therefore are not so easily suitable for repair, replacement, or rework.

Chemical Bonds. Adhesives for the structural joint and conductive cements for the electrical joint are examples of chemical bonding materials. Here repair is practically impossible.

Table 1-1 shows a comparison of these three major bonding methods. This table is mainly oriented for the electrical connection because structural bonding is more a function of the type of application and the strength necessary than of the considerations listed in the table.

Let us summarize by saying that soldering is probably best used when high density and volume of joints are to be made at a low cost, possibly in automation. The best example is a simultaneous connection of multiple terminals on a printed-circuit board or other uniformly shaped parts which are soldered in one dip or in one pass through the molten solder. In addition, soldering, being easily repaired, either during manufacturing or in field service, gives a complete metallurgical bond which is easy to reflow and makes later changes possible. Because of the possibility of 100 percent visual inspection, the highest reliability of bonding methods can be achieved through soldering. And finally, soldering offers a large degree of flexibility in an assembly with low equipment investment and relative simplicity of manufacture.

1-3 The Role of Wetting[1]   From the definition of soldering, we see that the basic solder process depends on wetting for the formation of a solder–to–base-metal contact. It is this physical wetting of the work by the molten solder that generates the joining interface. This process is really not different from the wetting of any solid by a liquid. The solidification of the molten solder after wetting, however, results in a permanent bond, described in detail later in this chapter.

The theory of the mechanism of wetting was developed many years ago in a general form. Many physics textbooks trace individual contributions to their rightful originators. In soldering, the mechanism of wetting has special significance, and so we shall discuss a general outline of the process and apply it to the specific conditions of soldering. This will help us understand the atomic structure of the bond and its source of strength. We shall also use the wetting characteristics of the bond for a quality-control and inspection system, both here and in other chapters.

When two metal parts are joined by solder, a metallic continuity is established as a result of two interfaces where the solder is bonded to both metallic parts. There are at least two such bonds in each solder joint. Let

[1]Parts of this chapter were presented as a paper, *65th Annu. Meet. ASTM, New York, 1962.*

us follow the metallic continuity from base metal A to base metal B (see Fig. 1-1): *metal* A to *solder-to-*A *interface to solder* to *solder-to-*B *interface* to *metal* B. One can see that the solder serves both as a bond maker and as a link in the metallic continuity.

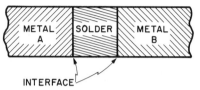

INTERFACE

Fig. 1-1 Two metals bonded by solder.

Actually we need to consider the formation of only one such interface in order to understand the fundamental process of soldering. In this chapter we shall talk only of the solder, flux, and base-metal system which produces the critical interface.

Under standard conditions, both the base metal and the fusible solder alloy have a thin film of tarnish on the surface, which interferes with the formation of the interface. Such thin films are the result of the environmental attack on the metal and may consist of oxides, sulfides, carbonates, or other corrosion products, depending on the environment and the base metal. These nonmetallic corrosion products act as a barrier to the formation of metallic continuity in a solder bond and must be removed. The removal of these films is one of the primary functions of fluxes. However, even if we could force the solder to bond to these layers of tarnish, the strength of the overall bond would be weakened by their presence because they lack the ductility or strength of the parent metal.

1-4 The Solder–Flux–Base-Metal System Assuming for a moment that we have perfectly clean metallic surfaces and that no reoxidation or other environmental attack can occur, let us consider the system involved in soldering. In order to obtain the basic interface between the solder and base metal A, we need to have three elements: the solder metal itself, the base metal, and either the flux or the atmosphere in which the process takes place (air or vapor). At soldering temperatures, the base metal, having a relatively high melting point, is solid, the fusible solder alloy is liquid, and the flux is usually a liquid although it can be a gas.

Let us consider a drop of liquid solder resting on a flat horizontal metallic surface in the flux atmosphere. In effect, we have a three-component system of a solid S, a liquid L, and a vapor V, as shown in Fig. 1-2.

At the point of total thermodynamic equilibrium, with no further diffusion or chemical reaction, the system will have a border line at which all three phases meet (*A* in Fig. 1-2). If we consider the plane perpendicular to the border, we see that the three phases meet at a certain angle to each other. This angle θ formed between the liquid and the solid is called the *dihedral angle*.

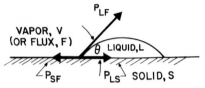

Fig. 1-2 Schematic of thermodynamic equilibrium in wetting.

Let us consider the balance of vectors at point *A* in Fig. 1-2. The three forces are actually the surface energies that brought this system to equilibrium. Vector γ_{LV} is the surface tension between the liquid and its vapor phase acting at a tangent to the liquid curvature. In effect, this is a force that would tend to minimize the surface of the liquid solder in the particular atmosphere of the flux. Every liquid, in an effort to minimize its surface and to satisfy the requirements of the lowest state of energy, has this force known as *surface tension*. In the absence of other forces, the surface tension of the liquid would draw it into a sphere, which has the smallest surface of any geometric configuration with an equal volume. Gravity and interfacial tensions between the liquid and its surroundings usually act against this surface tension, so that the liquid assumes other shapes.

The interfacial tension γ_{LS} is the force between the liquid solder and the base metal, and γ_{SV} is the interfacial tension between the solid base metal and the vapor phase. Both γ_{LS} and γ_{SV} act along the solid surface but in opposite directions. The magnitude of the interfacial energies, which are discussed later, is the result of the inherent properties of the materials involved. Both interfaces involve a rigid surface which requires large energies to change its dimensions. Unlike the surface tension of a liquid, these interfacial tensions when alone therefore do not express themselves by any physical changes in the system.

From the vector diagram we get

$$\gamma_{SV} = \gamma_{LS} + \gamma_{LV} \cos \theta \qquad (1\text{-}1)$$

Here γ_{SV} is the force which spreads the liquid on the solid, i.e., the spreading or wetting force. In other words, spreading or wetting will occur if γ_{SV} is larger than the combination of γ_{LS} and $\gamma_{LV} \cos \theta$. A measure of the relation between these forces can be obtained from the size of the dihedral angle θ. The two extreme conditions would be total nonwetting, where θ is equal to 180° (*a* in Fig. 1-3), and total wetting, where θ is equal to 0° (*b* in Fig. 1-3). Partial wetting will occur when θ is smaller than 180° and larger than 0° (*c* in Fig. 1-3). This concept of partial wetting needs further consideration, especially if we remember that during soldering the system seldom reaches true equilibrium. Normally the soldering time is too short, and the system is frozen before equilibrium is reached. In this case, the wetting angle reveals additional information. Let us divide this range of 0 to 180° into three ranges as follows:

1. $\theta > 90°$. If the system was allowed to freeze after reaching full thermodynamic equilibrium, the condition of $\theta > 90°$ indicates the lack of wetting affinity between the liquid surface and the solid surface. We probably have a situation where the liquid solder never did wet the surface but froze on it in a configuration dictated by the action of various forces such as surface tension and gravity. Only the physical presence of the solder droplets prevents us, at this stage, from determining whether θ was equal to or smaller than 180°. If a cross section can be made of a specimen of this nature, the true interpretation of the situation can be made by assessing the degree of wetting.

If the system froze before achieving thermodynamic equilibrium, $\theta > 90°$ indicates a condition termed *dewet*. The solder froze in the process of dewetting the surface to be soldered; hence the driving force was in the direction of nonwetting, and the dihedral angle was a direct function of the rate of dewetting at that specific time. In this case the surface was first wetted and for some reason (normally an intrafacial tension change due to solid-liquid interactions) the solder withdrew and the metal dewetted. The term dewet should therefore be qualified by saying that the dewetted surface should show signs of wetting; otherwise we have a nonwetted ($\theta =$ 180°) or a poorly wetted surface ($90° < \theta < 180°$).

2. $90° > \theta > M$. This indicates a condition of marginal wetting. Usually $M \leqslant 75°$, and unless special conditions exist, this type of wetting is

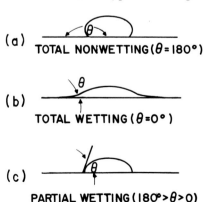

(a)   TOTAL NONWETTING ($\theta = 180°$)

(b)   TOTAL WETTING ($\theta = 0°$)

Fig. 1-3 Relation between dihedral angle θ and degree of wetting.

(c)   PARTIAL WETTING ($180° > \theta > 0$)

not acceptable. M is a purely arbitrary limit set by experience and fulfills the specific requirements of the individual solder system.

3. $\theta < M$. This indicates the condition of good wetting. M has the same value as in 2 above. If extremely high quality is required, the value of M can be taken as less than $75°$ (see also Table 8-2).

1-5 Anchorage Area and Other Corrections Another important factor which contributes to the balance of forces in the wetting system is the amount of *anchorage area*. This factor accounts for the actual percentage of metallic area available for wetting because of either a nonuniform surface or a partially contaminated surface. The nonuniform surface can be the result of the presence of inclusions and occlusions (metal oxides, sulfides, and so on) in either the solder or the base metal, nonmetallic foreign particles, or adsorbed vapor on the surfaces which are not displaced during the wetting process. No method is known at present to measure or evaluate the anchorage area of a surface. This accounts for variations in the reproducibility of comparable wetting in systems with the same materials under the same wetting conditions. However, as long as the surface-preparation methods and the cleaning procedures are kept the same, the anchorage-area factor does not seem to vary and the results are quite uniform.

It has been shown that the dihedral angle is solely dependent on the components of the system and is constant for any set of three phases. In actual soldering, the liquid-flux phase replaces the vapor phase, and another important property of fluxes becomes apparent. The flux should have the effect of decreasing the dihedral angle and thus enabling the solder to wet the solid metal better. In addition, the flux should be able to wet the base metal, replacing any vapor phase adsorbed to the surface and making the surface available for total contact with the solder, thus promoting wetting (100 percent anchorage area).

Several factors affect the energy balance of wetting. The following is a list of the more important corrections to our basic considerations.

1. γ_{LV} represents the surface tension between a liquid and its vapor, but this can be substituted by γ_{LG} to represent a liquid and a gas or by $\gamma_{L1,L2}$ to represent the interfacial tension between two immiscible liquids. In the soldering process, where flux is applied, the wetting equation can be presented as

$$\gamma_{SF} = \gamma_{LS} + \gamma_{LF} \cos \theta$$

where γ_{SF} is the interfacial tension between the metal (solid) and the flux and γ_{LF} is that between the solder (liquid) and the flux. The fact that the liquid solder wets the base metal, displacing the flux, indicates that $\gamma_{LS} < \gamma_{SF}$ and that no equilibrium will occur as long as there is enough solder.

This is not surprising since the free surface energy of solder, like that of most metals, is above 200 ergs/cm² while that of nonmetals like fluxing agents is below 50 ergs/cm².

2. γ_{LS} represents the interfacial energy between the base metal and the solder and is constant for the forward spreading of liquid. A receding liquid has a different contact angle because the surface left behind is not pure base metal but has some liquid molecules adsorbed to it (in some cases an alloy layer is left behind), changing thereby the solid phase and the interfacial tension. This phenomenon is labeled the *molecular hysteresis* of the contact angle. In tin–lead solder–copper systems, for example, it was empirically observed that the contact angle of spreading is greater than that of receding.

3. The dihedral angle is also dependent on the surface condition of the solid. The spreading over a rough surface exceeds the spreading over a smooth surface. This phenomenon is labeled *capillary hysteresis* because the grooves of a rough surface act like capillary tubes by virtue of their increased surface. Until now we have considered a homogeneous solid surface. Where two or more materials make up the solid phase, we must consider the composite. Here

$$\cos \theta = \sigma_1 \cos \theta_1 + \sigma_2 \cos \theta_2$$

where σ is the area (partial) and $\cos \theta_1$ and $\cos \theta_2$ are the intrinsic contact angles. Accordingly a rough surface of a composite will follow a homogeneous phase only if the individual components are metallic. When one constituent of the surface is nonmetallic, by virtue of the decreased free surface energy reduced wetting will result. In essence, then, we have a reduction of anchorage area, and in real life we would have a decrease in solderability. Lack of solderability of a metallic surface then would be due either to unreacted surface tarnishes (a result of a weak flux) or insoluble organic contamination (dirt), which would give a composite surface of insufficient free energy for wetting.

4. The temperature of the system affects the wetting and the surface tension. The rate of wetting increases with the rise of temperature. At a certain temperature, unique for each system, a sudden increase of spreading occurs. This point is labeled the *critical temperature.*

5. The hydrostatic pressure of the molten solder depends on the size of the liquid drop. It is of minor importance in soldering, as wetting and spreading proceed and outrun the bulk flow.

6. In the above considerations, the cohesive force of the liquid was neglected. This force, the measure of the affinity of the liquid for itself, is a restraining force to spreading because it works toward decreasing the surface area. For practical purposes, we can include the cohesive force in γ_{LV}.

7. In soldering, all the previous considerations are vital, but it should be remembered that we do not reach absolute equilibrium. We freeze the liquid phase before it has even approached chemical and physical equilibrium. This works to our advantage since the intermetallic compounds which would be formed at total equilibrium might embrittle the joint and also deplete the alloy of one of its components. In addition, short working times and smaller heat expenditures are desirable both from an economic standpoint and for minimum heat distortion or heat deterioration.

1-6 Wetting, Solubility, and Intermetallic Formation   A word about the effect of intermetallic on wetting is in order. From the foregoing it is clear that wetting is independent of any interaction in the solid-to-liquid phase. Obviously no intermetallic is formed in nonmetallic systems; neither is there any similar interaction. Thus the claim that intermetallic crystal formation is a prerequisite to soldering is without foundation. It was observed empirically by Bailey and Watkins[1] that molten metal will wet solid metal better if mutual solubility or intermetallic formation exists. They attributed this phenomenon to added alloying energy.

Klein Wassink,[2] however, showed that a simple relationship exists between wetting behavior and mutual solubility. He concluded that the alloying energy does not actually provide the driving force but that alloying energy and the wetting capacity both depend upon the same atomic properties. Jordan and Lane[3] confirm this in their work; they state that while the wetting of the liquid-solid metal pairs is aided by the formation of either intermetallic compounds or solid solution, it is not a necessary condition because liquid sodium is capable of wetting solid tantalum and niobium without these reactions. While their work in the area of atomic energy concentrated on unusual metals, a review of solder alloying elements and the engineering base metals normally found in industry reveals many more systems where none of these reactions exist.

1-7 The Solder Bond   A simplified explanation of the solder bond is now in order. Without going into the relation of the various energies holding a metallic lattice together, it is possible to explain the strength of a metal simply. Let us consider a base-metal surface that has just been wet by a liquid solder where the solder has solidified before or after chemical equilibrium is established. The same type of surface energy that promoted the wetting of the metal by the solder is now making the bond. This energy is due to the unsaturated bonds of the surface atoms, which have a less than average number of nearest neighbors surrounding them.

[1]G. L. J. Bailey and H. C. Watkins, *J. Inst. Met,* vol. 80, p. 57, 1951–1952.
[2]R. J. Klein Wassink, *J. Inst. Met,* vol. 95, pp. 38–43, 1967.
[3]D. O. Jordan and J. E. Lane, *Liq. Met. Coolants,* sec. 1, pp. 197–200, 1968.

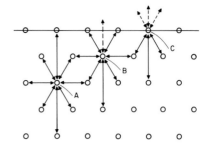

Fig. 1-4 Simplified diagram
explaining surface energies.

From Fig. 1-4 we can see that atom A has a balanced array of nearest neighbors surrounding it, whereas atoms B and C each have an unbalanced array of nearest neighbors and therefore have unsaturated bonds which generate the surface energies. This orderly arrangement of atoms in a metallic lattice is a fairly accurate representation of the internal structure of metals. There are regions of disorder and dislocation in every lattice; however, their overall effect on a metallic surface is negligible in area and can be disregarded in the solder bond. The unsaturated surface bonds, which are the major contributors to the surface energies, play an important part in the balance of vectors that determines wetting or nonwetting for a specific system. Once the solder has wet the surface, we get a mutual saturation of the surface bonds, and these surface energies are on an atomic level, giving the interface great strength and stability.

1-8 The Sequence in Solder-Bond Formation In the following chapters, we shall discuss the chemistry of fluxes and the metallurgy of solder. It is now time to fit all these into a whole picture to show the sequence of occurrences required to make the solder joint. By following the process step by step, we shall better understand the mechanisms described individually and be able to interpret our specific requirements more fully. A thorough treatment of the mechanism of corrosion and its effect on the solder joint will also be included.

1-9 The Base Metal In Fig. 1-4 we show a simplified diagram explaining surface energies. This is an acceptable schematic representing a metallic lattice. Let us consider such a base-metal structure. As long as there is absolute vacuum around the metal and the surfaces are absolutely clean, we get a configuration like Fig. 1-5.

However, when this clean metallic surface is exposed to the air, *adsorption* occurs. The molecules of the air will be attracted by the same surface energies and will orient themselves close to the surface. The rest of the molecules in the air will be free to move at random. This is depicted in Fig. 1-6. Since the oxygen in the air will attack most metals, let us consider the general case of oxidation as representing the general process of tarnish formation in environmental attack on metallic surfaces.

Figure 1-7 depicts the metallic surface, which is now covered with a nonmetallic layer of tarnish. To this layer of tarnish, another layer of air

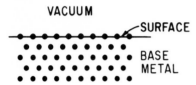

Fig. 1-5 Schematic of metallic lattice in vacuum.

molecules is again adsorbed, with the balance of the air scattered at random throughout the space above the metal. At this point, both the adsorbed air and the tarnish layer would have to be displaced before metal-to-metal continuity could be established with the base metal.

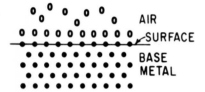

Fig. 1-6 Schematic of metallic lattice in air. Note adsorbed layer of air.

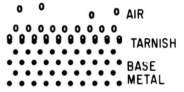

Fig. 1-7 Schematic of tarnished metal in air.

1-10 The Flux Action   Once the flux is introduced into the system, it will first displace the adsorbed gas from the surfaces as depicted in Fig. 1-8. This activity of the flux is called the *wetting characteristics.*

Once the liquid flux has wetted the tarnished metal, it is available to restore the metallic surface and remove the oxides. This can be done either by reduction of the oxides to restore the metallic surface or by removing the oxides from the surface. Figure 1-9 depicts the situation after the tarnishes have been removed.

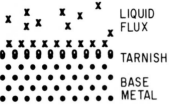

Fig. 1-8 Schematic of tarnished metal wet by flux, no chemical action yet.

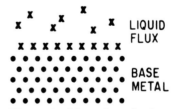

Fig. 1-9 Schematic of clean metallic lattice wet by flux.

1-11 Introducing the Solder When the solder is introduced on the flux–base-metal system, and if all the requirements spelled out earlier for wetting are fulfilled, the solder will displace the flux easily and bond with the base metal. This is depicted in Fig. 1-10. Note that the density of the air and the liquid flux was not so large as that of the metallic solder and that in Fig. 1-10 we have not yet introduced the results of the high temperature on the surface.

However, as we shall show in Chap. 3, there is a large area of thermal agitation between the solder and the base metal where the solder and base

Fig. 1-10 Schematic of clean metallic lattice wet by liquid solder.

Fig. 1-11 Schematic showing mutual diffusion of solder (liquid) and base metal (solid). This is the stage when intermetallics, if any, are formed.

metal diffuse mutually into one another. Various side effects are recorded such as intermetallic alloy formation and solution hardening. This condition is depicted in Fig. 1-11. The size of the alloyed region and the nature of the side effects are functions of the temperature and materials under consideration.

Once the solder is frozen on the surface, we obtain a good metallurgical bond between the base and the solder. Metallic continuity is established, which is good for both electrical and heat conductivity as well as for strength and other properties we seek in the soldered bond.

Let us remember, however, that more than one base metal is always going into the solder bond and that the same mechanism with the same degree of efficiency must therefore be carried out on more than one surface if the bond is to fulfill requirements.

1-12 Cleanliness of Soldered Surfaces and Corrosion Since we have shifted the surface from the base metal to the surface of the solder fillet, this solder surface requires some further consideration. In the presence of air, the same conditions depicted in Fig. 1-6 will occur. The air will be adsorbed to the surface and will concentrate in the same area as before. The mutual saturation of the bonds will bring the air molecules closer to the surface, and Fig. 1-12 describes this condition. In Fig. 1-13, we see the condition of the surface after the air has reacted with the surface molecules. A thin layer of lead oxide will be formed in the case of lead-containing solder. This lead oxide is a tenacious film which will protect the

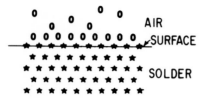

Fig. 1-12   Schematic of solder lattice in air (not tarnished yet, similar to Fig. 1-6).

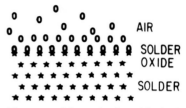

Fig. 1-13   Schematic of tarnished solder lattice in air (similar to Fig. 1-7).

metal from further environmental attack. Since there is a strong possibility that some flux residues are left on the surfaces, however, the picture might have an entirely different aspect. It is possible to demonstrate the importance of cleanliness on the solder surface by explaining the corrosion cycle of a tin-lead solder in the presence of chlorides.

1-13 The Corrosion Cycle in Tin-Lead Solders when Chlorides Are Present   The metallic lead is usually protected from environmental attack by a dense well-adhering tenacious layer of lead oxides, as shown on the top left-hand side of Fig. 1-14. In the presence of chloride ions we get the reaction shown in Fig. 1-14. For simplicity we use hydrochloric acid in Fig. 1-14 because it is easily obtained from chloride ions and water:

$$H_2O \rightleftharpoons H^+ + OH^-$$
$$H^+ + Cl^- \rightleftharpoons HCl$$

The lead chloride which is formed is a rather loosely adhering compound. It is not stable in a moist atmosphere containing carbon dioxide, CO_2. As we can see on the lower left-hand side of the corrosion cycle, the lead chloride is easily converted into the more stable lead carbonate, releasing in the process another chloride ion, which is free to attack the lead oxide layer again. The lead carbonate, which is the end product of this transformation, is a very porous white layer of material which does not protect the metallic lead. As a result, atmospheric oxygen reaches the metallic lead, and there is renewed oxidation of the metallic surfaces. The lead oxide is again converted into lead chloride by the presence of the free chlorine ions. The chlorine ions are regenerated upon further conversion of the

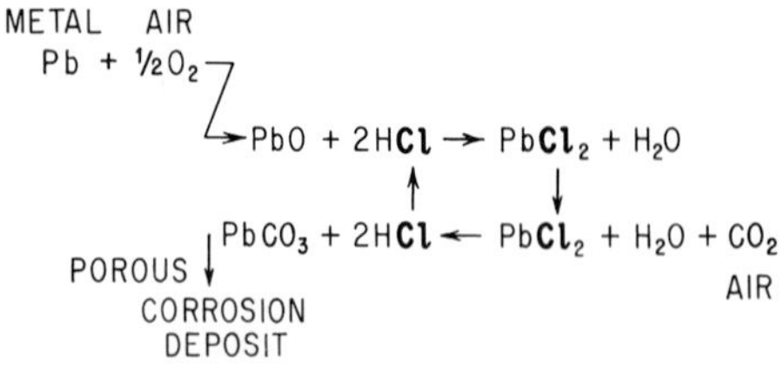

METAL   AIR
Pb + ½O₂

→ PbO + 2HCl → PbCl₂ + H₂O

PbCO₃ + 2HCl ← PbCl₂ + H₂O + CO₂
POROUS   AIR
CORROSION
DEPOSIT

Fig. 1-14   Simplified cycle for solder corrosion.

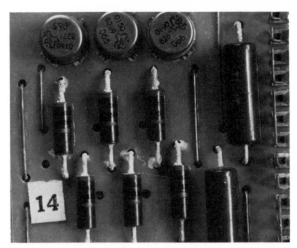

Fig. 1-15 Printed-circuit board showing white PbCO$_3$, which developed on pretinned leads contaminated with chlorides and exposed to 100 percent relative humidity for 24 h.

lead chloride into lead carbonate, and this process is a continuous cycle which does not terminate until all the lead is consumed, provided there are water and carbon dioxide in the environment.

Figure 1-15 shows the results of traces of chloride contamination left after improper removal of activated rosin flux. In cleaning only nonpolar solvents left activator exposed. Accelerated aging reproduced the same corrosion results on lead-rich resistor-lead coating as were noted in field service.

1-14 General Cleanliness Considerations The question is often raised: How clean is clean? When the equipment and its projected use are known, the environment in which it will operate helps to determine the degree of cleanliness. A home appliance like a radio or television is entirely different from an airplane's navigational system or a radar tracking station upon which human lives depend. These contrasting examples help establish the fact that no one clear rule can be applied to all equipment.

Besides consuming the solder fillet, corrosion has other dangers (see also Sec. 2-12). Corrosion can damage conductors. It can increase circuit resistance, and high resistance is undesirable. It can cause physical failure of conductors by weakening and embrittling them. In addition, corrosion products themselves can cause current leakage. Current leakages are particularly bad because they are not consistent. Humidity changes in the atmosphere will cause intermittent variations in the amount of current leakage.

Corrosion products can also cause contamination throughout a whole system in the form of nonconductive deposits on relay surfaces, mechanical contacts, etc. (Not only corrosion products are dangerous in this respect. Using rosin-base fluxes or similar materials which give off nonconductive fumes in the close vicinity of the contact surfaces might deposit from the vapor insulation layers which will be barriers to the electric contacts.)

Although flux and flux residues are usually the first materials to be blamed for corrosion in electronic assemblies, there are many sources of corrosion. Assembly contamination can occur in presoldering operations, because many assemblies are subjected to strong, chemically active plating solutions. Sometimes cleaning procedures are not complete or perfectly controlled. This solution carry-over can cause corrosion. Furthermore, there is always a danger of contamination due to storage. Parts can be packed in material that may contribute to contamination. And, of course, corrosive materials can always settle out of the air. Soldering fluxes themselves contribute to corrosive materials if they are improperly chosen and if cleanliness procedures are not followed. Furthermore, we must never forget that postsoldering operations, like final storage, can also be a source of contamination.

Let us take a quick look at some of these other sources of contamination.

1. *Processing solutions,* such as plating and etching materials.

2. *Human perspiration.* Corrosive chlorides deposited on an assembly through various handling operations can contribute more corrosive materials than any other single factor.

3. *Environmental sedimentation.* This is particularly important in industrial environments. Sulfur in the air, for instance, can attack a silver surface.

4. *Contamination from handling equipment.* This equipment is usually loaded with cutting oils, lubricants, and other kinds of solutions which contribute to contamination.

5. *The packaging materials.* Where corrosion is a problem, it is always wise to talk to packaging experts who can recommend a type of material that should be used in order to avoid contamination.

Once the danger of corrosion is recognized, there are various solutions to this problem. Since soldering is usually the last operation in the sequence of an assembly, violations of cleanliness can be corrected before actual service. Cleaning procedures are not complex and are described in detail in Chap. 5. Once we have cleaned an assembly, however, we must make sure that it is not recontaminated by improper handling, packaging, or storage.

TWO

The Chemistry of Fluxes

THEORETICAL CONSIDERATIONS

2-1 Introduction   The word *flux* comes from Latin and means "flow." In soldering, this definition is not adequate. We have shown that the flux has other functions besides helping the flow. It is clear that the flux itself does not enter into the bond formation. In that respect, it is similar to a catalyst in a chemical reaction, which triggers and promotes a process without entering into the end product. The flux also affects the rate and degree of completion. Given clean metallic surfaces in the right atmosphere, we obtain a solder bond without flux. To obtain such conditions in industrial applications would prove quite a burden, and here the addition of flux will facilitate the joint formation without adding excessively to the cost.

As shown in the first chapter, the flux has two major functions: (1) to provide tarnish-free surfaces and keep the surface in a clean state and (2) to influence the surface-tension equilibrium in the direction of solder spreading by decreasing the dihedral angle. Both these functions can be further broken down into more specific properties, leading to a series of broader statements. Let us list them more specifically and given them names.

2-2 Chemical Activity Before a good solder joint can be made, all the products of environmental attack (which are normally referred to as *oxides* but which for the purpose of this book are called *tarnish films*) have to be removed. These films are not soluble in any of the conventional solvents and cannot be removed like grease. They are reacted upon chemically with materials which result in compounds soluble in the liquid flux. Thus they are removed, exposing the metallic surface, which is now ready for bonding.

The chemical reaction can be one of several. It can be a reaction whereby the flux and the tarnish combine to give a third compound, which in itself is soluble in either the flux or its vehicle, or it can be a reaction whereby the metallic tarnish is reduced to its original form, restoring the metallic surface. Variations and combinations of these reactions are also found.

The reaction between a water-white rosin and a copper oxide serves to illustrate the first type of reaction. The water-white rosin consists mainly of abietic acid and other isomeric diterpene acids which are soluble in a number of organic solvents. When used as a flux, water-white rosin is usually in an isopropanol vehicle. When the oxidized copper surface is covered with this flux and heated up, it will be cleaned by the following reaction. The copper oxide combines with the abietic acid to give a copper abiet. The copper abiet is a green transparent rosinlike material, which mixes easily with the unreacted rosin, leaving a bare metallic surface for solder wetting. The abietic acid does not attack any of the copper underneath the copper oxide. When the liquid solder displaces the rosin flux, it displaces the copper abiet at the same time. The now exposed metallic copper surfaces can easily be wetted by the solder. When the flux residues are removed with the aid of organic solvents, the copper abiet is removed, together with the other rosin-type materials.

The soldering of tarnished surfaces under a blanket of hot hydrogen is an example of the second type of reaction. The hydrogen at elevated temperatures will reduce the metal oxides on the surface, forming water and restoring the metallic surfaces. This type of operation is very common in semiconductor-device soldering.

Nearly every organic or inorganic compound which upon heating removes the tarnish in one of the two mechanisms described above can act as a fluxing material. However, most chemicals have limited usefulness because of the operating conditions required in soldering. The range of materials is limited even further because a good flux material should also be applicable to many types of metallic surfaces simultaneously. Additional considerations discussed on the next pages eliminate a large number of materials as good fluxes on the basis of physical and chemical characteristics. Thus only a small number of the potential chemicals can be used for fluxes. Let us consider the other properties required to make a chemically useful material into a useful fluxing material.

2-3 Thermal Stability   Once the chemical reaction has taken place, the flux must provide a protective coating on the cleaned metallic surface at the soldering temperature. Otherwise the freshly exposed metallic surfaces will easily reoxidize in the atmosphere, accelerated by the elevated soldering temperatures. The flux must therefore be capable of withstanding soldering temperatures without evaporating or breaking down. The breakdown of the flux material would cause undesirable deposits of materials which are difficult to displace with a molten solder and hard to remove. Rosin will decompose and char at 545°F (285°C) even under a nitrogen blanket.

Here a distinction should be made between the flux and the solvent that serves as a vehicle for the flux. In many cases, the vehicle has a boiling point below the soldering temperature, while the flux material itself is heat-stable. Let us use the example of water-white rosin flux to illustrate this point. The alcohol (usually isopropanol) will evaporate long before soldering temperatures are reached, but the rosin material itself is stable for higher temperatures as long as it is not exposed to heat for long periods of time.

The reverse case may also be desirable. A flux vehicle could provide the thermal stability as well as the wetting ability for the surfaces and have no chemical activity. A rather corrosive activator would then be added to give the flux the chemical activity necessary to clean the surfaces. If this highly corrosive flux is formulated in such a manner that it evaporates to a great extent during soldering temperatures, the residue will not be so dangerous as if it contained only the activator. We can use the activated-rosin fluxes as an example of this. Here the rosin and the isopropanol both really serve as a vehicle to the more active organic chlorides which are added as activators (hence the name *activated-rosin fluxes*). The activators (mostly some form of amine hydrochloride) have a dissociation temperature below the soldering temperatures. At that point, the corrosive hydrogen chloride is released to do the chemical cleaning on the tarnished surfaces. As the temperature is raised to soldering levels, some activator may decompose and volatilize, but the major part will recombine upon cooling, leaving the residues as harmless as possible. For further information on this, see Sec. 2-4.

The heat stability we have been talking about should not be confused with the stability of the flux during storage before use. Such properties come under a different heading. It is also important to remember that when the heat stability of a flux is discussed, the soldering processes we have at our disposal cover such a wide range of soldering temperatures that the specific flux might be adequate to clean a certain surface in one operation and not suitable for the same surface using different equipment.

2-4 The Activation and Deactivation Temperatures of Fluxes

Thermal stability is not the only temperature characteristic required by a good flux. Various other temperatures are just as important in the overall consideration of a material for proper fluxing. These temperatures fall into two major groups which are interrelated as follows.

Temperature of Activation. We have seen that a portion of the flux which possesses the chemical activity has to remove tarnishes. This action usually takes place at a certain temperature range which is labeled the *temperature of activation.* It is at this temperature that the flux mechanism is triggered and/or reaches optimum reaction conditions. In the case of activated-rosin fluxes, for instance, unless the temperature is reached where the activators dissociate to give ionic halides for the cleaning of the surface tarnishing, no wetting will be possible. It is essential that this temperature range lie within the working range of the solder system used. Another good example is the use of hydrogen gas as a soldering flux. Unless a certain elevated temperature is reached, the hydrogen will not reduce the surface tarnishes of the metal sufficiently for adequate wetting. This temperature is unique for every base metal, and the time of exposure to the hydrogen gas at that temperature depends on the thickness of the tarnish coat.

The Temperature of Inactivation. It is conceivable that organic materials (or inorganic materials for that matter) might change their characteristics because of intermediate chemical changes at elevated temperatures and become inactive. A good example here is the water-white rosin-type flux, whose ability to clean copper is greatly reduced by overheating. If such a flux is heated rapidly to 600°F (315°C) and kept there, hardly any chemical reaction will take place. If this type of flux has to be used for such high temperatures, sufficient dwell time should be allowed in the active range to prepare the surfaces chemically. At the same time, the surface activity must still decrease the dihedral angle once the elevated temperature is reached.

The inactivation temperature is also referred to as that temperature at which the breakdown of the flux supposedly is complete and at which, if enough time at that temperature is allowed, the residues of the chemicals used for fluxing are supposed to be chemically inert and noncorrosive. This, once more, is a function of the uniformity of reaction. Claims like this should be viewed with suspicion, for only a few methods of flux application will guarantee total heating of the flux for long enough period to the right temperature to assure this inactivation.

2-5 Wetting Power In order to be able to clean the metallic surfaces chemically and influence the surface energies of both solder and base metal, the flux should be able to come into intimate contact with both surfaces. In other words, it should be able to wet them properly. This stresses the fact that the flux has to displace the vapor phase on the metallic surfaces to be soldered. Because of the same surface energies described in Chap. 1, a layer of gas—mostly air—is adsorbed to the surface of any solid. Before intimate contact between the metal and the flux is possible, this layer of gaseous molecules has to be displaced. This is a wetting action which is similar to that described in Chap. 1, except that here the flux wets the metal and only a temporary bond is formed.

If the system is based on aqueous solutions, the addition of surface-active agents can be of great benefit in a flux. However, a word of caution is in order here. The fluxes are usually associated with a cleaning operation and are expected to remove all grimes and greases. This is not the case, because most fluxes are rendered useless by the presence of insoluble layers of dirt on the workpiece. A thorough cleaning to remove oils and solid particles contributes largely to efficient fluxing and proper soldering. Unless a flux comes into intimate contact with the metallic surfaces, displacing all surface contaminations, it will be unable to react with the surface tarnishes and provide metallic surfaces for bonding.

2-6 Spreading Activity The presence of the flux in the soldering region should influence the surface-energy equilibrium in the direction of solder spreading. This also implies easy displacement of the flux by the liquid solder. As stated earlier (Chap. 1), the wetting and spreading of solder are directly related to the dihedral angle. This contact angle between the solder and the work surface can be used in practice as a measure for fluxing efficiency. The spreading activity can therefore be described as the power of the flux to reduce the contact angle of a solder–base-metal system.

If a flux has good wetting of the base metal and the solder, and has the chemical activity necessary, it still does not follow that it has a spreading activity. This is discussed further in the next section.

2-7 The Relative Importance of the Various Factors In order to illustrate the relative importance of the above-mentioned flux functions, a number of tests were made in the laboratory.

1. A material that removed tarnish at soldering temperatures was used as a flux. The material, however, evaporated at the same time. The

molten solder started spreading but was checked by a layer of freshly formed tarnish because there was no more flux present to provide a protective coating. The material used was oxalic acid. It has a melting point around 360°F (182°C), and the liquid volatilizes rapidly upon further heating without charring. (Oxalic acid is one of the few organic compounds that does not char but sublimates.) As long as the oxalic acid was present, the solder spread; but after the rapid evaporation, tarnish was formed again and the spreading was checked.

2. A material that removed tarnish but remained liquid at soldering temperatures, charring slowly, was used as a flux. The molten solder balled up in this liquid because of the surface-tension equilibrium. In spite of the clean surface, there was no spreading; but upon cooling and flux-residue removal, the solder ball was found to be firmly bonded to the copper at the point of contact. The same experiment was repeated, and this time the solder was moved around mechanically. Upon cooling and cleaning, the trace of the ball movement was visible as lines of tinned copper. This indicated that clean surfaces upon contact formed a solder bond. The material used here was regular glucose. The well-known ability of sugar to reduce copper oxide caused the cleaning action. The viscosity of the molten sugar and the surface activity did not favor spreading. The balling up of solder indicated that the dihedral angle was greatly increased.

3. A material that had the right surface activity but was unable to clean the tarnish was used as a flux. The solder, unable to wet the metal because of the tarnish layer, stayed balled up. A chemically active material was then added, removing the tarnish, and there was instant spreading. This demonstrated that surface activity alone does not promote solder bonds. The material used was abietic acid, one of the main constituents of rosin. The base metal was a heavily tarnished copper surface. The flux flowed freely over the tarnished metal but did not change its color. The molten solder remained balled up. When a drop of amine hydrochloride in alcohol was added, the color of the tarnish changed and the solder spread rapidly over the surface.

Thus we see that, although each factor in itself is very important, a combination of all three factors is vital before good wetting can occur. It is therefore possible when formulating fluxes to combine various materials to fulfill the various functions required, and it is not necessary for each component in the flux to fulfill all the demands imposed on the total formula. Thus an activated rosin will contain a vehicle which acts only as a diluent and makes it possible to apply the activated rosin uniformly over the surfaces. It contains the rosin itself, which has good wetting power and the thermal stability required, and finally it has the ability to influence the surface tension of the system in the direction of wetting. The chemical activity of this flux is mostly due to additions of activators, which give it the cleaning power required, although the rosin itself has some cleaning power, as shown earlier.

2-8 Electrochemical Activity   An additional function attributed to fluxes is worth mentioning here. Several previous workers have reported an electrochemical activity of fluxes whereby the flux is supposed to deposit solder ions on the surface to be joined, thus aiding the process of bonding. The author's work in this field showed that this process is true only for certain inorganic materials, mainly the zinc chloride-ammonium chloride fluxes and other fused salts. It also turned out that not only solder ions were deposited out of the fluxes but also some of the metals in the salts contained therein. Two different types of emf sources are available for this process as follows: (1) A replacement deposition of the solder or flux ions on the base metal. This is the case with *reactive fluxes* on aluminum, for instance. (2) A plated deposition due to an external emf on the base metal. This is much the same as plating out of fused salts and depends on the voltage. The work described by Bailey and Watkins[1] has such an emf in the form of a thermocouple, and their results fall in this group.

Electrochemical deposition actually has the same effect that pretinning would have. In most cases, much faster spreading of solder occurs on a pretinned surface because solder will wet itself faster than most foreign-base metals.

It appears that the action of pretinning by fluxes is mostly due to the replacement deposition, because no obvious external emf source is found in the solder–flux–base-metal system.

INDUSTRIAL AND ENGINEERING CONSIDERATIONS

2-9 The Practical Flux   To make soldering an economical connecting method, fluxes must meet some additional engineering and industrial requirements. These additional points will help to determine whether the flux has industrial value or just theoretical potential. Because the relative importance of these additional parameters varies from job to job, no attempt has been made to list them in order of importance.

2-10 Time   In industry, short soldering times are important, particularly in automatic setups and in electronics, where heat-sensitive assemblies make fast soldering essential. Therefore, a well-selected flux should

[1]G. L. C. Bailey and H. C. Watkins. Surface Tension in the System Solid Cu–Molten Pb, *Proc. Phys. Soc. Lond.*, vol. 63B, p. 350, 1950.

affect the solder system rapidly, allowing a maximum number of joints to be made per unit of time.

2-11 Temperature   The soldering temperature required depends largely on both the application conditions and the solder alloy used. It is therefore important to match the temperature characteristics of the flux to the overall temperature picture of the assembly. However, if a specific flux must be used because of some unique consideration, it is also possible to use a flux at temperatures outside its usual range provided that enough dwell time is allowed in the active region (Sec. 2-4) and the flux has enough thermal stability. Such practices are not recommended, however, and the safe way would be to use the flux at a temperature high enough to energize its active parts for the chemical cleaning function and not too high for breakdown of the materials.

2-12 Corrosion   The corrosion discussed here stems directly from the flux. For a general discussion of corrosion in electrical assemblies, see Secs. 1-12 and 1-13. This corrosion must be controlled or the solder joint may weaken and fail. In structural parts, the failure can occur by reduction of the cross section of the metallic parts, which weakens the metal below its load capabilities, or by weakening through preferred corrosion (along grain boundaries, etc.), which causes a fatigue-type failure. A good example of such a failure is brass when ammonia-containing fluxes are used. The ammonia will cause intergranular corrosion of the brass, weakening the material drastically although the overall amount of corrosion is small compared with the volume of the brass. Special non-ammonia-containing fluxes are therefore required for soldering brass parts.

In electric connections, these troubles may be compounded by changes in electrical characteristics (increase in resistance due to decrease of conductor diameter). In addition, corrosion products can lower the insulation resistance or bridge the components, causing short circuits. This is a function of the ionizable corrosion products formed, and the current leakage depends on the absorption of humidity to form thin layers of electrolytes.

The flux-related corrosion itself might originate in several places:

1. The corrosivity of the flux itself, which might be lodged in crevices and difficult-to-reach places.

2. The corrosivity of the fumes liberated during the soldering process. These fumes condensing on the assembly parts away from the solder bond may not be removed when the area of the joint is cleaned after bonding.

3. The corrosivity of the flux residue, which, even after cleaning, may be left in inaccessible places.

Consequently a good industrial flux should have the following properties:

1. Low corrosivity in raw flux at room temperature if any.
2. Low-corrosivity fumes given off during soldering.
3. Low corrosivity in residue at room ambient if any.
4. If corrosivity of flux cannot be controlled or avoided, its products must be adequately and easily removable after soldering.

Flux chemistry will dictate the color of the residues. The interpretation of the colors must be tied in with the chemistry of the system; otherwise it is very misleading.[1]

2-13 Safety   Safety must be considered for the personal handling of the material, the plant itself, and the ecology.

Personal Safety. The flux should be a nonirritating material with harmless fumes given off during soldering. An example of dangerous materials will help to demonstrate this point. Hydrazine-base fluxes can easily cause dermatitis, and the fumes of some fluxes contain poisonous metal compounds, e.g., indium chlorides.

Because of the Occupational Safety and Health Act (OSHA) flux vendors must pay close attention to their ingredients. Detailed safety sheets (Form OSHA-20) can be requested by the user for reference in case of need. These data also come in handy when drawing up purchasing specifications.

Plant Safety. The flux and/or its vehicle are often combustible, and some are even flammable. This requires thoughtful handling and storage to prevent fires or explosion: normal standard industrial pratices should suffice. While most people associate the fire hazard with the flash point of the material, this is not the entire story. The flash point is actually only a good indicator, but fume concentration is also important. Thus, with adequate air movement, fluxes containing low-flash-point alcohols like methanol (closed-cup flash point 54°F, 12°C), ethanol (closed-cup flash point 55°F, 13°C), and isopropanol (closed-cup flash point 53°F, 11.7°C) have been used in the electronic industry's formulations without difficulty for over 20 years. The autoignition temperature, on the other hand, is a very meaningful variable for slowly evaporating high-boiling fluxes such as the glycol-ether-base materials. Because the vapors are heavy, they tend to concentrate in low enclosed areas. Should this happen along hot surfaces like a preheater or the solder reservoir itself, if the autoignition temperature is within range of these surfaces, spontaneous fires will start.

---

[1] H. H. Manko, Color, Corrosion and Fluxes, *Electron. Packag. Prod.,* vol. 9, no. 2, February 1962.

This is true also for other soldering chemicals, e.g., solder blankets and solder oils.

The Ecology. The important work done by the Environmental Protection Agency has had an impact on the formulation of fluxes too. One must consider the quality of the fumes, decomposition products, residues, and discarded material in the light of their effect on the ecology. To be specific, fumes must not be noxious or photosensitive, breaking down into harmful materials. The fumes, the decomposition products, and the residual fluxes must be biodegradable, or specific instructions for their disposal must be given by the vendor.

Directions for proprietary fluxes are usually printed on the label and in data sheets supplied by the vendor. When the specified precautions are followed, no serious hazards are encountered with the fluxing materials available today.

2-14 Economics   This is a self-explanatory factor though it should be kept in mind that a cheaper flux is seldom the most economical one. More costly materials can often bring savings later in the process through their reliability and the elimination of rework. Soldering times and cleaning expenses usually outweigh the price of the fluxing material itself.

FLUXING MATERIALS

2-15 Materials Available   As seen earlier, one of the inherent properties of flux materials is the chemical activity which enables them to clean the tarnished surfaces, making a solder bond possible. Therefore, every fluxing material is corrosive to some degree. Dividing fluxes into corrosive and noncorrosive types, a widely used specification system, is inadequate and often misleading. It seems to suit the advertising purposes of flux manufacturers more than it pleases the customer. For the purpose of this book, therefore, fluxes are classified by their chemical origin. Table 2-1 identifies and compares the major flux types in terms of the properties discussed above except for properties which depend on the total flux–solder–base-metal system and cannot be tabulated. The degree of corrosivity is included under the properties of these materials.

The two major flux groups are inorganic and organic materials, the latter being subdivided into nonrosin materials and the rosin-base fluxes. Table 2-1 identifies the fluxing materials with the vehicle type normally used with them. Flux vehicles will be discussed separately in Chap. 5 in connection with cleaning.

2-16 Group I, Inorganic Materials   This group contains three kinds of compounds.

Inorganic Acids. These are fast-cleaning and will remove any common tarnish. They are stable and active at soldering temperatures. They are very corrosive before and after soldering, and their condensed fumes must be removed. Most of them are easily removed in aqueous solutions and can also be neutralized.

The most common members of this group are hydrochloric and ortho-phosphoric acid. In general, any acids or combination of acids used for etching of metals can be used provided the requirements mentioned earlier are met. The acids are seldom used by themselves but are a vital part of inorganic solder-flux combinations. Further details and formulations are given later in this chapter.

Inorganic Salts. These are fast-cleaning and good tarnish-removing materials which become active when molten and are stable at soldering temperatures. They are less corrosive in salt form unless they are in humid atmospheres. The fumes must be carefully removed. They are mostly water-soluble and can be neutralized easily. However, sometimes it is necessary first to soak them in a slightly acidic or basic solution in order to form soluble complex salts and then to continue with the normal aqueous rinsing procedure.

For example, when zinc chloride is used as a flux, the following reaction will take place:

$$ZnCl_2 + H_2O \rightarrow Zn(OH)Cl + HCl$$

This reaction of zinc chloride in water frees the HCl, which is now ready to clean the oxides off the surfaces, as in the case of copper oxide:

$$CuO + 2HCl \rightarrow CuCl_2 + H_2O$$

Thus the copper chloride, which is very soluble in water, can easily be removed from the surfaces. However, the zinc oxychloride which was left on the partial decomposition of the zinc chloride is a relatively insoluble material and has to be treated with an excess of hydrochloric acid in order to restore the zinc chloride, which is water-soluble:

$$Zn(OH)Cl + HCl \rightarrow ZnCl_2 + H_2O$$

Unless the zinc oxychloride is removed in a dilute acid, a layer of dull white may remain in the vicinity of the joint made with the zinc chloride.

The inorganic salts are less dangerous than acids in fluxes. A single salt is seldom used, and the flux formulas usually contain either additional acids or other salts in their makeup. When the melting point of the solder is below the melting point of the salt used, the solid material left after the vehicle of the flux is driven off because of the soldering heat is not easily displaced from the surfaces and actually retards wetting. Thus it is normal practice to use a salt with a melting point lower than that of the soldering alloy. In order to achieve this, it is often necessary to use a eutectic

TABLE 2-1 Comparison of Flux Materials in Order of Decreasing Chemical Activity

	Typical fluxes	Vehicle	Use	Temperature stability	Tarnish removal	Corrosiveness	Postsolder cleaning methods
Inorganic							
Acids	Hydrochloric, hydrofluoric, orthophosphoric	Water, petrolatum paste	Structural	Excellent	Excellent	High	Hot-water rinse and neutralize; organic solvents; degrease
Salts	Zinc chloride, ammonium chloride, tin chloride	Water, petrolatum paste, polyethylene glycol	Structural	Excellent	Excellent	High	Hot-water rinse and neutralize, 2% HCl solution, hot-water rinse and neutralize; organic solvents; degrease*
Gases	Hydrogen, forming gas; dry HCl	None	Electrical	Excellent	Very good at high temperatures	None normally	None required
Organic: nonrosin base							
Acids	Lactic, oleic, stearic, glutamic, phthalic	Water, organic solvents, petrolatum paste, polyethylene glycol	Structural, electrical	Fairly good	Very good	Moderate	Hot-water rinse and neutralize; organic solvents; degrease*
Halogens	Aniline hydrochloride, glutamic acid hydrochloride, bromide derivatives of palmitic acid, hydrazine hydrochloride or hydrobromide	Water, organic solvents, polyethylene glycol	Structural, electrical	Fairly good	Very good	Moderate	Hot-water rinse and neutralize; organic solvents; degrease*
Amines and amides	Urea, ethylenediamine, mono- and triethanolamine	Water, organic solvents, petrolatum paste, polyethylene glycol	Structural, electrical	Fair	Good	Moderate	Hot-water rinse and neutralize; organic solvents; degrease*

Organic: rosin base

Rosin, superactivated	Rosin or resin with strong activators	Alcohols, organic solvents, glycols	Structural, electrical	Fair	Very good	Moderate	Water-base detergents; isopropanol; organic solvents; degrease*
Activated (RA)†	Rosin or resin with activator	Alcohols, organic solvents, glycols	Electrical	Fair	Good	Only to critical electronics	Water-base detergents; isopropanol; organic solvents; degrease*
Mildly activated (RMA)†	w/w rosin with activator	Alcohols, organic solvents, glycols	Electrical	Poor	Fair	None	Water-base detergents; isopropanol; organic solvents; degrease*
Nonactivated (water-white rosin) (R)†	w/w rosin only	Alcohols, organic solvents; glycols	Electrical	Poor	Weak	None	Same as activated rosin but normally does not require postcleaning

*For optimum cleaning, follow by wash with demineralized or distilled water.
†Follows Federal Spec. QQ-S-571 or MIL-F-14265.

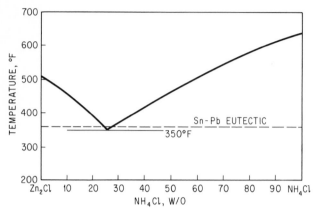

Fig. 2-1 Schematic zinc chloride–ammonium chloride eutectic system. Note dashed line (361°F) representing solidus of tin-lead solders.

combination of salts as in the case of zinc chloride–ammonium chloride and tin-lead solders.

Zinc chloride, which melts at 541°F (283°C), in itself would not be fluid at 361°F (183°C), which is the temperature at which the solders would be fluid. This salt would therefore not be suitable for many applications such as dip soldering, where the temperature differential between the soldering temperature and the melting point is relatively small. On the other hand, ammonium chloride by itself will melt at 662°F (350°C), which again is too high for this type of application. However, the zinc chloride–ammonium chloride eutectic with a ratio of 3 zinc chloride to 1 ammonium chloride will be very effective for soldering because its melting point is 350°F (176°C), which is lower than the melting point of the solder. Figure 2-1 shows a rough diagram of the eutectic composition of the zinc chloride–ammonium chloride system, and superimposed on it is the temperature of the eutectic tin-lead solders. The reader is reminded that the soldering temperature itself is not necessarily the melting point of the alloy and is usually higher. For details, see Chap. 6.

Inorganic Gases. These are chemically active only at elevated temperatures. They require clean surfaces, free from foreign materials. Because of their hazardous nature they require special equipment. In this group, we find dry hydrogen and hydrogen chloride. They are extensively used in the transistor industry.

2-17 Group II, Organic Materials (Nonrosin Base)   This category can be divided into three major subdivisions.

Organic Acids. These are slower-acting materials with a medium capacity to remove tarnishes. Being organic materials, they are temperature-sensitive. They are still corrosive after use, and any condensed fumes must be removed. Most are water-soluble, and organic solvents alone are not used for flux-cleaning procedures. In this group are lactic, oleic, stearic, phthalic, citric, and other acids.

Organic Halogens. These are similar to the inorganic-salt group in their activity and are used because of the easily available halogen ion. They are still temperature-sensitive by virtue of their organic radical. They are relatively more corrosive than the other organic fluxes, and their condensed fumes must be removed carefully. In this group, we find materials like aniline hydrochloride, glutamic acid hydrochloride, and bromide derivatives of palmitic acid.

Amines and Amides. This is a group of additives of many proprietary fluxes because they do not contain halogens. This group is still corrosive and very temperature-sensitive. It contains such materials as urea, ethylenediamine, and mono- and triethanolamine. Various derivatives of the amines and amides are also used for fluxing materials. The most common is aniline phosphate.

2-18 Rosin-Type Organic Fluxes   In the electronics industry rosin fluxes have been the standard for many years. When this industry was in its infancy, flux properties were simple to define. Because of the primitive nature of the components, cleaning was impractical and flux residues were left in place. As a result, these materials had to be nonconductive, nonhygroscopic, and noncorrosive. Researchers working on government specifications identified a particular grade of gum rosin as the best raw material for this purpose.

Considering fluxes from this point of view, the ideal material would be completely inert at ambient temperature and yet active at soldering temperatures. Furthermore, flux residues, including the reaction products with the metals, would be inert at assembly operating ambient temperature and the flux-fume condensates would behave likewise.

Although we have discussed fluxes in a decreasing order of strength, we shall reverse this procedure for the rosin-base formulations. It is easier to explain the differences especially, if we follow the historical sequence of development.

2-19 Water White (w/w) Rosin Flux   Being a solid at room temperature, pure rosin is both chemically inactive and electrically insulating. The same holds true for condensed rosin vapor and its metallic reaction compounds. Yet when heated to slightly above its melting point, the rosin becomes active and will react with some of the metals used in electronics applications (copper, silver, gold, etc.).

Rosins are a steam distillate of pine trees and are graded by their colors. The purest form of rosin by the ASTM designation is called *water white* (w/w) (Table 2-1). This material has been adopted as the mildest flux and is normally available in the core of solder or in liquid form dissolved in alcohols and designated liquid water-white rosin, also sometimes referred to as *nonactivated rosin*. The major disadvantages of water-white rosin are its inherent chemical weakness on many metals and the fact that a certain amount of presolder cleaning is mandatory in order to make it a reliable flux.

A large number of solder joints using this flux are being made year in and year out, mainly in the telephone and telecommunications industries. It has also found many good applications in the more modern microelectronics industry, where the chemical activity of the flux must be restricted to avoid changing any of the values of active and passive devices being soldered to hybrid circuitry.

2-20 Activated Rosin Flux   In an attempt to strengthen the fluxing properties of rosin, additives called activators are introduced. These materials fall into many categories; differences between various countries in theory and practice are quite evident (see Table 2-2). Instead of tackling this problem from a historical point of view and tracing the development of various national patterns, we shall discuss the rosins in the order of increasing activity.

This increasing activity, however, bears no correlation to the chemical and insulating properties of the fluxes and their residues after soldering. We shall discuss the activated fluxes in the order in which they appear in Table 2-2 and relate each of the concepts to the appropriate cleaning methods required. The full details of cleaning procedures can be found in Chaps. 5 and 8.

Two distinct types of soil exist. One type (salts, sugar, etc.) is soluble in polar solvents like water. Although not all examples of this type ionize, those which do may cause corrosion and, even worse, product malfunction due to electric-current leakage. The second type of soil comprises nonpolar soluble materials (oil, grease, rosin), normally removed in degreasing, which may be done by cold immersion or by vapor-phase solvent cleaning.

Low-Activity Rosin Fluxes. The first category of activated fluxes (second item, Table 2-2), although referred to as mildly activated by specification-issuing authorities, has been labeled low-activity rosin for this book. These materials are controlled mainly by limitations on the amount of activity permissible, so that removal of flux residues should not be required for most applications. The amount of activator (labeled chloride) is limited to 0.5 percent (the British DTD specifications). This, in the opinion of the writer, is a relatively weak specification since it has little bearing on the properties of the end result and the type of activator is not really specified. This would enable the unethical manufacturer of fluxes to incorporate

TABLE 2-2 Attributes and National Specifications Relating to Eight Categories of Solder Fluxes Listed in Order of Increasing Activity*

| Typical use | | Generic term (category) | Form† | Formulation | | | Cleaning | | | USA | | Germany, Europe, DIN | UK, DTD | Japan JIS-C | |
Electronics	Structural			Body	Activator	Vehicle	Need	Non-polar	Polar	MIL-F-14256	QQ-S-571			2519	2512
X		w/w rosin	L	Rosin	None	Alcohols	None	Yes	No	Type R	....	F-SW 31	None	None	......
			C	Rosin	None	......	None	Yes	No	......	Type R	F-SW 31	None	.....	None
X		Low-activity rosin	L	Resins	0.5% Cl	Alcohols	Optional	Yes	Yes	None	None	None	Yes	Yes	Type A
			C	Resins	0.5% Cl	......	Optional	Yes	Yes	None	None	None	Yes	.....	& AA
X		Halogen-free activated rosin	L	Resins	Yes‡	Alcohols	Optional	Yes	Yes	None	......	F-SW 32	None	None	......
		Mildly activated rosin	C	Resins	Yes‡	......	Optional	Yes	Yes	......	None	F-SW 32	None	.....	None
X			L	Rosin	Yes	Alcohols	None	Yes	Yes	RMA§	......	None	None	None	None
X		Activated rosin	C	Rosin	Yes	......	Optional	Yes	Yes	RA	RMA	F-SW 26	None	Yes	Type B
X			L	Rosin	Yes	Alcohols	Optional	Yes	Yes	......	RA	F-SW 26	None	.....	
X	X	Superactivated rosin	L	Resins	Yes	Solvents	Mandatory	Yes	Yes	None	......	None	None	None	None
	X	Organic fluxes (nonrosin)	C	Resins	Yes	......	Mandatory	Yes	Yes	None	None	None	None	None	None
X	X		L	No	Acids	Water	Mandatory	No	Yes	None	......	F-SW 24/25	None	None	None
X	X		L	No	Acids	Solvents	Mandatory	No	Yes	None	......	F-SW 24/25	None	None	None
	X		L	Yes	Bases	Solvents	Mandatory	No	Yes	None	......	F-SW 24/25	None	None	None
X			C	No	Acids	Solvents	Mandatory	No	Yes	None	Type OA	None	None	None	None
X		Inorganic fluxes	L	No	Acids	Water	Mandatory	No	Yes	None	......	F-SW 21/12	None	None	None
X			C	No	Acids	......	Mandatory	No	Yes	None	Type IA	None	None	None	None
X	X		L	Yes	Neutral	Solvents	Mandatory	No	Yes	None	....	None	None	None	None

*Am. Weld. Soc. 1st Int. Soldering Conf., Detroit, 1972.   †L = liquid; C = core.
‡Organic acids as activators.   §As forthcoming specification, Revision (e). Present specification has type A for RMA; no type RA.

relatively hazardous materials in the formula and still meet the chemical and electrical requirements of the specification. In essence, it eliminates whole families of truly safe activators, which, even at higher concentrations, would create less potential hazard to the soldered assembly.

Halogen-free Activated Rosins. A second approach to the composition of the flux is the German DIN specification. One of its two categories is again classified as mildly activated and thus is labeled as the type of material whose residues would not be dangerous to electronic assemblies. These fluxes are called *halogen-free,* and the type of activator here is limited to organic acids and similar materials which, according to the specification, are relatively harmless. Once again the author does not agree with this approach since it limits the manufacturer to a small variety of chemicals which can be proved to be rather dangerous, especially if the rosin is removed and they are left exposed behind by inadequate cleaning. Unlike rosin, organic acids used as activators require a polar solvent for removal.

Mildly Activated Rosins. The approach in the United States is exemplified in MIL-F-14256 and QQ-S-571 specifications which do not really relate to the chemical composition of the flux but put specific electrical and chemical requirements on the material before and after soldering. The manufacturer is therefore able to seek the most effective materials in this mildly activated category. According to this concept, any type of activator (halide base, organic acid, amines, amides, etc.) may be used. The end result is a flux formulation whose residues and recondensed fumes are noncorrosive and electrically insulating. In the opinion of the author, this type of flux can truly be considered nonharmful.

Although mildly activated fluxes are much stronger than the original water-white rosin (nonactivated) material, they are still relatively weak. They require solderability monitoring of all surfaces in order to achieve reliability and economy in soldering. These materials have found wide application in the computer, telecommunications, aerospace, and military applications in the United States and have also been successfully used in color televisions, where residues are left in place. They are even more widely used in industries throughout Europe.

Activated Rosins. For standard electronics manufacturing and for mass production, however, mildly activated fluxes have proved to be weak. Another category, called *fully activated rosin fluxes,* has been developed and is used throughout the industry. This group of materials constitutes probably 80 percent of all rosin fluxes sold in the United States and in many of the countries the author has visited. Activated rosins in liquid form have only recently been considered for inclusion in the latest revision of MIL-F-14256 specification. The category has been present in the core solder QQ-S-571 specification for many years under class RA. While high-quality activated rosins will leave residues which are not necessarily dangerous to many applications, e.g., radios and television sets, they are considered too dangerous for high reliability and long-life applications under many environmental conditions.

In the removal of fully activated rosin fluxes, the use of bipolar solvents is absolutely mandatory, i.e., first, a nonpolar solvent to remove the rosin, followed by a polar solvent such as water to remove activator traces and other ionizable materials. It is also possible to use a bipolar-solvent blend to clean both types of residue in one simple operation.

The German DIN classification F-SW-26 is similar to the United States group of materials, although, again, the degree of latitude in activator selection is restricted by specification and therefore these materials are not quite as strong as their United States counterparts.

Superactivated Rosins. Finally, it was necessary to develop a rosin-flux formulation labeled *rosin superactivated* for very specific applications. These fluxes would be outside the limits of the materials as specified by any national standard. Some of these fluxes are so strong, chemically, that they will solder bare Kovar, nickel, some stainless steels, and similar alloys even without descaling. The flux residues are very active, however, and thorough removal after the soldering operation is recommended. The need to use rosin in conjunction with the high activity is dictated by the general properties of the specific component manufacturing process and is therefore unique. These types of fluxes have limited applications and are not recommended for general use by the electronics or electrical industries unless thorough cleaning of flux residues will be performed.

Superactivated fluxes were originally designed for use on incandescent-lamp bases and seal beams, where their residues would be burned off by the soldering flame or torch. Their use expanded from there into the tinning of bare metal leads, the soldering of tin-nickel-plated printed-circuit boards, and other electronics applications. Their greatest advantage over acid fluxes come from the cleaning methods which can be used. Superactivated rosins can be removed by bipolar degreasing operations, which are water-free. This usually requires less floor space than water-wash equipment and leaves the parts dry. See Chap. 5 for a further discussion of cleaning.

2-21 A Word about Rosin   In the electronics industry, water-white gum rosin as a flux has been the subject of controversial claims. It has many uncontested qualities, but somehow it does not fulfill all the requirements for efficient fluxing. The author has spent a considerable amount of time studying the nature of this material.

From the chemical point of view rosin is a mixture of several compounds. It is the non-steam-volatile fraction of pine sap, and its specific composition varies with the source of the raw material. In general rosin is a mixture of several isomeric diterpene acids. The three major components are abietic acid, d-pimaric acid, and l-pimaric acid.

Abietic acid (also known as sylvic acid) is a heteroannular diene with a melting point of 174°C (345°F; $[\alpha]_0 = -140°$; $\lambda_{max} = 2375$ Å). At 300°C (572°F) the abietic acid rearranges into neoabietic acid, with a melting point of 169°C (336°F; $[\alpha]_0 = +159°$; $\lambda_{max.} = 2500$ Å):

Abietic acid   300°C   Neoabietic acid

Upon further heating the abietic acid undergoes disproportionation into a pyroabietic acid mixture, both components of which are chemically inert.

d-Pimaric acid is another primary rosin constituent. It is a nonconjugated diene which is stable to acid. It has a melting point of 219°C (426°F; $[\alpha]_0 = +73°$).

l-Pimaric acid, a homoannular diene, is also a constituent of rosin. It is easily isomerized by acid to abietic acid. It has a melting point of 152°C (306°F; $[\alpha]_0 = -275°$; $\lambda_{max} = 2725$ Å).

d–Pimaric acid l–Pimaric acid

In an average gum-rosin analysis abietic acid makes up 80 to 90 percent of the rosin, with the pimaric acids ranging between 10 and 15 percent. The commercial designation of water-white rosin refers to a grade of the material determined by colorimetric methods (see ASTM Designation D 509-70).

From the above we can see that the components of the rosin mixture are rather temperature-sensitive and that overheating will change the chemical structure of the components. A decrease in the efficiency of water-white rosin due to overheating reported in the literature was confirmed by the author in a group of tests. The conclusions from the tests are summarized as follows:

1. Overheated rosin turns dark and loses most of its tarnish-removal properties.

2. Overheated rosin retains its surface activity and protects tarnish-free surfaces (not quite so much as unheated water-white rosin).

3. Heavy tarnishes are penetrated only slightly by unheated water-white rosin.

4. Water-white rosin reacts with copper oxides and sulfides to give a green copper abiet, which can be separated.

5. Water-white rosin does not reduce the weight of pure metallic copper, regardless of time and temperature of exposure.

6. No other metal abiets except copper could be separated in the laboratory, although they are known to exist.

It is very easy to see from the foregoing that water-white rosin is not a universal and strong tarnish remover, but when relatively clean surfaces are involved, the surface activity of rosin causes good wetting and spreading. Therefore, rosin makes a good vehicle for more active tarnish removers by providing the proper surface activity and protecting the metallic surfaces from fresh tarnish formation. The residue of rosin flux is a hard transparent film of excellent electrical-insulation properties which does not absorb water.

2-22 Color, Corrosion, and Fluxes   By definition corrosion is the slow eroding of solids, especially metals, by chemical action. In other words, during the process of corrosion a metallic surface is consumed by the chemical attack of foreign materials. The results of the chemical reaction are the corrosion products. This natural process accounts for the rusting of iron, the tarnishing of bright surfaces like aluminum, and the formation of a green patina on copper and its alloys, the latter caused by a reaction between the exposed base metal and air containing moisture and carbon dioxide. The reaction is

$$2Cu + O_2 + CO_2 + H_2O \rightarrow Cu_2CO_3(OH)_2$$
$$\text{From the air} \qquad\qquad \text{Green patina}$$

Therefore it is not surprising that whenever a green substance appears in electronics assemblies, it is immediately assumed to be a corrosion product. This is misleading because (1) many corrosion products are *not* green; (2) many reaction products of a noncorrosive nature *are* green.

The main danger of corrosion in electronic equipment is seldom associated with actual deterioration of the metallic conductors. Only in rare cases does the corrosive action of the chemicals present cause mechanical failure in the components because the consuming agents usually are not present in large enough amounts to cause such damage. The danger actually originates from the presence of small amounts of the corrosion products themselves. The nature of the chemical reactions involved in corrosion means that a large amount of ionizable material is always present, ions which are found in both the corrosion products and the corrosive chemicals. These materials react with atmospheric moisture to form conductive films containing ions, in turn causing current leakage across insulation surfaces. Finally, this leads to erratic equipment behavior and may result in a short circuiting and a major assembly failure.

In addition, corrosion products, being nonmetallic, can cause a different problem on make-or-break contacts. In their dry form, they act as an insulation when deposited on mating contact surfaces. Thus, relays, mechanical switches, and similar devices are often rendered useless.

The current-leakage and contact-contamination problems are always connected with small amounts of corrosion. It might seem to the novice that after a short period of corrosive action and erratic behavior, a major failure will follow, depending on the type of attack. This is seldom the case. Corrosion mechanisms fall into three major categories:

1. The *continuous corrosion process,* whereby a small amount of corrosive material attacks the metallic surface. A porous corrosion product is formed which hydrolizes in the atmosphere, again releasing the corrosive agent, which is capable of attacking fresh amounts of metal. This mechanism is typified by the attack of chloride ions on lead (see Sec. 1-13).

2. The *self-limiting type of corrosion,* in which the corrosive material attacks the metallic surface and forms a nonsoluble product, rendering itself inactive. A typical example is the interaction between chromic acid and copper. This process, in the form of a *conversion coating,* is utilized in the finishing industry to protect copper surfaces. The insoluble copper chromate makes a nearly invisible protective coating. Its presence, however, is felt when one attempts to solder through it, the chromate drastically decreasing the solderability of the surface.

3. The *self-limiting but nonhydrolizing type of corrosion,* where the corrosion, although soluble, has no active agent and is therefore not available for further corrosive action. The reaction between copper oxides and rosin is a typical example (of which more later).

It becomes obvious from these basic types of corrosive mechanisms that small amounts of corrosive materials can cause electronic leakage without causing major failures. Only the first reaction described above can continue attack till failure.

2-23 Fluxes and Corrosion   For years the industry has classified fluxes into corrosive and noncorrosive categories. In the light of this chapter, it is

TABLE 2-3   Common Color of Some Metallic Salts*

Metal	Color of salts	Metal	Color of salts
Aluminum	Colorless	Gold	Yellow
Antimony	Colorless	Lead	Colorless
Arsenic	Colorless	Manganese	Red pink
Bismuth	Colorless	Nickel	Green
Cadmium	Colorless	Platinum	Many colors
Chromium	Blue, green	Silver	Colorless
Cobalt	Red	Tin	Colorless
Copper	Blue, green	Zinc	Colorless
Iron	Many colors		

*H. H. Manko, Color, Corrosion and Fluxes, *Electron. Packag. Prod.*, vol. 9, no. 2, February 1969.

apparent that this classification is unscientific. To be effective, a soldering flux must clean the base metal surfaces chemically. Only on such surfaces can a metallurgical bond between the base metal and molten solder be effected.

Actually, in the context of our discussion, the corrosivity refers to the chemical danger of the material and its residues on the assembly after it has performed its fluxing function. In this respect, it would be more accurate to talk about the possible detrimental effect of a particular flux on the assembly under a given set of conditions. Rosin-base fluxes, for example, might not be corrosive to a television assembly operating in a household environment, but they might be very corrosive in a missile computer.

We have stated that a flux is, by nature, chemically active; this is particularly true if the flux system was improperly selected, applied, or soldered. Flux removal, when required, must be thorough to avoid problems.

Most people associate color with corrosion. This adds confusion because the color of the corrosion products on metal surfaces is often misinterpreted and confused with harmless materials. Some corrosion products are colorless; others show color only on drying (see Table 2-3). The author has conducted a series of tests on the reaction of fluxes with various materials designed to highlight color changes and the importance of proper interpretation.[1]

2-24 Color Identities   From this discussion it is apparent that no direct correlation can be drawn between colors of residues appearing on a soldered assembly and the type of corrosion mechanism present, if any. It

[1]H. H. Manko, Color, Corrosion and Fluxes, *Electron. Packag. Prod.*, vol. 9, no. 2, February 1969.

is necessary to examine completely the full history of the parts. Only then can we draw a meaningful conclusion about the possible danger of surface residues and their coloration. Green residues do not necessarily mean a lowering of surface resistivity or the presence of ionizable corrosion products. On the other hand, green residues may mask the presence of blue residues, which could contain ionizable salts of copper. White residues may indicate the presence of chloride-containing salts, which continuously react with tin, lead, or alloys like solder. Nor does the absence of color indicate that the surfaces are free of ionizable materials; they might still conduct current.

The color of corrosion salts depends on the type of ions they contain. Not all metallic salts have color. This demonstrates once again that the presence or absence of color does not necessarily indicate the presence or absence of corrosion products. This is one reason why a correlation between color and corrosion products is difficult to make unless the complete history of the part is known. Table 2-3 refers only to inorganic salts; the color of organic materials is more difficult to classify. However, organic materials also contain metallic ions in their composition, tending to follow the same color pattern as inorganic materials.

THREE

The Metallurgy of Solder

THE THEORETICAL BACKGROUND OF
TIN, LEAD, AND THE PHASE DIAGRAM

3-1 An Introduction to Metallurgy The term *metal* is familiar to everyone, but many people would have difficulty defining it other than by specifying that a given material possesses some metallic characteristics. More accurately, a metal is "a chemical element which is lustrous, hard, malleable, ductile, and usually a good conductor of heat and electricity." Of all the elements, 73 are classified as metals. More sophisticated definitions of metals are available but are beyond the scope of this book.

The atoms in a metal are arranged in a definite order and pattern which repeat themselves in all three directions. This is typical for all crystalline materials. The orderly arrangement of the atoms in the crystal is also referred to as the *space lattice*. The distance between the atoms is measured in angstroms and depends on the individual material and its temperature; an angstrom (Å) is a unit of length equal to 1×10^{-10} m or 4×10^{-9} in. Most metals in everyday life consist of many crystals and are called *plolycrystalline*. Only under special conditions can a single crystal of metal be obtained. The polycrystalline structure is the result of how metals solidify from the molten liquid. As the liquid is cooled, many nuclei of crystals form and start growing simultaneously. As these solid nuclei grow, they consume the liquid metal until the walls of the expanding crystals (in all three dimensions) meet each other at *grain boundaries.* Because the metal nuclei had different orientations in space at the start of growth, they will not meet at orderly interfaces; hence the grain boundaries have an uneven configuration which is entirely unpredictable. By metallurgical processes such as annealing, it is possible to change the grain size and structure.

In everyday life, most metals are not used in the pure state but are combined with other metallic elements to form an *alloy.* Hence an alloy can be defined as a combination of elements which exhibits a property of a metal. By combining metals in various proportions (alloying them) it is possible to obtain a material with specific required characteristics. The properties of an alloy containing two or more elements differ appreciably from those of the parent metals. Alloys are generally divided into *ferrous* and *nonferrous* alloys, the first including iron-base materials and the second non-iron-containing alloys, which include soldering alloys. Other subdivisions of alloys according to their physical properties or prices are also found in literature, e.g., fusible alloys or precious alloys.

Alloys are usually formed by a solution process. The metals dissolve in each other very much like water and alcohol to form a single *phase.* A phase is a distinct homogeneous portion of the system, and regardless of where a sample is taken and analyzed, the composition will always be the same. In other cases, however, the metals might have only partial solubility. Then the two elements will be present in two separate layers, very much like water and oil, to give two distinct phases. Some alloys are soluble in the liquid state but segregate upon cooling and solidification to form a mixture which depends on the method of cooling. Finally, it is possible for two materials that dissolve in the liquid state to form *intermetallic compounds* upon cooling. These intermetallic compounds are unique inasmuch as they do not always follow chemical-valance rules like those followed by chemical compounds.

An intermetallic compound can be defined as a distinguishable homogeneous phase having a relatively narrow range of compositions with simple stoichiometric proportions. The intermetallic compounds can be metallic in nature or have an ionic atomic binding. The structure and nature of the intermetallic components depend on the atomic radii and the electronic activity of the two or more metals which form the compound. Intermetallic compounds can often be seen in metallographic examination of alloys and/or by defraction techniques. They will be discussed separately for alloy selection and design purposes as well as for their influence on the resistivity and strength of the solder.

When the elements combined in an alloy dissolve in each other in such a fashion that the atoms of one become part of the space lattice of the other element, we have a true *solid solution*. The solid solution differs from the intermetallic compound inasmuch as the ratio between the two metals is not fixed and a specific number. The range in which two elements form a solid solution is temperature-dependent. Often a material which solidified as a solid solution undergoes precipitation (the separation of a phase which can no longer stay in the solution) upon further cooling as it exceeds its solid-solubility limits. For further details, see the next section.

Two methods of strengthening of alloys are important in soldering and primarily in the tin-lead system. These are work hardening and solution hardening.

Work Hardening. When a metallic lattice is deformed because of some physical stresses, the alloy becomes stronger and harder. This is called *work hardening*. To relieve this stressed condition the metal has to be heated to a specific temperature for a period of time, a process called *annealing* or *recrystallization*. Once the stresses are removed, the alloy resumes its original strength.

For most soldering alloys, the annealing temperature or the recrystallization temperature is below or near room temperature. Thus, the stresses in the average solder systems do not cause work hardening.

Solution Hardening. When small amounts of alloying substances are added to a metallic lattice, they can cause internal stresses similar to those in work hardening (provided the atoms are different from the lattice atoms). Such strengthening of the alloy is called *solution hardening*. Neither annealing nor other treatment will affect the increased strength of the alloy. Antimony will affect the tin-lead alloys this way.

This hardening is a desirable effect in soldering. A certain amount of the base metal is usually dissolved in the solder, increasing its strength. For further details see Sec. 1-11.

3-2 The Phase Diagram   In order to understand the properties of an alloy better, let us look at some of the basic changes which occur within the system as the temperature and composition change. This type of information is usually presented in a *phase diagram* (see Fig. 3-1). Let us go briefly over the theoretical considerations to see its importance in the field of solder metallurgy.

For simplicity let us look at a two-component, or a *binary, alloy system*. Under any set of conditions, this binary system may have one or more homogeneous structures which are stable and unique for those conditions. For instance, it is possible that metal A will dissolve part of metal B in its lattice and metal B will dissolve part of metal A in its lattice; thus we have two solutions coexisting with a specific and uniform structure. Each

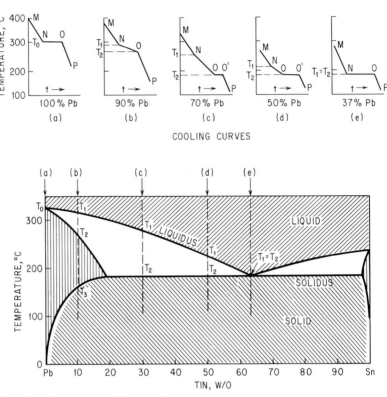

Fig. 3-1 Construction of tin-lead phase diagram.

of these portions of the material is called a *phase* and is physically a distinct entity. We say that the system has two phases. A multicomponent system might have one or more phases without any limitations. Thus a solid solution might coexist with an intermetallic compound and a second solid solution to form a three-phase system. The only requirement in all these cases is that the phases be at equilibrium under the specific conditions.

One additional term should be clarified, namely, the eutectic composition (or eutectic point or eutectic temperature). In binary alloys, for instance, it has been noted that certain compositions of the two elements have a distinct melting point lower than those of the parent metals. This composition is called the *eutectic composition,* and the temperature at which this specific mixture of metals goes from a complete solid to a complete liquid or from a complete liquid to a complete solid is termed the *eutectic temperature* or the *eutectic point.* Each system can have one or more eutectic compositions, and there are also eutectic compositions for multiphase alloy systems.

The eutectic composition behaves as an independent homogeneous phase and has a unique metallographic structure, shown in Fig. 3-3. In effect, a eutectic composition can be described as a mixture of two or more phases which have a distinct melting point where the two separate phases are transformed into one liquid solution. Upon cooling, the reversible process will give the two distinct solid-solution phases in an intermixed structure which is unique for eutectic compositions.

As we raise the temperature of a crystalline metal, the distance between the atoms in the lattice increases. The atoms are said to be *thermally activated*. Once this activation energy of the atoms exceeds the forces holding them in the solid lattice, the metal undergoes a change in its state of aggregate from a solid to a liquid. Thus the liquid metal consists of a nonorderly arrangement of thermally activated atoms, which follow different physical rules from the solid crystalline metal. Once the thermal activation is removed from the liquid metal (cooling), the atoms will coalesce again to form a lattice structure as described previously.

The behavior of metals in cooling can be recorded in the form of cooling curves. The method involves the slow cooling of various composition alloys in the binary system at a controlled rate. The temperature changes against time are recorded. The rate of cooling for both the liquid and the solid alloy will be represented in a curve by a line with a specific slope which depends on both the rate of cooling and the specific material employed (this is not truly a straight line). At the point of solidification, a noticeable change in the slope of the curve is recorded because the heat of fusion has to be dissipated before the alloy can lower its temperature.

For pure metals where the temperature of solidification is constant, the curve looks like Fig. 3-1a. Here the section of the curve MN represents the cooling rate of the pure liquid lead, the horizontal section NO represents the freezing- or melting-point temperature of lead (327°C, or 620°F), and the sloping section OP represents the cooling of the solid lead. We have referred to the term *heat of fusion* of lead as the number of calories which have to be dissipated out of liquid lead before it solidifies. The heat of fusion is therefore also the amount of heat necessary to change solid lead at the melting temperature of 327°C into liquid lead at the same temperature. For lead, the heat of fusion is 11.3 Btu/lb. For tin, the heat of fusion is nearly double, namely, 26.1 Btu/lb. This is an appreciable amount of heat, and it accounts for finding both liquid and crystalline lead coexisting at equilibrium.

Figure 3-1b shows a cooling curve for an alloy of 90 percent lead and 10 percent tin. Here MN and OP are the cooling curves for the liquid and the solid alloy, respectively. The curve NO represents the temperatures at which solid crystals of lead-rich composition form. The process starts at temperature T_1, and at temperature T_2 all the alloy has solidified. Note that another transformation occurs at a lower temperature. The information which can be derived from this change is not significant enough to be used later and is therefore neglected.

Figure 3-1c shows an alloy of 70 percent lead composition. This alloy upon cooling has the typical MN and $O'P$ cooling curves for the liquid and solid alloy, respectively. However, the curve NO is similar to that of Fig. 3-1b discussed previously. Here the lead-rich alloy starts crystallizing out of the liquid melt, and by the time we reach point O at temperature T_2 a unique change occurs. At that temperature, all the remaining alloy crystallizes out in the eutectic form, as discussed earlier in the chapter. This crystallization of the eutectic closely resembles the crystallization of the pure lead. Here the line OO' is horizontal. This fact is even more pronounced in Fig. 3-1d, where the horizontal line OO' is much larger, indicating more eutectic-alloy formation.

Figure 3-1e is exactly like Fig. 3-1a inasmuch as it has MN and OP curves representing the cooling of the alloy of 37 percent lead and 63 percent tin with a horizontal line NO indicating the formation of a material which we have already established as the eutectic. The eutectic behaves exactly like a pure metal having a definite solidification temperature and a specific heat of fusion.

The behavior of the alloys from this point on as we proceed to the tin-rich region is very much a mirror picture of the alloys that we have seen in Fig. 3-1a to e (the lead-rich region).

The information to this point has been presented in individual cooling-curve form. It is possible, however, to present all the temperatures where a change in the cooling rate has occurred in a composite picture of the alloy composition vs. the temperature in the form of a *phase diagram* as presented on the bottom of Fig. 3-1. If we take the points in Fig. 3-1a, we shall find that there had been a specific cooling-curve change at the melting point of pure lead at 327°C. This was presented on the cooling curve as a horizontal line and is presented as a point on the phase diagram. Let us go to Fig. 3-1b. Here we had a temperature change at point N corresponding to T_1 and at point O corresponding to T_2. Both these represent the complete liquefaction and the beginning of melting, respectively, and can be presented on the phase diagram as shown in the lower part of the figure. A third transformation at temperature T_3 was noted, which corresponds to a change in the crystallographic structure of the solid state. Let us plot this temperature too. Figure 3-1c yields complete liquefaction temperature T_1 and a lower solidification temperature T_2, which are plotted on the phase diagram, as does Fig. 3-1d. Finally, in Fig. 3-1e we find the eutectic-temperature transformation of the system from a complete liquid to a complete solid at one temperature. Note that the temperature of the eutectic transformation is the same temperature as in Fig. 3-1d and c. We are now ready to complete the left side of the phase diagram by connecting all the T_1 temperatures and all the T_2 temperatures in the graphs as presented in the lower part of Fig. 3-1. More information is required to get the phase diagram in exactly the form shown in Fig. 3-1. In practice, it is customary to have cooling curves of many more alloys in order to have all the transformation temperatures which would give us an exact phase diagram. The right-hand side of this phase diagram can be obtained in a similar fashion.

Let us discuss in detail the various areas and lines which are found in the phase diagram. Starting from the top, we have the liquid region. The line which makes up the lower boundary of the liquid region is called the *liquidus*. The two triangular areas immediately beneath the liquid region represent the mushy stage, where the material has solid crystals floating in a liquid at equilibrium for that particular temperature. The lower border of this mushy region is the *solidus*, which in the tin-lead diagram lies between the melting points of the pure elements and 361°F (183°C). This is the temperature at which all the material changes into solid upon cooling or upon heating starts forming a liquid phase. The area under the solidus is completely solid. The two triangular areas on both sides of the phase diagram are the regions of solid solution. The lower boundary of the solid-solution region is the temperature at which the specific composition of the alloy changes from one solid homogeneous phase into two phases. A detailed account of the changes in the phase diagram and their corresponding importance to the tin-lead system is given in Sec. 3-7.

3-3 Tin, Lead, and Their Alloys   Both tin and lead have been known for thousands of years. *Oferet* (Hebrew for lead) and *bedil* (Hebrew for tin) are mentioned in the Bible. The Romans seemed to have been confused about the difference between tin and lead. They referred to tin as *plumbum album* in order to separate it from *plumbum nigirum*, which stood for lead. Whereas lead was used by itself and in various alloyed forms for the manufacture of artifacts, tin was mainly used as an alloying addition to copper to make the alloy bronze, which was fairly widespread in ancient times. Today both these metals are used in an ever-increasing tonnage throughout industry, and the tin-lead alloys are the most widespread solder alloys.

Tin ores are found in Australia, Czechoslovakia, Bolivia, Cornwall, Malaya, Nigeria, East Germany, South Africa, etc. Tin ores are usually found in what is called *tinstone* or *cassiterite,* which is a major source of commercial tin. This mineral, SnO_2, is usually found in crystals and is divided into *lode* or *vein tin,* which is cassiterite obtained from veins or lodes in primary deposits, and *stream tin,* which is cassiterite from alluvial secondary deposits which are usually rounded lumps. Small quantities of metallic tin are reported in various parts of the world, but these deposits are not large enough for production purposes.

Lead ores are usually found in nature as galena, PbS, which is the most common ore, although there are ores of lead carbonate, $PbCO_3$, called cerussite, and lead sulfate, $PbSO_4$, called anglesite. Lead ores are found in Australia, Belgium, Canada, Germany, Mexico, Russia, Spain, and the United States. The last is the world's largest producer of lead.

The extraction of tin from its ores is relatively simple. The ore is crushed and then concentrated by simple flotation methods, which are fairly easy because of the high specific gravity of tinstone (6.8 to 7.0). Unwanted impurities such as arsenic and sulfur are removed by an oxidizing roast and a dilute acid leach, while other impurities such as lead, bismuth, antimony, and silver are removed by a chloridizing roast and an acid leach. The ore is now ready for a reduction in either a blast furnace or a reverberatory furnace. The furnace is charged with the concentrated tinstone mixed with coal. The reduction reaction is

$$SnO_2 + 2C \rightarrow 2CO + Sn \qquad (3\text{-}1)$$

The molten tin collects on the bottom of the furnace and is drawn and cast directly into block tin. The average purity of the tin in this form is 99.5 percent. The tin is further refined by various methods.

Lead can be extracted by open-hearth smelting, by reverberatory-furnace smelting, and by blast-furnace methods. The most common today seems to be the last method, where the ore is subjected to preliminary roasting for the purpose of sintering the previously concentrated ore. During this roasting operation, most of the ore is converted into oxides. The oxidized ore is then charged together with the fuel into the furnace, where it is reduced by the carbon as well as by some of the unconverted lead sulfite. The quality of the lead obtained in this process is relatively poor, and further refining is necessary to remove the antimony, tin, copper, and other impurities which make it hard and brittle. The presence of relatively large quantities of silver makes this additional refining profitable because of the recovery of the silver.

3-4 The Element Tin (Stannum)   Tin is a silver-white lustrous metal. It resists oxidation well and maintains its luster on exposure to air. The metal is harder than lead but is soft enough to be cut with a knife. The ductility and/or malleability of tin is great, and it can be rolled and extruded as well as drawn into wire. Tin has a relatively large crystal structure. When a bar of tin is bent, it emits a unique sound called *tin cry*, assumed to be the result of friction of the crystal interfaces on each other.

When cooled to low temperatures, pure tin transforms from a white metallic material into a gray amorphous powder. In the past this transformation was called *tin pest,* because tin articles for no explainable reason crumbled into dust. The transformation temperature has been scientifically established at 13.2°C (55.7°F).[1] This transformation is still under study, but it has been shown to have a heat of transformation of 4.2 cal[1] and other specific properties of its own, e.g., entropy and specific heat, which can be measured precisely. In solder, this transformation of tin into gray tin has been always considered a potential danger, especially as it appears that the transformation is contagious. In other words, nuclei of gray tin will accelerate the formation of gray tin in untransformed materials. The addition of various alloy elements greatly reduces the danger of the tin pest, and the federal government specification QQ-S-571 calls for the addition of a minimum of 0.25 percent antimony as a preventive for the formation of gray tin.[2]

Tin is not attacked by water and/or air separately or combined. This makes it desirable for protective coatings. However, the presence of chlorides in marine water will facilitate the formation of stannous chloride. For further information on tin, see Sec. 6-13.

3-5 The Element Lead (Plumbum)   Lead is a bluish-gray metal with a bright metallic luster when the surface is freshly exposed. In ordinary air the surface deteriorates rapidly, taking on a dull gray appearance. This tarnish is very tenacious and protects the metallic surface from further environmental attack, which is why lead articles were preserved for thousands of years in the ground or other relatively corrosive environments. Lead is a very soft metal with great ductility and can be easily formed. It is so soft that when it is drawn across a smooth surface like paper, it leaves a trace behind. Hence the name *lead* is used for the core of a pencil, although today it is essentially made of graphite.

Although lead is very soft and in its pure state can be scratched with a fingernail, small additions of impurities such as antimony, arsenic, copper, zinc, etc., make the lead much harder. Small traces of these elements are enough to increase the hardness and strength appreciably. In addition to being soft, lead is one of the heavier metallic elements with a relatively low melting point and a high density. This high density is best utilized in shielding applications of radiation such as x-ray and nuclear energy. However, its most unusual property is its high resistance to corrosion under a wide variety of conditions.

As indicated earlier, the metal surfaces under normal environmental conditions tend to tarnish rapidly, giving a tenacious protective layer, which prevents further attack and gives lead its unique resistance to many

[1]"The Properties of Tin," The Tin Research Institute, 1962.
[2]For specific information see Jack Spergel, The Transformation of Tinned Copper Wire, *ASTM 65th Annu. Conv. 1962;* W. Lee Williams, Gray Tin Transformation in Soldered Joints Stored at Low Temperature, *ASTM Spec. Tech. Publ.* 189, 1956.

chemical and environmental corrosive atmospheres. It would be more accurate, therefore, to say that the metallic lead itself is prone to corrosion. Once a layer of corrosion products, whether chromate, carbonate, oxide, sulfate, or any other corrosion product, is formed, the attack on the metallic lead ceases and the corrosion layer acts as a protective coating. As long as these coatings are tenacious and nonsoluble in the liquids around them, the metal surfaces underneath can be preserved for prolonged periods. Most tarnish layers on lead are not water-soluble; hence lead can be satisfactorily used in most environments and in water. Depending on the exact compositions of the marine atmospheres, lead can be used with a minimum amount of loss. Lead is attacked, however, by water containing air, nitrates, carbon dioxides, and ammonium salts. For the mechanism of corrosion with chlorides see Sec. 1-13.

3-6 The Tin-Lead Alloy System   To many people the term *solder* has become synonymous with the tin-lead alloy system. There are many other alloy systems and alloying additions to the tin-lead system itself, but since most soldering alloys actually are binary alloys of tin and lead, the tin-lead alloy system is discussed separately and thoroughly in this section. Figure 3-34 shows the variety of soldering alloys which can be used, and the shaded portion indicates the small temperature range in which all the tin-lead alloys fall. This narrow temperature range satisfies the requirements of most everyday soldering needs. Another important factor in their popularity is the relatively low price of tin-lead solders in relation to their wetting and strength properties.

Although a large amount of work has been done in the general investigation of the tin-lead system and the properties of the various alloys, it is difficult to find correlation between the figures published by the various workers. This can generally be attributed to the degree of purity of the alloys and materials used by these investigators. Most of this work was carried out when little or no attention was paid to minor alloying constituents and to the general condition of the sample. Recent advances in analytical methods and especially in spectrographic techniques make it possible today to gain much more information about the specimens under investigation than was possible when most of these studies were performed. Vacuum-melting techniques for the degassing of alloys and the removal of included oxides and nonmetallic impurities have also been developed only in recent years.

In a research study aimed at measuring certain physical parameters such as resistivity of tin-lead alloys and the effect of impurities on them the author discovered the following interesting facts.[1] Calibration of the

[1] H. H. Manko, The Effect of Some Metallic Contaminations on the Tin-Lead Solder System, IBM TR 00.717, Poughkeepsie, N.Y., May 2, 1960; also The Effect of Some Metallic Additions on the Tin-Lead System, thesis, New York University, December 1959.

equipment for resistivity measurements on pure lead and pure tin proved difficult because the results were not reproducible. The raw materials used were of 99.999 percent purity and were the best obtainable at the time. Only after these metals were vacuum-melted for degassing and also for oxide and inclusion removal was it possible to get repeatable measurements of the samples.

The inclusions and oxides present in the lead were removed by repeatedly vacuum melting the lead specimen of 99.999 percent purity and cropping the head of the ingot. After three treatments, the alloy developed no further scum on the surface. The volume of the inclusions constituted nearly 2 percent of the volume of the specimen.

Thus it becomes apparent that although many technical data are available on the tin-lead system, the exact figures reported by various observers cannot be used without further investigation of the purity of the samples they used and correlation with the type of material to be used. In addition, the method of sample preparation and sample testing causes appreciable differences in results. For further information see the discussion of metallurgical and physical testing in Chap. 8.

3-7 The Tin-Lead Phase Diagram   As described earlier in this chapter, the phase diagram enables us to study the interrelations at equilibrium of the various phases of the alloy system for different alloy compositions at various temperatures. It is therefore the first step in the fundamental understanding of the specific characteristics of the tin-lead alloy system. The components of this system are completely soluble in the liquid state but only partially soluble in the solid state. This is the most common type of binary system and is the same as the systems of copper-silver, copper-tin, copper-zinc, aluminum-copper, aluminum-magnesium, etc.

Let us first discuss a simple diagram showing the phases only. Figure 3-2 shows the phases, which are identified by a Greek letter and by regular metallurgical nomenclature. A certain amount of confusion in the way this phase diagram is presented in the literature originates from the two terms used for this important phase relation. Because the metallurgical convention dictates that the elements be written in their alphabetical order the system should be called the *lead-tin system* if the diagram is presented in the scientific manner. However, in industry, solders are called *tin-leads,* the opposite of the metallurgical convention. The solder phase diagram is therefore a mirror image of the scientifically correct diagram. The reader is urged to disregard a specific familiar arrangement and always to check the bottom of a phase diagram to make sure that the tin-lead ratios are correct.

An additional point should be clarified here concerning the units used for the percentage ratio of one metal in the other. The metallurgical convention dictates that the phase diagram should be presented in atomic percent, which is important in the interpretation of the diagram using the lattice structure. In solders, however, the relation is usually presented in weight percent. Many books compromise by showing the atomic percent on the bottom and the weight percent on the top of the diagram or vice versa (see Fig. 3-2). Also, it is relatively simple to convert atomic percent into weight percent according to the following formula for a binary system:

$$y = \frac{100x}{x + (A/B)(100 - x)} \tag{3-2}$$

where y and x represent the atomic or weight percent of a metal with the atomic weight A. B is the atomic weight of the second metal.

Starting from the top of the diagram in Fig. 3-2, we see that the region marked I is completely liquid. In other words, any composition of the tin-lead system will be totally liquid and in complete solution at the temperatures above line ABC. This ABC line is referred to as the *liquidus* of this system and will go from the melting point of tin (449°F) at point A to the eutectic temperature of the tin-lead system (361°F) at point B to the melting point of lead (621°F) at point C. Regions II and III in the diagram marked β (beta) and α (alpha), respectively, are areas of solid solution. When an alloy of a given composition and a given temperature falls within the regions of α and β, it will have a single phase. Lines AD and CF, which form the upper limits of these regions, are called the *solidus*, as is line DF. Any material underneath solidus A, D, F, C will be a complete solid. The material in area IV consists of a combination of β and α in various ratios. The curvature of lines FG and DE indicates that the amount of solute

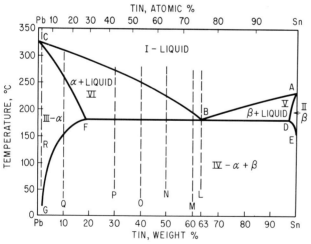

Fig. 3-2 A simplified tin-lead phase diagram.

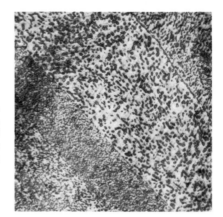

Fig. 3-3 Eutectic tin-lead solder (63/37) ×200 magnification. Note the fine intermixed structure of dark lead-rich α particles and the light tin-rich β particles. This is a typical eutectic structure.

(primary element) in the solid solution of regions II and III diminishes with cooling and that area IV consists of a double phase where α and β coexist in various configurations, as we shall discuss later. Regions V and VI, as shown in the diagram, lie between the solidus and the liquidus of the system and therefore consist of β plus liquid and α plus liquid, respectively. These regions are usually labeled the *pasty range*. As solid crystals of β or α are dispersed in a liquid solution of tin and lead, and as the temperature is lowered, the liquid can no longer dissolve the large quantity of the second phase and more crystals of β or α are formed until the eutectic temperature is reached, where the balance of the material freezes totally in a mixture of α and β.

Let us follow the cooling characteristics of various tin-lead alloys and see how the phase changes with temperature (Fig. 3-2). Starting from line L, we can cool an alloy of this eutectic composition from a complete liquid solution into a solid material where β and α are intermixed in the typical eutectic structure (see Fig. 3-3). Here the transformation of the single liquid phase into two coexistent phases takes place at the eutectic temperature of 361°F (183°C).

Let us consider another alloy M with a composition of 60 percent tin and the balance lead. Here, upon cooling of the single-phase liquid solution to a temperature of approximately 372°F (189°C), a segregation of solid crystals of α (lead solid solution) occurs. This, in the overall picture, leaves a more tin-rich liquid behind. Upon further cooling, additional α is segregated, and the liquid which is left tends to become more and more tin-rich until the eutectic composition is reached at a temperature of 361°F. This eutectic tin-lead liquid freezes completely with time. At this point, we go from the pasty range, region VI, into the solid range, region IV, where crystals of α are suspended in a matrix of eutectic α plus β in the alloy as it solidifies (see Fig. 3-4). The same will hold true

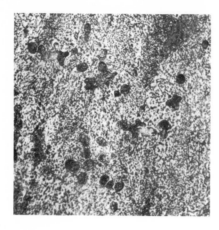

Fig. 3-4   Tin-lead 60/40 solder, ×200 magnification. Note particles of dark lead-rich α scattered in background of eutectic matrix.

for 50/50, 40/60, and 30/70 lines N, O, and P, respectively (see Figs. 3-5 to 3-7).

Let us go to even higher lead concentrations, as in line Q in Fig. 3-2, where we have a 90 percent lead, 10 percent tin alloy. Here we go upon cooling first into the pasty range, where the alloy at a temperature of approximately 570°F (299°C) starts forming crystals of α. Upon further cooling at a temperature of approximately 527°F (275°C) the alloy totally freezes in the form of α phase, and in the temperature range of 527 to 294°F (275 to 195°C) we have a single phase of α solid solution. However, at this low temperature of 294°F, the solid solution no longer can hold all the tin, and upon further cooling we get a formation of the β phase from the α and go into the two-phase region IV of β and α coexistence.

Only in the case of line R in Fig. 3-2 with a 98 percent lead, 2 percent tin composition would we have a single phase of α even at room temperatures because the amount of tin (2 percent) can stay in solid solution even at room temperatures.

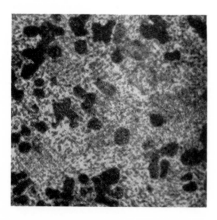

Fig. 3-5   Tin-lead 50/50 solder, ×200 magnification. Note more lead-rich particles than in Fig. 3-4 dispersed in eutectic matrix.

Fig. 3-6 Tin-lead 30/70 solder, ×200 magnification. Note large amount of α phase and the less refined eutectic structure.

Several points should be stressed. The temperature of 361°F (183°C) is not the only one which was reported for the eutectic transformation. Serious problems with supercooling and superheating have made the determination of this temperature difficult. In addition, the exact composition of the eutectic, although it is considered for practical purposes at 63 percent tin and 37 percent lead, has never been accurately established as such. However, for the purposes of this book and all practical engineering considerations, these two figures for temperature and composition can be used successfully without any noticeable errors.

If lines M to R on Fig. 3-2 are on the other side of L through regions V and II rather than regions VI and VIII, respectively, the situation will be the same because the second side of the phase diagram is similar to the first. Figures 3-5 to 3-7 show the corresponding micrographs.

The micrographs shown (Figs. 3-3 to 3-7) were prepared as follows. A 1-lb bar was poured at 100°F (37°C) above the liquidus temperature into

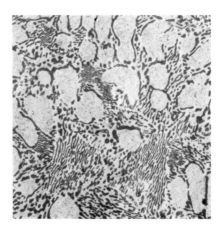

Fig. 3-7 Tin-lead 70/30 solder, ×200 magnification. Note light β phase of tin-rich particles in eutectic matrix.

room-temperature molds. The new bar was then annealed at 300°F (149°C) for 8 h to eliminate any internal variation. The sample was then cut into sections, which were ground first on a 180-grit belt sander and then on 240-, 320-, 400-, and 600-grit wet silicon carbide papers with intermediate etching with a 50/50 acetic and nitric acid etch solution. Polishing was carried out in two stages using alpha aluminum no. 2 (0.3 μm) and then Linde-B (0.1 μm). The final etchant was a glycerin acetic acid (1 part nitric acid, 1 part acetic acid, and 8 parts glycerin).

The phase diagram described in Fig. 3-2 gives only the theoretical background for understanding the tin-lead system. For engineering purposes, this diagram lacks essential details which would give the information vital for soldering considerations. Figure 3-8 takes the place of this diagram for all engineering considerations. The melting temperature of the solder alloy is very important in many selection considerations, as shown elsewhere in this book. However, an alloy cannot necessarily be soldered at its melting point. At or slightly above its melting temperature the alloy is still sluggish, does not flow very easily, and seems to have some restraint in its wetting characteristics. It is therefore a good rule of thumb to use a range of approximately 60 to 160°F (15.5 to 71°C) above the melting point for the soldering operation. This range is suitable for most alloys. To be more specific in the case of tin-leads, the use of an approximately 100°F thermal gradient between the melting point and the soldering temperature of the alloy is recommended. The smaller temperature differentials are meant for low-melting-point alloys.

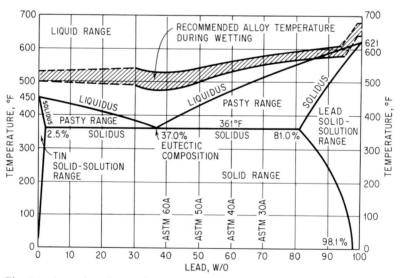

Fig. 3-8 An engineering version of the tin-lead phase diagram.

Figure 3-8 shows a band of temperatures recommended for soldering in the tin-lead system. It can be seen that for the solder compositions normally used for electrical soldering, the temperatures fall within the recommended rule of thumb. The higher-melting-point solders with a higher lead content are used not only for electrical joints but also for mechanical constructions. Wiping or body solders are used close to the melting point, where the mushy and pasty properties of the alloys are utilized for the manufacturing process.

It is usually good practice to start with as low a soldering temperature as possible and raise it only if it is not suitable for the soldering operation. In addition, it should be kept in mind that the temperature recommended for soldering still does not always give the temperature of the soldering tool itself. The temperature the joint will reach during the soldering operation is the mean temperature, which lies between the temperature of the tools and that of the work and is the temperature that should be considered.

This higher temperature is necessary to ensure proper flow of the solder and also to take care of all heat losses to the surfaces being wetted. In special cases soldering can be accomplished near the liquidus temperature, but time of soldering and flux activity usually dictate higher temperatures.

Let us discuss simply how we can use the information on the specific regions in the tin-lead solder phase diagram. In soldering we complete a cycle in the solder alloy from a solid to a complete liquid and back again to a complete solid. Depending on the composition and the phase changes, we get various properties in the joint. For the eutectic solder, as we have shown earlier, we get at a specific temperature of 361°F a total change from solid solder to liquid solder. Again, once the solder cools after the joint has been properly made, we get the reverse, a complete change from a liquid solder to a solid solder at the same temperature. However, in the phase diagram the important element time has not been considered. There is no place to indicate on this diagram how long a transformation from one phase to another will take at a specific temperature. While it is possible to develop a time relationship for a specified mass under controlled heat-loss conditions, it would be of little practical use because the mass and general conditions of heating and cooling in real life depend so much on outside factors that they cannot be included in a scientific diagram. To demonstrate this point, consider a very hot soldering iron (large heat source) and a drop of solder (small mass) going through a heating cycle. It is apparent that the solder will melt and heat up to high temperatures within a short time and that once the iron is withdrawn the small quantity of solder will cool down in a relatively short time. However, a big solder pot loaded with hundreds of pounds of solder will take much longer to heat up past the solidus and liquidus temperatures. Once the heat source is cut off, the solder pot will take a long time to cool.

In the heating cycle of the solder joint, the major factors are the temperature of the parts before soldering; their heat content and conductivity, or in other words their heat-sink effect on the joint; and finally the amount of caloric heat required to activate and heat the flux, heat and melt the solder, and effect good wetting. These temperature requirements must be at equilibrium with the amount of heat supplied by the soldering tool.

In the cooling cycle the picture is slightly different. The soldering tool has been removed, and the heat dissipates into several places, e.g., the assembly itself (this is similar to the heat-sink effect) or the room environment. This overall condition will determine the rate of cooling. The net result is a temperature-time relationship for each specific configuration.

Let us consider a solder alloy with a pasty range. Not very much can go wrong if the heating cycle is prolonged and the soldering temperature is reached after unduly long times. Cooling presents a different picture, however. It is absolutely necessary for the surfaces to be joined to be fixed in space relative to each other. Any movement of the surfaces will disturb an otherwise homogeneous joint. The first crystals of the phase, which are formed in the pasty range, usually settle close to the metallic surfaces to be joined because a rather sizable heat loss occurs along these surfaces. Should the surfaces move relative to each other, the liquid staying behind can no longer freeze in a continuous fashion and a jagged array of crystals

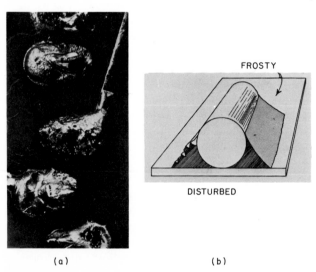

(a)   (b)

Fig. 3-9   (a) Typical appearance of disturbed, or cold, joint shows grainy solder and rough surface. (b) Artist's conception of the same condition.

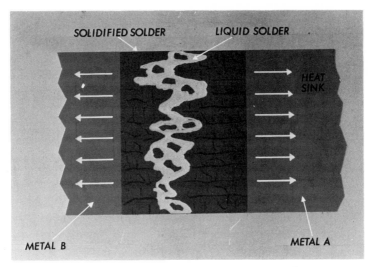

SOLIDIFIED SOLDER   LIQUID SOLDER

HEAT
SINK

METAL B   METAL A

Fig. 3-10   Schematic illustrating the formation of a disturbed joint when the two surfaces of metal A and B vibrate while the solder joint in the center freezes.

which segregated in the pasty range will result. When total freezing occurs, the eutectic liquid that stayed behind is no longer capable of bridging all the intercrystalline spaces which have been formed, and the result is a frosty-appearing solder connection. In its ultimate condition, such a connection can conceivably have extended cracks and discontinuities which might cause it to be a fractured joint. These connections are called *disturbed joints,* a very appropriate name. In the past, they were erroneously termed *cold joints,* which is very misleading. Figure 3-9 shows such a connection, while Fig. 3-10 is a schematic of the cooling surfaces and the formation of the final interface, which may not be disturbed by vibrations.

In wiping, the situation is completely different. The solder joint requires a solder which is in the pasty range, where it can be applied and wiped along the work. The longer the alloy stays in this physically workable condition, the better the alloy composition for the wiping operation.

The tin-lead phase diagram indicates only the actual changes in pure tin-leads. It does not give any information about alloying conditions and/or impurities in the binary system. Further information can be acquired only with three-phase diagrams, which are not suitable for a book of this nature because of their complexity. However, we shall discuss some contaminants of the tin-lead solder system later in this chapter.

Before closing this discussion, we should stress that all the conditions represented on the diagram are conditions of total equilibrium. It is sometimes possible, by extremely fast heating or extremely fast cooling, to have superheating and supercooling. These are the scientific terms for the possibility, because of the speed of the operation, of having an alloy reach a temperature not shown in the phase diagram before a phase change. However, these temperature and time conditions do not usually prevail in soldering and are simply nonequilibrium conditions of little importance.

FOREIGN MATERIALS IN THE SOLDER ALLOY

3-8 The Types of Contamination in the Tin-Lead System Contamination in the tin-lead solder system refers to metallic alloying additions which are found in the material because of its origin rather than artificial additions introduced to obtain certain properties. The impurities can result from the ores used as raw material or can come from contamination picked up during the soldering process. A solder pot can be contaminated by the work which is passed through it. Many contaminants, e.g., silver and bismuth, are not specifically detrimental in small quantities and therefore are only an indication of the purity of the solder. Other contaminants, e.g., aluminum and zinc, are very detrimental to the performance of the alloy and are definitely an indication of how well the material would perform in service. Nonmetallic impurities such as sulfur are also dangerous to the solders since they can cause certain dewetting conditions of both the tool and the work. Finally one class of contaminants has received more attention recently, namely, oxides and sulfides of tin and lead, which are present in the form of inclusions in the bulk of the material.

3-9 Grades of Solder as a Function of Purity To protect the consumer, many government and industrial organizations have issued specifications to cover the purity of solder in order to set up safeguards. The level of contamination of the solder is an indication of the source of the material, and the price increases with the purity of the alloy. Let us look at the three basic grades of solders and the specifications covering them.

Reclaimed and Refined Alloys. The recycling of tin and lead is important in the solder industry. Scrap tin and lead are collected and segregated to be reclaimed or refined metallurgically. Dross, drippings, clippings, and discarded contaminated solder have an intrinsic metal value and should be sold for a substantial fraction of the new-metal cost. To obtain the best price for the solder scrap, the user must keep it segregated from copper-wire clippings, other metallic scrap, and foreign materials, since scrap is sold by weight.

Refining must be differentiated from reclaiming. Scrap metal is normally melted down and analyzed for its impurity content. Contaminating elements are classified for their intrinsic metal value. Thus low-grade scrap containing only such impurities as copper, zinc, and iron is destined for reclamation, while high-grade used solder containing precious metals, e.g., gold and silver, is earmarked for refining.

In the process of reclamation, little if any metallurgical purification takes place. Although the purchase price of scrap metal is extremely favorable, it is not economically feasible to refine such alloys because the refining costs are higher than the value of the combined materials. Some impurities, e.g., copper, can be removed by treating the melt with elemental sulfur. After the chemical reaction is complete, the copper reaction products and excess sulfur are skimmed off with the dross. This process requires extensive pollution-control equipment, however, and cannot be carried out by the solder user. In addition, we must remember that solder is poisoned by the presence of sulfur, which should be kept below the 7 ppm range. Thus special sulfur-removal treatment of the tin-lead alloy is required if the solder is not to lose much of its wetting power. This sulfur-removal treatment is expensive and not widespread. As a result of these and other difficulties with contaminating elements, metallurgical reclamation of solder is not attractive, and the common industrial practice is to dilute the scrap with virgin metal, bringing it within permissible contamination levels.

During World War II, when the tin supply of the United States was greatly curtailed and the country had to control its use of tin, reclaiming scrap became one of the important sources of tin. Federal specification QQ-S-571 (Table 3-2) and ASTM B-32 (Table 3-1) reflect the level of contamination permissible in reclaimed material. The use of reclaimed material for critical applications is not encouraged because the unpredictable impurity content may cause serious problems in automated mass production by introducing unforeseen variables in solder behavior. Reclaimed materials are perfectly acceptable for plumbing, mechanical-structural joining, and many other nonsensitive applications.

When valuable impurities are present, the economic picture changes dramatically. The intrinsic metal value of noble metals is often higher than the cost of pure tin and lead. With gold prices in the range of $220 per troy ounce, a contamination level of 0.04 percent justifies the use of high-efficiency electrolytical refining processes. This level of gold contamination, often referred to as the *break-even concentration,* is justified not only because of the precious metal retrieved in the process but also because refining yields a tin-lead matrix in the 99.999+ purity range. This solder is suitable for critical applications like anodes. Solder scrap containing gold should therefore be segregated from the regular scrap and sold accordingly.

Virgin-Grade Solder. The term *virgin grade* refers to solders made of tin and lead extracted from the ore. The reader is reminded, however, that not all virgin metal is of high purity or uniformity. Here again the value of

TABLE 3-1 Chemical Composition^{a-c} from ASTM Specifications for Solder Metal B 32–76

Alloy grade	Sn desired	Pb nominal	Sb Min	Sb Desired	Sb Max	Ag Min	Ag Desired	Ag Max	Bi	Cu	Fe	Al	Zn	As
												Maximum		
70A	70	30	...	...	0.12	...	...	...	0.25	0.08	0.02	0.005	0.005	0.03
70B	70	30	0.20	...	0.50	...	...	...	0.25	0.08	0.02	0.005	0.005	0.03
63A	63	37	...	...	0.12	...	...	...	0.25	0.08	0.02	0.005	0.005	0.03
63B	63	37	0.20	...	0.50	...	...	...	0.25	0.08	0.02	0.005	0.005	0.03
60A	60	40	...	...	0.12	...	...	...	0.25	0.08	0.02	0.005	0.005	0.03
60B	60	40	0.20	...	0.50	...	...	...	0.25	0.08	0.02	0.005	0.005	0.03
50A	50	50	...	...	0.12	...	...	...	0.25	0.08	0.02	0.005	0.005	0.03
50B	50	50	0.20	...	0.50	...	...	...	0.25	0.08	0.02	0.005	0.005	0.03
45A	45	55	...	...	0.12	...	...	...	0.25	0.08	0.02	0.005	0.005	0.03
45B	45	55	0.20	...	0.05	...	...	...	0.25	0.08	0.02	0.005	0.005	0.03
40A	40	60	...	...	0.12	...	...	...	0.25	0.08	0.02	0.005	0.005	0.02
40B	40	60	0.20	...	0.50	...	...	...	0.25	0.08	0.02	0.005	0.005	0.02
40C	40	58	1.8	2.0	2.4	...	...	...	0.25	0.08	0.02	0.005	0.005	0.02
35A	35	65	...	...	0.25	...	...	...	0.25	0.08	0.02	0.005	0.005	0.02
35B	35	65	0.20	...	0.50	...	...	...	0.25	0.08	0.02	0.005	0.005	0.02
35C	35	63.2	1.6	1.8	2.0	...	...	...	0.25	0.08	0.02	0.005	0.005	0.02
30A	30	70	...	...	0.25	...	...	...	0.25	0.08	0.02	0.005	0.005	0.02
30B	30	70	0.20	...	0.50	...	...	...	0.25	0.08	0.02	0.005	0.005	0.02
30C	30	68.4	1.4	1.6	1.8	...	...	...	0.25	0.08	0.02	0.005	0.005	0.02
25A	25	75	...	...	0.25	...	...	...	0.25	0.08	0.02	0.005	0.005	0.02
25B	25	75	0.20	...	0.50	...	...	...	0.25	0.08	0.02	0.005	0.005	0.02
25C	25	73.7	1.1	1.3	1.5	...	...	...	0.25	0.08	0.02	0.005	0.005	0.02

20B	20	80	0.20	...	0.50	...	...	...	0.25	0.08	0.02	0.005	0.005	0.02
20C	20	79	0.8	1.0	1.2	...	...	...	0.25	0.08	0.02	0.005	0.005	0.02
15B	15	85	0.20	...	0.50	...	...	...	0.25	0.08	0.02	0.005	0.005	0.02
10B	10	90	0.20	...	0.50	...	...	...	0.25	0.08	0.02	0.005	0.005	0.02
5A	5[d]	95	...	...	0.12	...	...	...	0.25	0.08	0.02	0.005	0.005	0.02
5B	5[d]	95	0.20	...	0.50	...	...	...	0.25	0.08	0.02	0.005	0.005	0.02
2A	2[e]	98	...	...	0.12	...	...	...	0.25	0.08	0.02	0.005	0.005	0.02
2B	2[e]	98	0.20	...	0.50	...	...	...	0.25	0.08	0.02	0.005	0.005	0.02
2.5S	0[f]	97.5	...	...	0.40	2.3	2.5	2.7	0.25	0.08	0.02	0.005	0.005	0.02
1.5S	1[g]	97.5	...	...	0.40	1.3	1.5	1.7	0.25	0.08	0.02	0.005	0.005	0.02
95TA	95	0.20 max	4.5	5.0	5.5	...	...	...	0.15	0.08	0.04	0.005	0.005	0.05
96TS	96	0.20 max	...	...	0.50	3.6	4.0	4.4	0.15	0.08	0.02	0.005	0.005	0.05

[a] Analysis shall regularly be made only for the elements specifically mentioned in the table.

[b] The chemical requirements of SAE specifications 1A, 2A, 2B, 3A, 3B, 4A, 4B, 5A, 5B, 6A, and E-07 conform substantially to the requirements for alloy grades 45B, 40B, 40C, 30B, 30C, 25B, 25C, 20B, 20C, 15B, and 2.5S, respectively.

[c] Federal specifications are similar to the above alloy grade 70B, 63B, 60B, 50B, 40B, 35C, 30C, 20C, 2.5S, 1.5S, and 95TA.

[d] Permissible tin range, 4.5–5.5%.

[e] Permissible tin range, 1.5–2.5%.

[f] Tin maximum, 0.25%.

[g] Permissible tin range, 0.75–1.25%.

TABLE 3-2 Chemical-Composition Requirements of Solder, Weight Percent, from Federal Specification QQ-S-571E

Composition	Sn, %	Pb, %	Sb, %	Bi, max, %	Ag, %	Cu, max, %	Fe, max, %	Zn, max, %	Al, max, %	As, max, %	Cd, max, %	Total all others, max, %	Approximate melting range, °C	
													Solidus	Liquidus
Sn96	Remainder	0.10, max	....	.....	3.6–4.4	0.20	.....	0.005	.....	0.05	0.005	.....	221	221
Sn70	69.5–71.5	Remainder	0.20–0.50	0.25	0.015	0.08	0.02	0.005	0.005	0.03	0.001	0.08	183	193
Sn63	62.5–63.5	Remainder	0.20–0.50	0.25	0.015	0.08	0.02	0.005	0.005	0.03	0.001	0.08	183	183
Sn62	61.5–62.5	Remainder	0.20–0.50	0.25	1.75–2.25	0.08	0.02	0.005	0.005	0.03	0.001	0.08	179	179
Sn60	59.5–61.5	Remainder	0.20–0.50	0.25	0.015	0.08	0.02	0.005	0.005	0.03	0.001	0.08	183	191
Sn50	49.5–51.5	Remainder	0.20–0.50	0.25	0.015	0.08	0.02	0.005	0.005	0.025	0.001	0.08	183	216
Sn40	39.5–41.5	Remainder	0.20–0.50	0.25	0.015	0.08	0.02	0.005	0.005	0.02	0.001	0.08	183	238
Sn35	34.5–36.5	Remainder	1.6–2.0	0.25	0.015	0.08	0.02	0.005	0.005	0.02	0.001	0.08	185	243
Sn30	29.5–31.5	Remainder	1.4–1.8	0.25	0.015	0.08	0.02	0.005	0.005	0.02	0.001	0.08	185	250
Sn20	19.5–21.5	Remainder	0.80–1.2	0.25	0.015	0.08	0.02	0.005	0.005	0.02	0.001	0.08	184	270
Sn10	9.0–11.0	Remainder	0.20 max	0.03	1.7–2.4	0.08	.....	0.005	0.005	0.02	0.001	0.10	268	290
Sn5	4.5–5.5	Remainder	0.50 max	0.25	0.015	0.08	0.02	0.005	0.005	0.02	0.001	0.08	308	312
Sb5	94.0 min	0.20 max	4.0–6.0	.....	0.015	0.08	0.08	0.03	0.03	0.05	0.03	0.03	235	240
Pb80	Remainder	78.5–80.5	0.20–0.50	0.25	0.015	0.08	0.02	0.005	0.005	0.02	0.001	0.08	183	277
Pb70	Remainder	68.5–70.5	0.20–0.50	0.25	0.015	0.08	0.02	0.005	0.005	0.02	0.001	0.08	183	254
Pb65	Remainder	63.5–65.5	0.20–0.50	0.25	0.015	0.08	0.02	0.005	0.005	0.02	0.001	0.08	183	246
Ag1.5	0.75–1.25	Remainder	0.40 max	0.25	1.3–1.7	0.30	0.02	0.005	0.005	0.02	0.001	0.08	309	309
Ag2.5	0.25 max	Remainder	0.40 max	0.25	2.3–2.7	0.30	0.02	0.005	0.005	0.02	0.001	0.03	304	304
Ag5.5	0.25 max	Remainder	0.40 max	0.25	5.0–6.0	0.30	0.02	0.005	0.005	0.02	0.001	0.03	304	380

the impurities will determine the extent to which they will be removed from the raw material. The smelter must equate the cost of refining with the value of the impurities. The end product also reflects the original composition of the ore which was processed.

Virgin-grade metals are pretty much the standard in the electronics solder industry, especially where automatic mass-production techniques are used, as in the wave soldering of printed-circuit boards. The consistency of this grade of solder in impurity content is relatively predictable. Unfortunately, no industry or government specifications exist to cover virgin-grade materials in the critical area of contamination content. It is up to the individual user to hammer out a specification in conjunction with the suppliers, who cannot control the fluctuations in impurity content in the material they receive from the prime manufacturer. Reasonable virgin-grade-solder specification allowing for these variations is insurance enough against the use of diluted reclaimed materials with their potential production difficulties.

Antimony-containing Solder. The allotropic transformation of metallic tin into a gray amorphous state poses a potential danger to solder (see Sec. 3-4). Deliberate additions of antimony to the solder has been found to arrest tin pest.[1] Solders are commercially available with and without these antimony additions. Federal specification QQ-S-571 and ASTM B-32 grade B both reflect these additives; antimony-free solders are normally designated as ASTM B-32 grade A alloys. In the past the solder manufacturer has not charged for the deliberate addition of antimony and supplied either alloy on request. The author is inclined to favor this addition because of the added safety it imparts under future-use conditions. More about the benefits of antimonial tin-lead solders and their properties can be found in Sec. 3-20.

3-10 The Effect of Metallic Contamination on the Physical Properties of Tin-Lead Solders   The effect of four types of alloying additions on the eutectic tin-lead solder system was studied by the author.[2] The four contaminating metals which were introduced represented different cases of metallographic systems. They were selected to show what happens when the metallic addition is soluble (1) in tin and lead, (2) in tin only, (3) in lead only, and (4) in neither tin nor lead. These four situations gave an indication of the influence of small metallic additions on the tin-lead system. The additions were made in quantities calculated from the individual solubility in tin and/or lead and constituted, in one case, 5 percent

[1]Tin Disease in A. Bornemann, Solder Type Alloys *ASTM Spec. Tech. Publ.* 189, 1956.
[2]H. H. Manko, The Effect of Some Metallic Additions on the Tin-Lead System, thesis New York University, December, 1959; also The Effect of Some Metallic Contaminations on the Tin-Lead Solder System, IBM.

over and, in the second case, 50 percent of the total solid solubility. Additional levels of contamination were run as necessary. The eutectic composition was chosen for this study to eliminate additional variables, as it affords the greatest homogeneity in the tin-lead alloy system. Once the alloys were prepared, resistivity measurements and metallographic examination were utilized to study some of the physical properties. The following specific conclusions on the four metallic additions of copper, bismuth, magnesium, and zinc were drawn on the basis of the work described.

1. The copper addition was made in order to study the case of an alloying metal that has practically no solid solubility in tin or lead. The effects were as follows:
   a. *Resistivity*. The addition of copper lowered the overall resistivity because of the lower resistivity of the tin-copper intermetallic compound that was formed (see Sec. 3-12).
   b. *Microstructure*. The typical eutectic structure of pure tin-lead was retained. The tin-copper intermetallics were found aligned at what were probably as-cast grain boundaries which disappear with annealing.
2. The bismuth addition was made in order to study the case of an addition with large solid solubility in both tin and lead. The effects were as follows:
   a. *Resistivity*. The addition of bismuth raised the overall resistivity because of the formation of solid solution in both the tin and the lead phases. The effect was rather large, owing to the large amount of bismuth that was added.
   b. *Microstructure*. The large additions of bismuth changed the microstructure from a eutectic binary to a noneutectic two-phase ternary structure.
3. The magnesium additions were made to study the case of additions which go into solid solutions with lead only. The effects were as follows:
   a. *Resistivity*. The overall resistivity was raised because of the formation of a solid solution in the lead phase.
   b. *Microstructure*. No effect on the typical eutectic structure of the tin-lead was found.
4. The zinc additions were made to study the case of additions which go into the solid solution with tin only. The effects were as follows:
   a. *Resistivity*. The overall resistivity was lowered. This indicates that the zinc did not go into solid solution but formed in small zinc-rich regions in the eutectic.
   b. *Microstructure*. No effect on the typical eutectic structure of eutectic tin-lead was found.

The pickup of copper, magnesium, and zinc as contaminants in solder is not harmful from the point of view of resistivity. The amount of contamination for which this statement holds is limited to alloys containing 5 percent over the full solid solubility. However, these quantities are several orders of magnitude larger than those allowed in virgin solder (see Table 3-2).

To summarize, the effect of alloying additions to solder depends on the solid solubility of the pure metal added in the tin and/or lead phase and the formation of intermetallic compounds, if any. Thus the formation of a solid solution raises the resistivity (bismuth and magnesium additions), while the formation of an intermetallic compound lowers the overall resistivity (copper additions). If a third phase is formed, the overall resistivity is theoretically the weighted sum of the resistivity of the three phases (zinc additions).

3-11 The Effect of Metallic Contamination on the Wetting Properties of Tin-Lead Solders   As can be anticipated, relatively small additions of metallic contaminants to the tin-lead solder will change the surface energies of the resulting alloys and thus influence the wetting characteristics. Figure 3-11 shows the result of a test run by the author. Contaminated alloys were prepared by adding predetermined amounts of the contaminant into a 63/37 eutectic tin-lead alloy. The mixture was heated up to about 100°F (38°C) over the melting point and stirred magnetically for 20 min. The alloys were then chill-cast in order to avoid segregation. The alloying additions were made disregarding any solubility ranges. The flux used consisted of stearic acid and small quantities of ammonium chloride, thus eliminating the addition of any metallic ions from the flux. After the dross and flux were removed, the alloys were extruded in a special die. The extruded alloy wire was metallurgically checked for uniformity and examined by a wet analysis for exact composition.

Using the spread-rate analyzer, described elsewhere,[1] the alloys were evaluated for their spread factors. The time-temperature cycle was selected to ensure that the system would reach equilibrium, although no time was allowed for extensive mutual diffusion of the solder and the base metal. Obviously a test of this kind is not designed to get a complete evaluation of the effect of these impurities on the tin-lead system. The results obtained from this test were not intended to be used to establish the maximum allowable level of contamination in applications where bulk molten solder is applied with soldering pots, wave-soldering equipment, and the like. The results are indicative only when solder spreads, for

[1] H. H. Manko, "Solders and Soldering," 1st ed., McGraw Hill, New York, 1964, pp. 266–272.

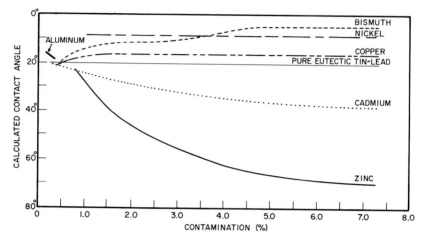

Fig. 3-11   Effect of contamination on the spread of eutectic solder. Remember that this simulates only self-spreading operations. (The aluminum did not dissolve more than 0.042 percent in this test.)

example, in preforms or pastes or rises in a plated-through hole of a printed-circuit board.

The six[1] alloy additions which were made have the following metallurgical properties:

Bismuth. The bismuth addition to the tin-lead goes into solution with both tin and lead. The spread rate was improved with increasing amounts of bismuth, and the highest spread rate of all alloys tested was obtained with this addition. The solid solubility of bismuth in lead is 18.9 atomic percent at 100°C (212°F). The solubility of bismuth in tin is 0.5 atomic percent at 25°C (77°F).

Nickel. The addition of nickel also improved the spread rate of solder. There is no solid solubility of nickel in tin, although three intermetallic compounds form at high temperatures (several orders of magnitudes higher than soldering temperature). There is a slight solid solubility of 0.08 atomic percent of nickel in lead. No intermetallic of lead is formed.

Copper. The addition of copper gave a slight increase in the spread rate of the solder. The solid solubility of copper is negligible in either tin or lead. There are two copper-tin intermetallic compounds at room temperature.

Aluminum. The addition of aluminum could be studied only over a narrow range because the aluminum does not dissolve in the solder at the temperatures of the test. There is no solid solubility of aluminum in tin or of aluminum in lead. This was borne out by the tests.

[1] For similar information on antimony alloying additions see Fig. 3-32.

Cadmium. Cadmium goes into solid solution with tin (1.1 atomic percent at 100°C). The solid solubility of cadmium in lead is 0.7 atomic percent at 100°C. The addition of cadmium lowered the spread rate on the solder sample.

Zinc. The addition of zinc lowered the spread rate of the solder to the largest extent. The solid solubility of zinc in tin is 2 atomic percent at 198°C (388°F), and there is no solid solubility of zinc in lead.

In conclusion, therefore, we can say that the addition of bismuth, nickel, and copper to the eutectic tin-lead solder seems to improve the wettability of the contaminated solders. The addition of cadmium and zinc decreases the wetting power of the eutectic solders.

The addition of aluminum was limited to a very small composition range. Even when large quantities of aluminum were added to the alloy, only a small amount up to 0.042 percent of the aluminum was found in the contaminated alloy. It is assumed that the rest was removed with the dross on top of the melt.

3-12 The Effect of Large Alloying Additions on the Microstructure of Eutectic Tin-Lead   In an effort to get a clear view of the intermetallics that form and their effect on the microstructure of the solder alloy, a study was made in which 15 percent by weight of the more important alloying and contamination additions were introduced into a eutectic 63/37 tin-lead solder. The following microstructures and short descriptions will help readers identify any intermetallics or changes in the structure found in their own work. The addition of 15 percent across the board was chosen because it exceeds the solid solubility at room temperature and therefore causes the formation of a third phase and/or intermetallics with all the materials selected for this work.

1. *Antimony.* The microstructure (Fig. 3-12) clearly indicates the cubic tin-antimony intermetallic crystals in a matrix of α-rich eutectic.

2. *Cadmium.* The microstructure (Fig. 3-13) shows a typical dendritic structure of the cadmium phase (containing tin in solid solution) with a tin-lead eutectic matrix containing some α.

3. *Copper.* The light-colored crystalline orthorhombic copper-tin intermetallics are shown in the micrograph (Fig. 3-14) in a matrix of a fine eutectic tin-lead with distinct lead-rich α regions. The needle-shaped crystals can also be seen with the naked eye when a copper-rich solder pot is cooled to 370°F (187°C) and the sludge is removed.

4. *Gold.* In the micrograph (Fig. 3-15) we can see a distinct gold-rich phase which takes a laminar crystalline shape and can be seen as the gold-tin intermetallic. Then there is a secondary eutectic phase apparently between the gold and the lead which freezes around these gold-rich

Fig. 3-12   Antimony (15 weight percent) in eutectic tin-lead, ×200 magnification (5% nital etch).

Fig. 3-13   Cadmium (15 weight percent) in eutectic tin-lead, ×200 magnification (glycerin acetic etch).

crystalline sections, and finally there is a lead-tin phase which is the darker portion of the picture.

5. *Silver.* Crystalline shapes of hexagonal close-packed silver-tin intermetallic crystals are shown suspended in a eutectic tin-lead matrix with lead-rich areas dispersed in it (Fig. 3-16).

6. *Zinc.* The dark zinc-rich phase (Fig. 3-17) of a much higher melting point than the rest of the composition froze in a quasi-crystalline shape. It contains mostly tin, and therefore the eutectic tin-lead left

Fig. 3-14   Copper (15 weight percent) in eutectic tin-lead, ×200 magnification (5% nital etch).

Fig. 3-15   Gold (15 weight percent) in eutectic tin-lead, ×200 magnification (ferric chloride etch).

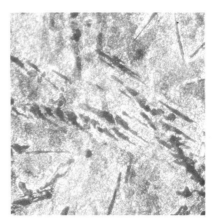

Fig. 3-16 Silver (15 weight percent) in eutectic tin-lead, ×200 magnification (glycerin acetic etch).

Fig. 3-17 Zinc (15 weight percent) in eutectic tin-lead, ×200 magnification (glycerin acetic etch).

behind is rich in the α phase, which came out in a dendritic structure usually found close to the zinc-rich phase.

3-13 The Practical Approach to Contamination in Solders   In our discussion of the effect of various contaminants on the tin-lead solder system, our remarks have of necessity been confined to theoretical considerations. Let us see how this can be carried over to practical applications. In general, all soldering methods can be divided into two major categories as related to contamination. In the first the solder is applied in wire or bulk form, and the metal is molten during a short soldering period and freezes immediately thereafter. In the second the solder is kept in its liquid form and applied as such to the work surfaces.

When the solder is applied in solid form, usually in wire, as with soldering irons, for instance, there is very little chance for it to pick up contamination other than from the working tool and the surfaces to be wetted. In most of these cases the solder is physically placed over the surfaces to be soldered and is given little chance to spread by itself and/or to pull back of its own accord from the wetted surfaces. Thus the effect of contaminants is less noticeable, especially when good inspectability procedures are not maintained and large fillets are acceptable.

A good example of this can be seen in the high-copper-loaded soldering alloys used for iron soldering. These are soldering alloys, usually tin-leads, with as much as 5 percent copper added. The purpose, of course, is to maintain the shape of the soldering tip, which is made of copper. This type of solder has found little use in the United States because the ironclad tip has replaced copper tips (see Chap. 7).

Because of the short time that the solder is actually molten in this type of application, the effect of the impurities on the surfaces is rather limited, although with gold a frosty solder joint appears even on short exposure times. If we did not pile up solder in most of these short applications and drained off excess solder, letting the system come to equilibrium, we would find many unsatisfactory conditions not noticed under present conditions. This is best illustrated by the following demonstration.

Two identical ¼-in stainless-steel disks are placed side by side. The first is not fluxed, and the second is coated with an appropriate flux. With a hot soldering iron, enough solder is melted onto both disks to cover the surfaces completely (Fig. 3-18).

In the first case, no wetting will occur, yet the solder will physically cover the disk. Even with prolonged heating, this solder cannot draw back from the nonwetted surfaces because it actually has no place to go.

In the second case, the solder can adequately wet the disk. If the same amount of solder were applied as in the first case, no difference could be seen. However, if both samples were allowed to drain off by inverting them or shaking the hot solder away, the difference between the samples would be apparent.

The second major group of applications is exemplified by dip soldering in a solder pot. Here the solder is kept molten for prolonged periods and has a definite chance of being contaminated by various metallic constituents which come into contact with it. In many of the applications where molten solder is applied to the work, we rely on the wetting forces and capillary and similar mechanisms to deposit small shallow fillets which are

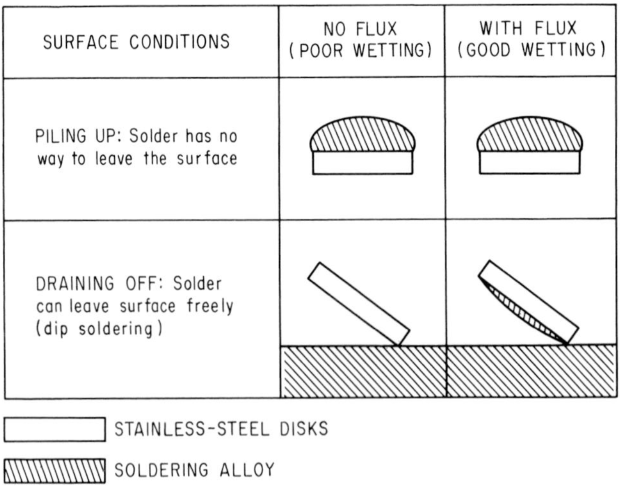

SURFACE CONDITIONS	NO FLUX (POOR WETTING)	WITH FLUX (GOOD WETTING)
PILING UP: Solder has no way to leave the surface		
DRAINING OFF: Solder can leave surface freely (dip soldering)		

STAINLESS-STEEL DISKS

SOLDERING ALLOY

Fig. 3-18   Masking wetting conditions.

reliable through inspectability. In this type of application drain-off is mostly inherent in the removal operation (see Fig. 3-18). Here the presence of contaminations greatly influences the work. For more discussion see Sec. 5-14.

Let us look at the various contaminants and their effect on solder as it has been reported in literature and as taken from the author's own experience. The materials are presented in alphabetical order. When a mechanism is available to remove parts of the contaminations by some kind of simple treatment usually performed by the user, it is mentioned together with the contaminants. It is normally not feasible for the solder users to reclaim and refine their own scrap material because of the equipment and techniques involved.

Aluminum. As indicated earlier, less than ½ percent aluminum can be introduced into tin-lead solders at soldering temperatures. No work has been performed on the solution of aluminum at higher temperatures in the tin-lead solder system. Aluminum does not have any solid solubility in either tin or lead at room temperatures. Only small amounts of aluminum are dissolved in the liquid tin at elevated temperatures, mostly above those used for regular soldering. Aluminum in molten solder usually causes sluggishness in the melt, with a considerable amount of grittiness on the dipped parts. Amounts as small as 0.001 percent aluminum are reported[1] to cause "lack of adhesion, grittiness, or liability to hot-short cracking."

It is unusual to find aluminum in electronic assembly surfaces which are exposed to the molten metal. Very little aluminum is used in electronic soldering because of the large galvanic potential present between the aluminum and the tin-lead (1.53 V, see Table 4-1). Because aluminum does not wet using ordinary fluxes acceptable for the electronics industry, there is a tendency to make soldering fixtures out of aluminum. Because of the bad effects listed above, however, this is not a recommended practice. The continuous erosion of the solder against the aluminum surface will eventually introduce aluminum into the solder, making it necessary to replace the solder with pure material. Should it be absolutely necessary to have some aluminum surfaces exposed to the liquid solder even for short periods, it is recommended that these surfaces be anodized beforehand. The author has shown that anodized surfaces withstand molten liquid solder for longer periods than aluminum surfaces covered by their normal oxides.

Antimony. The solubility of antimony in tin at room temperature is between 6 and 8 percent, whereas little antimony is dissolved in lead at room temperatures. As mentioned earlier, small additions of antimony up to 0.3 percent seem to improve the wetting of solder, whereas larger

[1]W. R. Lewis, "Notes on Soldering," Tin Research Institute, 1961.

additions seem to deteriorate the wetting slowly (Fig. 3-32). The reader is reminded that antimony is used to retard tin pest. For various government specifications, the presence of antimony is mandatory.

Addition of antimony in large quantities does not seem to affect the properties of the molten metal greatly. In many formulations of special solders, antimony is present in relatively large quantities. Contamination problems with antimony are minor because it is not likely to be introduced into the molten solder.

Arsenic. There seems to be no solid solubility of arsenic in either tin or lead. Two intermetallic compounds, Sn_3As_2 and $SnAs$, appear as long needles in the microstructures. Since there are no sources of arsenic to contaminate a liquid solder in electronic assemblies, this element has not presented any problems as a pickup contaminant. The amount of arsenic in the raw material should be closely controlled.

Bismuth. Bismuth has a large solid solubility in lead, up to 18 percent at room temperature, and little solid solubility in tin, around 1 percent at room temperature. Actually bismuth is not a contamination in solder but is usually an alloying addition. Bismuth as such improves the wetting characteristics of solders. It is not used more extensively only because it has some unique lattice changes after solidification. For further details, see Sec. 3-23.

Cadmium. The solid solubility of cadmium in both tin and lead is practically negligible. There is an intermetallic phase at elevated temperature. At around 130°C (266°F) however, a transformation takes place and the material decomposes. Cadmium will occur in special solder in many of the alloys used for low-temperature soldering. However, when the solder is applied from a molten bulk, the cadmium seems to impart a sluggish property to the material, and upon slow cooling a sludge containing most of the cadmium can be found on the bottom of the solder pot. The material apparently segregates because of the temperature difference between its melting point and the melting point of pure tin-lead. In addition, it appears that the presence of cadmium promotes the development of tarnishes and heavy oxide layers in the solder pot.

Unfortunately, cadmium is considered by many people a good solderable surface, and cadmium platings, which are relatively cheap, are used extensively throughout industry. Provided that the surfaces are really maintained in their solderable condition, cadmium can be used safely with soldering methods where no molten bulk is applied. It is strongly suggested that cadmium not be used for dip-soldering applications or the like.

Copper. Solubility of copper in both the tin and lead phase is negligible, but two intermetallic compounds are formed between the copper and the tin (orthorhombic Cu_3Sn and Cu_6Sn_5). These materials exist at room

Fig. 3-19 Copper-tin intermetallic crystals, ×4 magnification. These crystals were dredged out of a highly contaminated solder pot (copper content 2.4 percent).

temperatures and can easily be seen (Fig. 3-19). Micrographic examination shows them as hexagonal needles floating in the solder (Fig. 3-14).

As described earlier, wire solder with high copper percentages is sold to preserve the life of a copper soldering tip. When the solder is applied from a pot or in liquid bulk, however, it rapidly loses its effectiveness with increase of copper content. It is very difficult to give any rule about the level of copper contamination that can be tolerated in a solder pot. As the copper content increases, the working temperature of the pot must be raised in order to overcome the grittiness and sluggishness of the liquid metal. This increases the rate of solution of additional copper from the surfaces to be soldered, and from that point on the deterioration of the soldering conditions is rapid until the solder is used at too high a temperature. Actually, the solution to this problem goes in the opposite direction. When the solder temperature is dropped to within 10 to 20°F ($\approx$ 5 to 10°C) of its freezing point, the copper-tin intermetallic will start precipitating out in the pot and can be removed by a special ladle. This method will remove a large percentage of the copper contamination. It is suggested that the solder pot be replenished at this point with a makeup solder of higher tin content because of the loss of tin in the copper-rich intermetallic crystals. Now when the solder temperature is raised back to working conditions, the material will perform as before. However, it is impossible in this type of treatment to remove copper contamination below 0.7 to 0.8 weight percent. When a high volume of high-quality parts is soldered automatically, the industry prefers to change the solder completely as soon as the contamination level reaches anywhere from 0.3 to 0.8 percent copper (see Table 5-5). The best rule of thumb for this type of application in less critical cases is to use a solder pot until the effect of the copper contamination becomes evident, at which time action must be taken.

In order to reduce the amount of copper dissolved from the surfaces to be soldered by bulk solder, two steps can be taken: (1) A solder resist can be applied to the surfaces in those areas where the copper does not have to be wetted. Only the areas needed for the solder-joint formation are left bare. These resist materials are usually organic polymers which are applied to the surfaces by screening and cured in a pattern which leaves exposed copper only in the selected areas to be wetted. These materials also have a tendency to reduce icicling and bridging between adjacent conductors. (2) The soldering time and temperature are reduced to the barest minimum in order to slow down the rate of solution of copper in the bulk solders.

Under most circumstances, the rate of copper solution into the solder can be controlled by the above means. Regular replenishment of the solder alloy to compensate for drag-out helps maintain the solder pot at a level of contamination which is not harmful to the operation. This is found in many industrial applications where the same solder pot is used for many months and additions of pure material are made to maintain the liquid level (see Fig. 5-13).

The amount of copper in solder alloys can be reduced by the application of sulfur. This is not a recommended practice for the consumer of solders, however, because of the harmful fumes given off during the treatment and the difficulty of removing all the sulfur once the copper has been completely removed. In many cases the sulfur left in the solder is more detrimental than the copper it removed.

Gold. The solid solubility of gold in tin-lead at room temperature is negligible. However, several intermetallic compounds are formed between gold and lead (Au_2Pb and $AuPb_2$) and tin (Au_6Sn, $AuSn$, $AuSn_2$, and $AuSn_4$). Unless soldering is done very rapidly and the gold intermetallics have no chance to rise to the surfaces of the fillet, an extremely dull and scummy surface is found in soldering to gold. In the solder pot a level of as low as 0.02 percent and as high as 0.2 percent gold has been reported to be the critical amount of gold contamination, after which the solder seems to become too sluggish and dull for further use. Solder is usually sold for its high scrap value before it ever reaches the danger level.

Although gold was considered an excellent solderable surface for many years, industry and the government have recognized serious problems in soldering to gold, and its use is now being discontinued.[1]

Iron. There is no solid solubility of iron in lead, but there is some solubility of iron in tin at elevated temperatures, and two intermetallic compounds are formed ($FeSn$ and $FeSn_2$). The presence of iron in solders

[1] F. Gordon Foster, Gold from Plated Surfaces May Embrittle Solder and J. D. Keller, Elimination of Gold Plating as a Surface Preparation for Printed Circuits, *ASTM 65th Annu. Conv.*, 1962.

at low levels such as 0.1 percent shows grittiness and is rather detrimental. However, iron as such does not readily dissolve in solder at soldering temperatures. Most solder pots are made of cast iron and do not present a problem. The iron, however, will go readily into the tin-lead at temperatures above 800°F (427°C). It is therefore very important to avoid hot spots in the construction of solder pots and heating elements in contact with molten tin-lead.

Wetting of iron can be avoided when ordinary fluxes are used with this simple technique. The iron surfaces, to be used as fixtures, for example, are either blued or heavily oxidized by flame. Stainless steel has not been reported to have any advantage over regular cast iron although it resists corrosion from fluxes better than regular iron and steel. Gritty intermetallic crystals, also called *hardheads*, float to the surface and can be removed with the dross. Small quantities, however, keep on floating in the molten solder and must be specifically removed.

Magnesium. Magnesium has the same effect as aluminum on solders. The solubility of magnesium in tin at room temperature is negligible although the intermetallic Mg_2Sn is formed. There is no solubility of magnesium in lead, and the intermetallic formed in this system is Mg_2Pb. Because of the rarity of magnesium parts in electronic assemblies, this contamination is seldom introduced into the solder system.

Nickel. There seems to be no solid solubility of nickel in either tin or lead. However, three intermetallics are formed with nickel (Ni_3Sn, Ni_3Sn_2, and Ni_3Sn_4). This material is seldom found in solders, and no detrimental effect on molten solder has been reported to date.

Silver. There is no solid solubility in either tin or lead, but there are two intermetallics between silver and tin (Ag_6Sn and Ag_3Sn). Silver is not considered a contaminant in tin-lead solders until it reaches several percent, when it might cause grittiness and give small pimples on the soldered surfaces. The addition of silver is usual where silver-fired ceramics or other thin layers of silver are to be soldered and the scavenging of silver from the surfaces is to be eliminated. When the silver content rises over 2 percent, the silver-tin intermetallic will segregate upon cooling and can be removed from the solder pot with the technique described earlier for copper.

Sulfur. A word about sulfur is in order. This is a contaminant which has a detrimental effect on wetting and is reported to be dangerous at levels of 7 ppm. However, a good grade of solder will not contain sulfur over and above a few parts per million.

Zinc. This material has small solid solubility in tin and no apparent solubility in lead. No intermetallic compounds are formed with either tin or lead. The addition of zinc is reported to be very detrimental to the solder alloy. As little as 0.005 percent of zinc is reported to cause lack of adhesion, grittiness, and/or susceptibility to failure during solidification. Little zinc occurs in electronics soldering to endanger the properties of solder pots.

USE AND PROPERTIES OF TIN-LEAD SOLDERS

3-14 Bulk Solder Compared with Solder Joints   Let us review the function of the solder metal in the joint. As we have seen in Chap. 1, the bond is made at the interface by wetting both base metals, and in that respect the solder is the bond maker. The solder adheres to the base metals, thus providing metallurgical continuity. The solder has an additional function, however; it is also the bridge between the wetted surfaces holding the joint together. When the solder serves as a link between the metals to be joined, the bulk-solder properties become important since they determine the property of the entire joint (see Fig. 3-20).

The definition of soldering was given as a metallurgical joining method using a filler metal (the solder) with a melting point below 600°F (315°C) for soft solder or 800°F (427°C) for hard solder. Soldering relies on wetting for the bond formation (see Chap. 1). In other words, the solder alloy is truly a fusible alloy, a fact with many implications. Metallurgically speaking, all alloys are very similar in their mechanical properties when they are measured at identical temperatures relative to the melting point. For example, cold-rolled steel has a melting point of 1536°C (2797°F), and at room temperature its strength is approximately 64,000 lb/in². Solder, on the other hand, has a melting point of 183°C (361°F), and at room temperature its tensile strength is 7250 lb/in². Yet both metals are actually similar in strength, since when the cold-rolled steel is heated to 1400°C (2552°F), which is close to its melting point, it will have a tensile strength of only 8000 lb/in². The conclusion we can draw from this is that while solder is a fusible alloy, its mechanical properties at room temperature are dictated by its proximity to the melting point; thus the metal is softer,

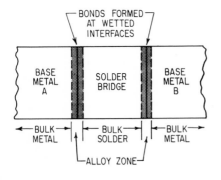

Fig. 3-20 Makeup of solder joint.

weaker, and more ductile than higher-melting-point alloys. On the other hand, the solder is metallurgically speaking a sound metallic matrix, capable of conducting electricity and heat and possessing all the other metallurgical properties such as ductility, luster, etc.

This proximity to the melting point gives the soldering industry an advantage which requires some elaboration. The ductility of solder and the ability of most alloys in this group to anneal at or near room temperature make the solder in the bond an excellent *stress coupler*. This ability of the solder to absorb stresses which are relieved with time enables the industry to use this material under adverse design conditions which could not be tolerated by the higher-strength filler metals used in brazing or welding. Vibrations and thermal excursions of many assemblies set up substantial stresses in the joint which would cause failure if the solder were less ductile. While a higher-strength filler metal in those cases would not fail, the stresses would be transferred to other weak elements in the design, e.g., the glass matrix of the laminate in a printed-circuit board or the metal-to-ceramic interface in metallized assemblies, causing disruption. Thus we see that the "weakness" of the solder is actually its strength.

As we discuss the physical properties of tin-lead solders, we should keep in mind that these data apply to the bulk solder as such and not particularly to the strength of the solder in the joint. As seen previously, the effect of solution hardening is to strengthen rather than to weaken the solder joint unless undesirable intermetallic compounds are formed at interfaces or in layers which will make the fillet brittle and weaken its overall strength. In the following sections we shall discuss the various physical properties of bulk solder and try to keep in mind how these properties change with the change of the composition of the material in use. The practical use of these data is discussed in Chap. 4.

3-15 Physical Properties, Accuracy of Data, and Sources of Additional Information   The physical properties of solder represent a field of investigation which has received unusually little scientific support. Very few serious investigations into the properties of the fusible alloys used in this industry have been conducted over the last 50 years. Work is scattered, often poorly reported (lack of background information), and very limited in the number of samples tested, accuracy, etc. It is not unusual to find major contradictions between workers on identical data, and this probably stems from some of the problems inherent in the research of these alloys. Solder properties are very sensitive to alloy purity, nonmetallic-occlusion content, grain size, time from casting, and surface finish. Many scientific reports omit these fundamental facts, making comparison and evaluation of results extremely difficult.

For example, the author conducted a computer literature search using the METADEX files compiled by the American Society for Metals. It revealed 138 publications concerned with the physical properties of solder from 1966 to 1977. The author has tried to compile the data and include them in this book with little success. Many publications oriented to users of solder address themselves to specific soldering problems encountered by their organization, and the data which resulted often conflicted with similar work in other sources. When no conflict existed, data were limited in value. It was therefore decided to keep the format of the first edition of this book, reporting on the best information available and listing references which appear to provide the best information without passing judgment on its quality. It is left to readers to select the data they determine to be the most appropriate for their situations. It is hoped that in the future an organized effort will be funded by various organizations to establish solder research on a more scientific basis.

Should it be necessary for the reader to run a set of experiments to support in-house work, it is absolutely imperative that the pitfalls described earlier be avoided. Only solders with known accurate analysis should be used. The test conditions should be selected carefully to consider the strain-sensitive nature of the alloys, and the thermal conditions of the experiment must be monitored and recorded. Grain size, time from casting, proper annealing, and similar factors need consideration. These and other precautions to be taken in testing are described in Chap. 8.

3-16 Properties of Bulk Solder at Room Temperature   We discuss these properties one by one and see the variations as they occur. The most interesting of the properties, which determines many characteristics of the solder joint, is the tensile strength of the bulk material. Curve 1 in Fig. 3-21 gives the ultimate tensile strength in pounds per square inch. A quick survey of the graph indicates that a maximum in tensile strength exists at the eutectic composition or slightly above it. The tin-rich alloys seem to be stronger than the lead-rich alloys, but the peak does not seem to be exactly at the eutectic composition. As mentioned elsewhere, the strength of a copper solder bond would be several times larger than the results reported here. However, these values of the bulk solder should be considered for all strength and design considerations because they involve a safety factor that applies when not enough solution hardening or possibly no solution hardening at all occurs and the solder in the joint is as strong as the bulk material. This is an extremely safe assumption to make. The strength of the pure tin and pure lead is much lower than that of the alloys, and its value falls off rapidly as the material becomes pure.

The shear strength reported in curve 2 of Fig. 3-21 falls below the tensile strength. Unfortunately no results are reported in the area of 65 percent or pure tin. Also, on the other side of the scale, no data are available between 85 percent lead and pure lead. As with the tensile strength, it can be assumed that the values fall off rapidly as the element

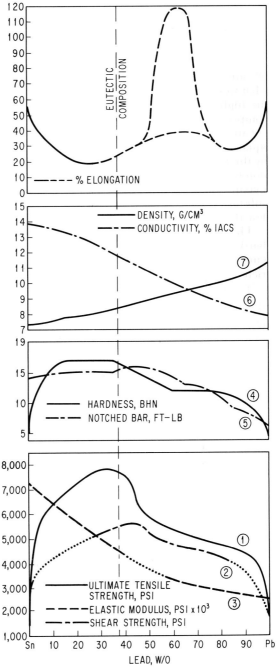

Fig. 3-21 Physical properties of tin-lead alloys. The elongation figures for these alloys depend on casting conditions; hence the difference in values in the 30 to 80 percent range. *(Adapted from L. T. Greenfield and P. G. Forrester, "The Properties of Tin Alloys," The Tin Research Institute, Greenford, England, 1961.)*

becomes pure, and the dotted line is therefore a good approximation of what would occur. It is also important to notice that the maxima shift from the high-tin area to the high-lead area on the other side of the eutectic composition.

Curve 3 in Fig. 3-21 gives the elastic modulus in million pounds per square inch. Here we get a curve which apparently is not greatly affected by the tin-lead composition. Unfortunately, not enough detailed information is available in the region between 40 and 60 percent lead to establish a maximum or minimum in the eutectic region, although their existence is substantiated by the change in pitch of the curve somewhere in this area. For that purpose, the dotted line was used between the two values.

The hardness of solder is seldom a factor in design. In order to obtain a harder material, various additions to the tin-lead system should be employed. The results as reported here have a plateau higher in the tin-rich alloys than in the lead-rich alloys.

The notched-bar impact strength is of interest, but solder is not really suitable for repeated impact applications. Because the values between 35 percent lead and a pure tin are not available, it is not possible to assume that there is any kind of maximum or minimum in this region.

The conductivity of the solder specimen as a percentage of copper (percent IACS) is a fairly straightforward linear relation, as can be expected (see curve 6 in Fig. 3-21). The results as far as the pure resistivity measurements are concerned depend on the purity of the alloys, as shown in Sec. 3-10. For practical purposes, however, and remembering that a safety factor is usually involved in any considerations, these figures are reliable.

The density of the tin-lead alloys when plotted against percentage composition gives a line as represented in curve 7 in Fig. 3-21, where the slope increases with the increased percentage of lead, which is the heavier of the two elements.

Curve 8 of Fig. 3-21 gives interesting results of elongation measure-

TABLE 3-3   Surface Tension and Viscosity*

Lead, %	Temp., °C	Surface tension dyn/cm	Viscosity, P
	290	545	0.0165
20	280	514	0.0192
37	280	490	0.0197
50	280	476	0.0219
58	280	474	0.0229
70	280	470	0.0245
80	280	467	0.0272
100	390	439	0.0244

*A. Latin, *J. Inst. Metals,* vol. 72, p. 265, 1946.

ments taken under two different conditions. Elongation figures for these alloys depend on the casting conditions, especially in the range of 30 to 80 percent lead. The figures given by different investigators therefore show large variations. The surface tension and viscosity of several tin-lead compositions are reported in Table 3-3. Note that the surface tension has a direct relation to the composition of the alloy and does not seem to have any maximum in the range reported. The same holds true of the viscosity.

3-17 Creep Properties of Solder Creep is sometimes a misunderstood term and requires some explanation before we consider the creep properties of solders. Although we measure the strength of solders and other metals by their ability to resist loads to failure (the *ultimate tensile strength*), it is possible to break a solder specimen or for that matter many soft and weak alloys with a small load that is only a fraction of the force required for failure. However, this low-stress failure does not occur instantaneously but is a function of time. This property is referred to as the *creep strength.* The material under small loads will actually rearrange its atom lattice because of the energy invested in it by the constant load. At first, the rearrangement of the atoms and the orientation of the dislocations and imperfections in the structure are rapid until there has been maximum reorientation for the specific load. This is usually reached in a relatively short time (see Fig. 3-22). From there on any change in the metal which accompanies such reorientation of the crystals requires higher energies; i.e., under constant load the rate decreases markedly. The metallic lattice, once it reaches this quasi-equilibrium state, will creep at a constant rate for a prolonged time until the joint is ready to fail. Deterioration from there on is rather rapid.

Figure 3-22 is a typical creep curve for a solder alloy. Here the change in strain is plotted against time. Strain is used because it is an indication of both the reduction of cross-sectional area and specimen elongation. Strain is defined as stress per unit area. In a constant-stress system, it is most meaningful.

In the schematic creep curve (Fig. 3-22) the curve section from point *O* to point *P* is the area of initial reorientation of the lattice under the steady stress. The random atoms and dislocations are oriented in the proper

Fig. 3-22 Schematic of creep-rate curve for solder.

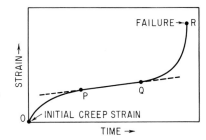

crystallographic planes because of the continued stress, and the change in the specimen (elongation and/or strain) is relatively rapid, as described earlier. From point P to point Q the elongation is in a quasi-steady state, where the changes are rather slow and the lattice pattern has the largest resistance to reorienting, the mechanisms involved in the elongation of the solder specimen. From point Q to point R, the change is rather rapid until at point R failure of the specimen occurs. Here the cross-sectional area of the joint has been reduced to the point where the creep stress that was set up originally is capable of destroying the joint by plain shear or tension as the case may be.

Figure 3-22, which is just a schematic, helps to demonstrate why area PQ is the important one that is usually determined by a creep test. The rate of creep in region PQ is the lowest (the pitch of the curve is the smallest), and the creep rate is practically constant. The results from creep tests are usually expressed as strain (extension per inch) per day. The next step involves carrying out creep tests at varying stresses and plotting their minimum creep rate against the stress (this is usually done by plotting the logarithm of the creep rate against the logarithm of the stress). Thus the stress which would give a particular creep rate can be determined by interpolation and gives a possibility of comparing creep properties of different materials. Also, it gives an indication of the stresses at which the rate of deformation is such that it can be tolerated in service. In the following tables and the literature, the creep characteristics of an alloy are expressed in pounds per square inch required to give a minimum creep rate of 0.0001 in/(in)(day).

The information given in Table 3-4 was part of an investigation to determine the effect of antimony on the creep properties of soft solder. Three separate series of alloys were run consisting of
1. High-purity materials.
2. Commercial-grade materials.
3. High-purity eutectic solders with the addition of copper and silver
The results in Table 3-4 are shown for both room temperature and elevated temperature. The bars for this test were chill-cast and were made under controlled conditions. The microstructure and macrostructure of the bars were examined, and the conclusions and the results are as follows.

The creep properties of chill-cast soft solders are determined principally by temperature, composition, constitution (e.g., proportion of eutectic material in the overall construction of the solder), and macrostructure. Generally, a coarse macrostructure is associated with high creep resistance,[1] but the kind of macrostructure, i.e., whether columnar or equiax-

---

[1]The term *creep resistance* refers to the stress required to produce a given small creep rate.

TABLE 3-4 Creep Properties of Chill-cast Soft Solders[a]

ND = not detected, HP = high purity, CQ = commercial quality, E = eutectic

Mark	Composition					Impurities							Stress, lb/in² for 10⁻⁴ strain per day	
	Sn	Pb	Sb	Other	% eutectic	Sb	Cu	Bi	Ag	Fe	As	S	Room temp.	176°F (80°C)
HP1	62.2	37.8	....	......	100	ND	<0.005	0.02–0.03	<0.001	ND	<0.001	0.001	335	68
HP2	55.0	41.4	3.6	......	100	....	<0.005	0.02–0.03	<0.001	ND	<0.007	0.001	480	125
HP3	49.5	50.5	....	......	80	ND	<0.005	0.02–0.03	<0.001	ND	<0.001	0.001	125	28
HP4	45.1	52.2	2.7	......	80	....	<0.005	0.02–0.03	<0.001	ND	<0.007	0.001	480	90
HP5	34.3	65.7	...	......	55	ND	<0.005	0.02–0.03	<0.001	ND	<0.001	0.001	140	
HP6	31.5	66.7	1.8	......	55	....	<0.005	0.02–0.03	<0.001	ND	<0.003	0.001	295	
HP7	30.4	69.6	...	......	50	ND	<0.005	0.02–0.03	<0.001	ND	<0.001	0.001	115	39[c]
HP8	28.1	70.2	1.7	......	50	....	<0.005	0.02–0.03	<0.001	ND	<0.005	0.001	245	52
CQ1	59.8	40.2	...	......	96	0.01	0.007	0.003	<0.001	0.01	<0.001	0.001	430[d] / 350[e] / 320[f]	
CQ2	60.2	39.52	0.24	......	97	....	0.01	0.015	<0.018	0.001	<0.002	0.001	800[c]	
CQ3	44.3	55.68	...	......	71	0.01	0.009	0.003	<0.001	0.01	<0.001	0.001	140	
CQ4[b]	44.3	55.68	...	......	71	0.01	0.093	0.003	<0.001	0.01	<0.001	0.001	400	
CQ5	40.0	57.8	2.0	......	71	....	0.09	0.10	<0.011	0.007	<0.010	0.001	420	
CQ6	31.6	68.3	...	......	50	0.01	0.004	0.005	<0.001	0.008	<0.001	0.002	130	
CQ7	27.8	70.35	1.6	......	50	....	0.10	0.15	<0.002	0.009	<0.008	0.001	460	
E1	62.2	37.8	...	......	100	ND	<0.005	0.002	<0.001	ND	<0.001	0.001	450[c]	45[c]
E2	54.4	42.2	3.4	......	100	....	<0.005	0.003	<0.001	ND	<0.001	0.001	660[c]	80[e]
E3	62.3	37.68	...	0.02 Ag	100	ND	<0.005	0.002	......	ND	<0.001	0.001	510[c]	
E4	53.8	42.68	3.5	0.02 Ag	100	....	<0.005	0.003	......	ND	<0.001	0.001	650	
E5	62.3	37.52	...	0.18 Cu	100	ND	......	0.002	<0.001	ND	<0.001	0.001	830	110
E6	54.5	41.71	3.6	0.19 Cu	100	ND	......	0.003	<0.001	ND	<0.001	0.001	1030[c]	110

[a] S. J. Nightingale and O. F. Hudson, Tin Solders, Br. Non-ferrous Met. Res. Assoc. Res. Monogr. 1, 1942.
[b] Prepared by adding copper to CQ3.
[c] Slight extrapolation.
[e] Machine, coarse-grained columnar bar.

[d] Machine, coarse-grained equiaxial bar.
[f] Machine, fine-grained equiaxial bar.

ial, also has an effect. Thus bars with a coarse-grained columnar structure were less resistant to creep than coarse-grained equiaxial bars. In any comparison, therefore, of the creep properties of solders due allowance should be made for the effect of macrostructure. One or two cases of inconsistent results observed in the course of the work were ascribed to different or nonuniform structures.

In the tin-lead alloys (nonantimonial solders), which covered a range of composition from about 30 percent tin (50 percent eutectic) to 62 percent tin (100 percent eutectic), the creep resistance showed little change from 50 to 80 percent eutectic (30 to 50 percent tin) but increased to a high value for the eutectic alloy. The eutectic alloy was also the most resistant to creep at 80°C (176°F) although not so markedly superior to the solders with lower tin content. The creep strength of the alloys at 80°C was of the order of 20 to 30 percent of their creep strengths at room temperature.

The antimonial solders were all much superior to their nonantimonial equivalents both at room temperature and at 80°C. The improvement due to antimony was much greater in creep tests than that observed in tensile tests. As in the case of the nonantimonial solders, the creep strengths of the antimonial alloys at 80°C were of the order of 20 to 30 percent of their creep strengths at room temperature.

The creep tests have shown that copper, the most common impurity in commercial solders, markedly improves the creep resistance of nonantimonial solders when present to the extent of 0.1 percent, and above 0.18 percent copper also benefits antimonial solders. Silver appears to have an effect similar to that of copper.

3-18 The Physical Properties of Solder at Other than Room Temperature   In the eyes of the metallurgist, any metal has physical properties which depend on the temperature difference between its melting point and the temperature at which it is used. In general, alloys of identical lattice structure have similar physical properties for equal temperature differences. In this respect the fusible alloys used for soldering derive their unique characteristics from their proximity to their melting point. Let us see how these properties change at other than room temperature.

In Cold Service. As expected, the yield point and tensile strength of solder increase and the elongation and reduction of area decrease sharply with lower temperature. In addition, the Izod impact strength decreases (see Table 3-5, which gives the results of a series of tests run on a 65/35 percent tin-lead alloy).

Low-temperature-strength studies of tin-lead solder alloys[1] indicate that the face-centered-cubic (fcc) structure of the lead phase remains ductile to cryogenic temperatures, while the body-centered tetragonal tin phase

---

[1]Jaffee, Minarcik, and Gonser, *Met. Progr.*, December 1948, pp. 843–845; Kalish and Dunkerley, *Trans. AIME,* vol. 180, pp. 637–659, 1949.

TABLE 3-5   Properties of 65/35 Tin-Lead Solder at −180°C*†

Temp., °C	Yield point, tons/in²	Ultimate tensile strength, tons/in²	Elongation in 2 in, %	Reduction of area, %	Izod, ft·lb
Room	1.75	3.63	28	54	15.0
−180	3.86	6.86	4	4	2.0

*On regaining room temperature from −180°C the Izod value rose to 12.0. Electrical conductivity of solders is slightly improved at lower temperatures.

† W. R. Lewis, "Notes on Soldering," Tin Research Institute, 1961.

loses ductility below −110°C. The exact temperature at which any particular alloy becomes brittle depends on the tin-lead ratio. A 50/50 tin-lead alloy, for instance, became brittle below −150°C while its ultimate strength increased to 18,500 lb/in². Solders with a higher lead content became brittle only at successive lower temperatures, until at about 70 to 80 percent lead the tin-lead alloys remained ductile down to temperatures near absolute zero.

The danger of tin transformation (tin pest) is described elsewhere in this book, but because it is always a serious consideration for low-temperature applications, pure tin or tin-lead is seldom used without antimony additives.

Solder alloys have also found many applications in superconducting devices. Superconducting behavior of the solders, which can affect the result of the assemblies, has been studied by Warren and Bader.[1] Their study of 18 solder alloys also gives valuable information on methods of application, useful fluxes, etc. A special solder for use with superconductors is mentioned elsewhere. More details of superconducting materials are beyond the scope of this book.

In Elevated-Temperature Service. At elevated temperatures the solder alloy approaches its melting point and thus loses strength while its ductility and elongations decrease. The author has developed for the electronics industry a general rule of thumb useful for deriving the maximum service temperature for a particular alloy:

$$T_{max} = \frac{T_{sol} - T_{room}}{1.5} + T_{room}$$

where T_{max} = maximum recommended temperature of alloy in use

T_{room} = ambient room temperature, normally taken as 65°F or 18°C

T_{sol} = solidus temperature of alloy = melting point of eutectic

[1] W. H. Warren, Jr., and W. G Bader, Super Conductivity Measurements in Solders Commonly Used for Low Temperature Research, *Rev. Sci. Instrum.*, vol. 40, no. 1, pp. 180–182, January 1969.

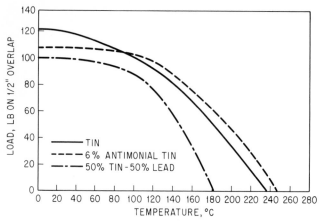

Fig. 3-23   Solder strength at elevated temperatures. *(From S. J. Nightingale and O. F. Hudson, Tin Solders, Br. Non-ferrous Met. Res. Assoc. Res. Monogr. 1, 1942.)*

For example, for eutectic 63/37 tin-lead solder which has a T_{sol} of 183°C (361°F) T_{max} would be

$$T_{max} = \frac{183 - 18}{1.5} + 18 = 118°C$$

or

$$T_{max} = \frac{361 - 65}{1.5} + 65 = 243°F$$

Obviously, soft solders are not intended for structural joining of materials for elevated-temperature service or for that matter at room temperatures. Only sketchy information is available in the literature on the strength of bulk solder at elevated temperatures. Figure 3-23 shows how the strength of the solder joint falls off with rising temperatures and compares it with antimonial lead and pure tin, which retain more strength

TABLE 3-6   Temperature Coefficient of Electrical Resistivity*

Sn	Pb	Sb	Temperature coefficient 0–100°C (32–212°F)
100	......	.....	0.0045
97.1	......	2.975	0.00366
95.0	......	5	0.00351
44.8	55.2	.....	0.00356
42.4	56.38	1.22	0.00380
40.0	57.55	2.45	0.00417
	100		0.00356

*S. J. Nightingale, The Joining of Metals, *Br. Non-ferrous Met. Assoc. Monogr.*, 1929.

TABLE 3-7   Linear Coefficient of Thermal Expansion and Thermal Conductivity*

Sn, %	Pb, %	Linear thermal expansion		Thermal conductivity	
		°C	Coefficient per °C × 10^{-6}	cal/ $(cm^2)(cm)(°C)(s)$	Btu/ $(ft^2)(in)(°F)(s)$
100	...	0–100	23.0	0.157	0.127
70	30	15–110	21.6		
63	37	15–100	24.7	0.121	0.0976
50	50	15–110	23.4	0.111	0.0895
20	80	15–110	26.5	0.089	0.072
5	95	15–110	28.7	0.085	0.068
...	100	17–100	29.3	0.083	0.067

*American Society for Metals, "Metals Handbook," 1961 ed., vol. 1, pp. 54–56.

under the same conditions. This information, however, should not be confused with creep strength, discussed earlier, which refers only to one single direct tensile measurement on the solder joint to failure. One should not take the information in Fig. 3-23 to show the superiority of pure tin and antimonial tin over solder without remembering that they are also higher-melting materials.

Table 3-6 shows the temperature coefficient of electrical resistivity for pure tin and a 45/55 tin-lead alloy. Several variations of the second alloy with the addition of antimony are also reported. This value is important in general consideration of the resistivity, but unfortunately few values are reported in the literature.

The linear coefficient of thermal expansion for tin-lead solders is shown in Table 3-7. A more detailed discussion of the contraction and expansion of alloys with temperature and after freezing is given in Sec. 6-20. Table 3-

TABLE 3-8   Brinell Hardness of Solders at Various Temperatures*

Sn, %	Pb, %	Brinell hardness†			
		19°C	65°C	115°C	140°C
100	0	4.5	3.3	1.5	1.0
97	3	7.8	4.6	2.7	1.7
87.6	12.4	7.3	4.3	2.4	1.7
80.4	19.6	6.0	3.7	1.6	1.3
76	24	4.9	2.2	1.0	
72.4	27.6	5.0	2.3	0.9	0.6
59.4	40.6	5.1	2.5	1.2	0.8
43	57	4.9	2.6	1.2	
9	91	7.0	4.9	2.9	2.3
0	100	2.6	2.1	1.1	0.7

*W. R. Lewis, "Notes on Soldering," Tin Research Institute, 1961.
†Specimens annealed and slowly cooled. Values obtained by measurement from curves.

7 also includes the thermal conductivity of solder alloys. The information is given in both the cgs system and the British system. The data seem to indicate that the change in thermal conductivity from pure tin to pure lead is a straightforward function. If the data are required for intermediate points, it appears that an extrapolation will give adequate information.

Table 3-8 shows the dependency of the hardness of the solder alloys on both composition and temperature. As in tensile strength, there is a rapid decrease with increasing temperatures.

3-19 The Properties of Solder Joints   In solder joints, the effect of the soldering parameters is the controlling factor. Solder joints made with tin-lead solders as well as some antimonial solders will be discussed here. Such parameters as time and temperature of soldering, the type of base metal and flux used, the mode of heating and cooling, and other considerations influence the characteristics of the joint enough to be recorded.

Unfortunately, however, in the situation we are now dealing with, perfection is rarely achieved. There is no way to assure that the heat characteristics of soldering tools can be reproduced in the laboratory or that they are representative of what happens in industry. The possibility of surface contamination causing nonuniform and noncomplete wetting becomes of primary importance, and the possibility of voids or imperfect joints is also evident.

Much of the work presented here was prepared by the British Non-ferrous Metals Research Association during World War II in an effort to establish some characteristics of solders and solder substitutes containing less tin because of the tin shortage in the overall wartime economy. Although additional and later information has been scattered through the literature, their work has been adopted for this book because it is the best documented and most appropriate. Much specific information on solder joints on printed-circuit boards can be found in federal documents.[1]

The Influence of the Quantity of Solder Used on the Shear Strength of Joints. Nightingale and Hudson[2] in an attempt to establish the importance of the

[1]Development of Highly Reliable Solder Joints for Printed Circuit Boards, *Westinghouse Def. Space Cent., Aerosp. Div., Baltimore, Final Rep. (Natl. Tech. Inf. Serv.),* NTIS-N69-25697, August 1969; B. D. Dunn, The Properties of Near-Eutectic Tin/Lead Solder Alloys Tested between Plus 70 and Minus 60 C and the Use of Such Alloys in Spacecraft Electronics, NTIS-N76-20264/74ST, September 1975; R. Gohle, The Significance of Strength for the Reliability of Solder Joints In Electronics, NTIS-N74-32686/95L, May 1974; F. L. Lane, Investigation of Gold Embrittlement in Connector Solder Joints, NTIS-N72-25522, Apr. 1, 1972; D. H. Liebenberg, Shear Strength of Soft-soldered Brass Joints as a Function of Age and Temperature History, NTIS-LA-3658 (CFSTI), Jan. 11, 1967.

[2]S. J. Nightingale and O. F. Hudson. Tin Soldiers. *Br. Non-ferrous Met. Res. Assoc. Res. Mongr.* 1, 1942.

quantity of solder used on the shear strength of joints, came to the following conclusions: "Since, in practice, large fillets usually conceal empty or partially empty joints, this occurrence is to be discouraged. It can be said, therefore, that there is nothing to be gained by the use of an amount of solder greater than that necessary adequately to fill the joint."

These results (Table 3-9), which clearly indicate that additional solder in the joint does not increase its strength, are extremely important and fortunate. Good inspection, as described later in Chap. 8, dictates that shallow fillets and inspectable fillets are absolutely necessary for high reliability in the solder joint. The fact that the amount of solder does not influence the strength is therefore welcome.

The Effect of the Time of Contact of the Molten Solder on the Shear Strength of Soldered Joints. As shown in Chap. 1, the mechanism of bond formation in soldering depends largely on the degree of wetting, and alloying itself is just a side effect. This conclusion was borne out by the work of Nightingale and Hudson in their effort to establish the effect of time of contact with molten solder on the shear strength of soldered joints. Their findings are that the results shown in Table 3-10, "though inclusive for the most part, suggest that the time of contact with molten solder within the limits of the experiment has little or no influence on the strength of the joint."

The mechanism of solution strengthening and the effect of alloying

TABLE 3-9   Effect of Quantity of Solder Used on the Shear Strength of Soft Soldered Joints*

Material joined	Soldering temp., °C	Joint thickness, in	Quantity of solder used, g	Shear strength	
				tons/in^2	kg/cm^3
Copper	300	0.003	24.8	2.60	365.6
			39.7	2.42	340.3
			62.2	2.14	300.9
			79.8	2.55	358.6
Brass	280	0.003	21.8	2.00	281.2
			41.5	2.22	312.2
			84.5	2.09	293.9
			92.0	2.37	333.3
Mild steel	280	0.005	9.7	2.39	336.1
			10.0	1.32	185.6
			26.3	1.17	164.5
			48.3	2.18	306.6
			62.7	2.66	374.0
			79.9	2.32	326.2

*Solder was 45/55 tin lead solder; flux $ZnCl_2$.
†S. J. Nightingale and O. F. Hudson, Tin Solders, *Br. Non-ferrous Met. Res. Assoc. Res. Monogr.* 1, 1942.

TABLE 3-10   Effect of Time of Contact with Molten Solder on the Shear Strength of Soldered Joints*

| Time of contact, min | Shear strength | | | | | |
| | tons/in^2 | | | kg/cm^2 | | |
	Copper	Brass	Mild steel	Copper	Brass	Mild steel
5	2.72	0.65	1.66	382.5	91.4	233.4
10	1.41	1.89	1.37	198.3	265.8	192.6
15	2.40	2.02	1.67	337.5	284.1	234.8
20	2.08	1.23	2.14	292.5	173.0	300.9
25	....	1.34	....	.....	188.4	
30	2.40	1.31	1.49	337.5	184.2	209.5
40	1.43	....	....	201.1		

*Solder was 45/55 tin-lead solder; thickness of joint 0.003 in; soldering temperature 300°C; flux $ZnCl_2$.

†S. J. Nightingale and O. F. Hudson, Tin Solders, *Br. Non-ferrous Met. Res. Assoc. Res. Monogr.* 1, 1942.

were discussed in more detail at the beginning of this chapter. However, it should be noted that for each base metal the fluctuations fall within the limits of accuracy of the test employed.

The Influence of Soldering Temperature on the Shear Strength of Joints. Under optimum soldering conditions, the strength, as indicated by shear, depends on the joining temperature. The type of flux used at the particular temperature is also extremely important. Not all fluxes are suitable for good wetting and good tarnish removal at all temperatures, and therefore there is an optimum temperature for each particular flux. The work of Nightingale and Hudson with zinc chloride indicated that below 260°C (536°F) the zinc chloride did not become fluid enough to leave the soldering area and was found in the form of inclusions on the surfaces. Around 280°C best results were obtained. Above that temperature, the fluidity of the flux as well as the solder seemed to be large enough to let the solder flow out from the joint area in spite of the capillary forces. Figure 3-24 shows a graph of the strength of two solders on three base metals in comparison with the temperature of joint formation. Note that the pretinned specimens are much stronger.

The Influence of Thickness on the Strength of Joints. It has been established that the strength of the solder joint is a function of the spacing between the soldered interfaces. The optimum conditions seem to be around 0.003 in when joint strength is at a maximum (much higher than the solder alloy itself). With this clearance, the flux and the solder can flow easily into the joint to give uniform wetting. When the gap is narrower, gases and fluxes are liable to be trapped, which results in a decrease of wetted areas and consequently less solder-fillet strength. Larger spacings have less capillary

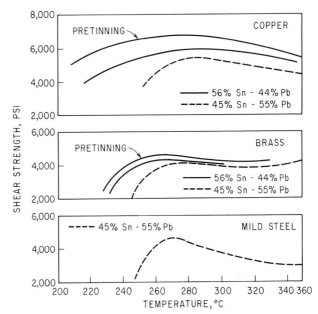

Fig. 3-24   Joint shear strength as a function of soldering temperature. Flux used ZnCl₂, joint thickness 0.003 in. *(From S. J. Nightingale and O. F. Hudson, Tin Solders, Br. Non-ferrous Met. Res. Assoc. Res. Monogr. 1, 1942.)*

force to aid wetting, and the mechanical strength of a thicker solder fillet is lower and usually approaches that of the solder alloy. Thus the chances of complete solution strengthening are reduced.

Figure 3-25 shows the effect of solder thickness on the shear strength of a 60/40 tin-lead solder joint. Here the base metal used was brass, and the soldering flux was eutectic zinc ammonium chloride salt. The samples were carefully prepared by the author and were inspected after destruc-

Fig. 3-25   Bond shear strength as a function of joint thickness. Flux used, eutectic ZnCl₂ and NH₄Cl, solder 60/40 tin-lead, base metal brass.

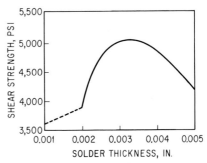

tion for the perfection of the solder fillets. The values obtained for the lower spacings were corrected for inclusions and imperfections.

Similar work with a lower peak was done by Nightingale and Hudson. Figure 3-26 gives the results of their work. They used only zinc chloride as a flux. The results indicate a peak similar to that found by the author. The solder used in these experiments had less tin (56/44 tin-lead). The purity of the alloy and the soldering parameters were not reported.

It is interesting to note that in spite of the difference in the base metals used and shown in Fig. 3-26, the peaks seem to lie in the same general area. It is also interesting to note that the solution strengthening of the material was definitely responsible for the change in the shear strength of the joints. The copper, which would be the major contributing or strengthening addition to the tin-lead system, gave the highest strength in shear. The brass came second, and the mild steel just approached the value of pure bulk material of the same tin-lead composition. As mentioned earlier, it is not quite sure that the work was carried out in such a way that full benefit of solution strengthening occurred. The reader is again reminded that it is important to know the time and temperature at which any specific set of experiments were run in order to be able to explain theoretically why certain conditions occurred.

This work led Nightingale and Hudson to a formula which can be used

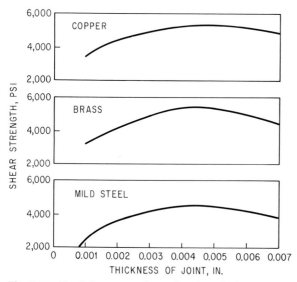

Fig. 3-26 Bond shear strength as a function of joint thickness. Flux used, $ZnCl_2$, solder alloy 56/44 tin-lead, soldering temperature 280°C. *(From S. J. Nightingale and O. F. Hudson, Tin Solders, Br. Non-ferrous Met. Assoc. Res. Monogr. 1, 1942.)*

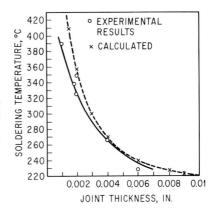

Fig. 3-27 Soldering temperature vs. joint thickness for maximum joint strength. *(From S. J. Nightingale and O. F. Hudson, Tin Solders, Br. Non-ferrous Met. Res. Assoc. Res. Monogr. 1, 1942.)*

to express the relationship between joint thickness and soldering temperature. A third series of specimens varying in thickness between 0.001 and 0.006 in were run at soldering temperatures varying between 215 and 500°C (419 and 932°F). The results were then plotted as in Fig. 3-27, and a smooth curve resulted for the maximum joint strength. The curve closely resembles that of a hyperbola; thus the following formula was found to be applicable:

$$(T - t)s = k \tag{3-3}$$

where T = soldering temperature
t = eutectic temperature of tin-lead
s = joint thickness
k = constant found empirically to be 0.34

Figure 3-27 shows the close resemblance between the theoretical curve, which is marked by the dashed line, and the empirical solid line.

The results of this test were so conclusive that it can be stated as a law connecting maximum joint strength and the thickness with the soldering temperature.

Whether the value of k is the same for all soft solders has not been determined. Considering for the given joint thickness the same soldering temperature gave the best results with two different solders and three different base metals [Fig. 3-24] and further that on the same three materials soldered at the same temperature the best results were obtained with the same joint thickness [Fig. 3-25] it would suggest that the constant k is the same for all solders containing tin-lead eutectic irrespective of the materials soldered. It is, however, likely that this relation is not so unassociated with the flux used. Those joints made at lower temperatures and the smaller spacing nearly always had flux inclusions.[1]

[1]S. J. Nightingale and O. F. Hudson, "Tin Solders," *Br. Non-ferrous Met. Res. Assoc. Res. Monog.* 1, 1942.

The reader is reminded that the higher temperatures and smaller thickness are possible because of the high fluidity of solders at elevated temperatures, whereas at lower temperatures a more sluggish solder will not fill in and penetrate the crevices. The theoretical curve should not be used beyond the limit at which the experimental results helped to establish the accuracy of the statement, however, because the solder does not wet readily at lower temperatures. When the temperature gets too high, oxidation and flux problems become extreme and erroneous values are obtained.

Change of Joint Strength with Solder Alloy and Base Metal. From the foregoing, it can easily be seen that the strength of the solder joint, as exemplified by the shear or tensile strength, depends on the base metals to which it is soldered because of their contribution as solution strengtheners. Hudson and Nightingale show the effect of the base metals on the solder joints in the useful range in Fig. 3-28. We can compare the change in the properties of the joint with those of the bulk solder (shown as the lower graph) by a quick review. It appears that the solution strengthening when it takes effect influences the solder strength in relatively the same fashion as the solder bulk when pure. However, the increase in strength obtained by making joints to mild steel is surprising and cannot be explained through solution effects. The phenomenon as such is therefore unexplainable. The solid line in these graphs indicates the tin-lead solders, and the dotted line is a lead-tin-antimony solder. The antimony additions correspond to those allowed under the British standards for those alloys.

SOME GENERAL REMARKS. Again we stress the importance of using the strength of the bulk solder itself for most design purposes, excluding any possible increase in the strength of the solder joint due to alloying and including this effect in the safety factor, because there is no assurance whatsoever that uniform solution strengthening will take place every time. It is also important to remember that the flux, the base metal and its condition, and the soldering parameters contribute largely to the joint characteristics.

It is also important to remember that although antimony imparts greater strength to solder joints and improves creep resistance and solder-joint strength at elevated temperatures, it produces a certain amount of sluggishness in the solder and does not improve its wetting characteristics. Figure 3-32 shows that the spread factor of the tin-lead alloys decreases with the increase in percentage of antimony added. This is not significant in dip and wave applications, a fact corroborated by practice (see Sec. 3-21). For preform and paste soldering, however, this may be an important consideration.

The Creep Strength of Solder Joints. We discussed the creep stength of solder alloys in Sec. 3-15, but the creep strength of solder joints is also of primary importance. It reflects the behavior of solder joints under small but constant loads. Baker[1] describes a method followed to prepare a series of special creep-test specimens with a lap joint of controlled thickness. The joints were made using zinc ammonium chloride eutectic flux in an aqueous solution. Various parameters were checked, such as soldering temperature, time, and joint thickness. Joints were made on steel, copper, and brass using three commercial solders of which the compositions are given in Table 3-11.

The results of the test are summarized in Figs. 3-29 and 3-30.

Because of the necessarily small dimensions of the joints tested, it was not possible to obtain actual creep curves, and hence creep rates could not be determined. The work by Baker therefore showed a comparison of

[1]W. A. Baker, The Creep Properties of Soft Solders and Soft Soldered Joints, *J. Inst. Metals,* vol. 65, pp. 277–297, 1939.

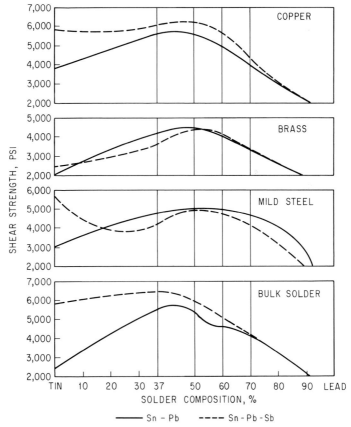

Fig. 3-28   Solder-joint strength as a function of alloy composition. *(From S. J. Nightingale and O. F. Hudson, Tin Solders, Br. Non-ferrous Met. Res. Assoc. Res. Monogr. 1, 1942.)*

TABLE 3-11   Solders Used for Thin Film Joints*

Mark†	Sn	Pb	Sb	Cu	Stress, lb/in² for 10⁻⁴ strain per day, room temperature
CQ3	44.3	55.68	...	0.009	140
CQ4	44.3	55.68	...	0.093	400
CQ5	40.0	57.8	2.0	0.090	420

*S. J. Nightingale and O. F. Hudson, Tin Solders, *Br. Non-ferrous Met. Res. Assoc. Res. Monogr.* 1, 1942.
†CQ = commercial quality.

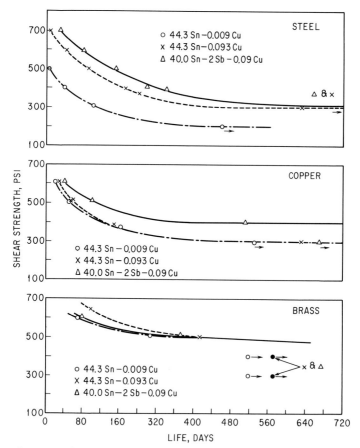

Fig. 3-29 Shear stress, duration of solder bond (at room temperature) for single lap joints. Arrow indicates unbroken joint. *(From S. J. Nightingale and O. F. Hudson, Tin Solders, Br. Non-ferrous Met. Res. Assoc. Res. Monogr. 1, 1942.)*

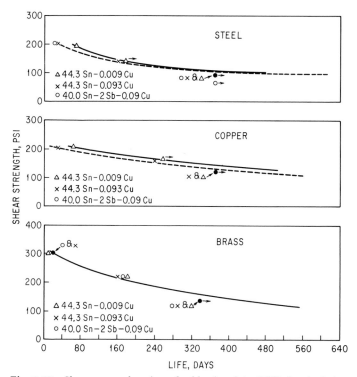

Fig. 3-30 Shear stress, duration of solder bond (at 80°C) for single lap joints. Arrow indicates unbroken joint. (*From S. J. Nightingale and O. F. Hudson, Tin Solders, Br. Non-ferrous Met. Res. Assoc. Res. Monogr. 1, 1942.*)

stress-life curves; in other words, stresses were plotted against number of days required to break the joint under that stress and, where possible, were also compared by deducing from the stress-life curve the stress corresponding to a life of 500 days under the continuous application of that stress.

When a fracture occurred, the total displacement of the members before failure was clearly defined on the solder film, and this quantity, which is a guide to the capacity for deformation of the joint under prolonged shear stress, was measured and recorded. In the case of joints prepared under the standard conditions described, this deformation varied from about 0.04 to a little over 0.1 in, being generally greater the longer the life of the joint. Cleaner film joints had a lower capacity for shear deformation.

A comparison of the shear strength of the bulk solder and the solder joint under similar conditions is reported in Tables 3-12 and 3-13.

When Figs. 3-29 and 3-30 and Tables 3-12 and 3-13 are carefully surveyed, it can be noted that the single lap joint on steel indicates a life of the joint in the prolonged shear stresses which is similar to that of the bulk solder. This can be explained by the fact that no solution strengthening with steel will occur. In addition, it can be seen that high-purity solders with no additions are weaker than those with minor alloying additions and therefore solution strengthening. In all cases, the addition of antimony to the alloy gives it superior creep-resistant properties.

With copper and brass the difference between the pure alloys and the less pure alloys is greatly diminished because of the solution of copper from the base metal which diffuses into the lattice of the solder, imparting

TABLE 3-12   Correlation between Creep of Solder and Creep of Joint at Room Temperature*

Solder used		Joint		
Composition, %	Stress, lb/in^2 for 10^{-4} strain per day	Material joined	Stress, lb/in^2 for 500-day life	Properties of solder film in joint
Sn 44, Cu 0.009	140	Steel	210	Apparently little alloying
		Copper	310	Marked alloying; solder film strengthened by absorption of copper
		Brass	470	Marked alloying; solder film strengthened by absorption of copper and zinc
Sn 44, Cu 0.09	400	Steel	310	Apparently little alloying but solder film strong owing to initial copper content
		Copper	310	Marked alloying but solder film not further strengthened by absorption of copper
		Brass	470	Marked alloying; solder film further strengthened by absorption of copper and zinc
Sn 40, Cu 0.09, Sb 2.0	420	Steel	325	Apparently little alloying but solder film strong owing to initial antimony content
		Copper	390	Marked alloying; solder film further strengthened by absorption of more copper
		Brass	470	Marked alloying; solder film further strengthened by absorption of copper and zinc

*S. J. Nightingale and O. F. Hudson, Tin Solders, *Br. Non-ferrous Met. Res. Assoc. Res. Monogr.* 1, 1942.

TABLE 3-13   The Effect of Soldering Conditions on Joint Strength*
Solder CQ4, 0.08% copper

Soldering conditions	Material joined	Stress, lb/in²	Life, days†		Displacement
Soldering temp. 320°C;	Steel	600	23	(42)	0.031
time 45–60 s; joint		400	215	(155)	0.043
thickness 0.0015 in	Copper	600	29	(38)	0.014
		400	119	(124)	<0.015
	Brass	600	94	(120)	0.035
		400	>384	(>492)	<0.001‡
Soldering temp. 320°C;	Steel	600	32	(42)	0.057
time 3 min; joint		400	193	(155)	0.112
thickness 0.006 in	Copper	600	48	(38)	0.056
		400	125	(124)	0.057
	Brass	600	153	(120)	0.050
		400	>350	(>492)	<0.001‡
Soldering temp. 400°C;	Steel§	600	13	(42)	
time 45–60 s; joint		400	105	(155)	
thickness 0.006 in	Copper§	600	35	(38)	
		400	93	(124)	
	Brass§	600	150	(120)	0.053
		400	197	(>492)	0.075

*S. J. Nightingale and O. F. Hudson, Tin Solders, *Br. Non-ferrous Met. Res. Assoc. Res. Mongr.* 1, 1942.
†Figures in parentheses refer to lives to be expected from joints soldered under conditions applicable to tests summarized in Figs. 3-29 and 3-30.
‡Unbroken joint.
§Joint slightly defective.

properties of its own. The solution of the zinc from the brass appeared to improve the properties of the joint even more than the copper, hence the stronger joints with brass. This makes it appear that the addition of zinc to the solder is beneficial, but we know that zinc is detrimental to other properties of the solder. For more details see Sec. 3-13.

ALLOY FAMILIES USED IN SOLDERING

3-20 Antimonial Tin-Lead Solders and Their Properties   Although antimonial tin-lead solders have not been discussed as such, information about these alloys was included in the previous section together with information on pure tin-lead alloys. Let us consider the tin-lead-antimony system for solders as a group and cover some of their properties.

The solid solubility of antimony in lead at room temperature is very small, whereas the solid solubility of antimony in tin at room temperature is around 7 weight percent. It is therefore possible by neglecting the solubility of the antimony in the lead to construct a similar phase diagram for the ternary alloy system tin-lead-antimony which would give us the same information as the tin-lead diagram discussed earlier. However, since it is a ternary phase diagram, the two-dimensional representation is no longer possible, and Fig. 3-31 gives a three-dimensional model of the tin-lead-antimony system.

Let us remember that the effect of the addition of antimony on the tin-lead system is beneficial if we keep the additions of antimony within the limit of solid solubility (hence 7 percent antimony to the percent tin present in the alloy). The small solid solubility of the tin in the lead at room temperature, which is less than 2 percent and will not affect the overall picture unduly, is neglected. When the amount of antimony exceeds this limit, it comes out of the solid solution as an intermetallic, the character of the alloy is altered, and the excess antimony becomes quite useless to the solder alloy (See Fig. 3-12).

There is considerable economic advantage in the use of antimony to replace tin because of the strength gained as well as price considerations (see Sec. 3-21).

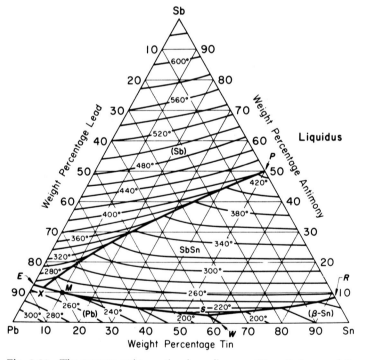

Fig. 3-31 The ternary antimony-tin phase diagram. *(From K. Iwase and N. Aoki, Kinzoku-No-Kenkyu, vol. 8, p. 253, 1931.)*

TABLE 3-14

	Weight percent			Temperature	
	Pb	Sb	Sn	°C	°F
X, eutectic	85.0	11.5	3.5	240	464
S, peritectic	40.0	2.5	57.5	189.0	372
M, pseudobinary eutectic	80.0	10.0	10.0	245.0	473

Figure 3-31 is the ternary tin-lead-antimony diagram with antimony, Sb, at the peak, lead, Pb, on the left and tin, Sn, on the right. Since most of the data leading to this equilibrium diagram were collected early in this century, the analytical methods used do not match presently available techniques, and it is hoped that these data will be rechecked soon. The points of interest in this diagram are listed in Table 3-14. Areas of primary crystallization have the following reactions. At point X we have a ternary eutectic, where three phases crystallize together

$$L \rightleftharpoons (Sb) + SbSn + (Pb)$$

Where (Sb) is the antimony-rich solution and (Pb) is the lead-rich solution. At point S we have a peritectic reaction, where SbSn reacts with the liquid to form

$$L + SbSn \rightleftharpoons (Pb) + (\beta\text{-}Sn)$$

where (β-Sn) is the tin-rich solution. At point M we have a coincident with the intersection of lines XS [where there is eutectic separation of (Pb) and (Sb)] and the pseudobinary line from Pb to SbSn.

Representation of ternary alloys in equilibrium diagrams is rather complicated and has limited usefulness in soldering. From Fig. 3-31 we can see that the properties of the tin-lead-antimony system (at the base of the diagram) will be similar to those of the pure tin-leads with a slightly higher melting point as long as we do not exceed the solid solubility. In practice, it is customary not to exceed 6 percent by weight of antimony in the tin because of the safety margin which might be required owing to variations in production of alloys. If the 7 percent antimony addition is exceeded, the saturation point of the solid solution is past and the antimony forms intermetallic cubical compounds, as shown in Fig. 3-12. This causes grittiness in the alloy and lowers the viscosity, making it more difficult to work as well as more brittle and more difficult to manufacture.

For the physical properties of tin-lead-antimony solder alloys in bulk, the reader is referred to Table 3-15. The physical properties of the tin-lead-antimony solder alloys are listed both as percent antimony in tin and

TABLE 3-15 The Physical Properties of the Tin-Lead-Antimony Solder Alloys*

Assayed composition wt %			Sb % of Sn content	Strength, lb/in²		Elongation in 4 in, %	Izod impact strength, ft·lb	Electrical conductivity, % IACS
Sn	Sb	Pb		Tensile	Shear			
100	0.0007	.....	0	1880	2560	55	14.2	13.9
97.1	2.975	.....	3.0	4080	4080	40	17.8	12.1
95.0	5.17	.....	5.5	5300	5360	38	20.4	10.9
94.5	5.59	.....	6.0	5620	5720	42	21.2	10.7
93.7	6.2	.....	6.5	5800	6020	32	21.5	10.4
66.1	.....	33.9	0	6860	5540	31	14.7	12.2
63.9	0.985	35.1	1.56	7780	5720	20	15.0	11.4
59.9	3.4	36.7	5.5	9000	6140	17.5	11.5	10.3
59.35	3.55	37.1	6.0	8840	6140	17.5	10.7	10.3
58.85	3.85	37.36	6.5	9100	6160	12.5	9.9	10.2
56.1	.....	43.9	0	5900	5580	38.7	15.7	11.2
52.82	1.62	45.56	5.0	6720	5840	31.2	15.7	10.2
50.67	2.75	46.58	5.5	7360	5980	30	12.2	9.8
50.15	3.00	46.85	6.0	7500	6120	29	11.1	9.6
49.65	3.13	47.22	6.5	7520	6020	27.5	10.2	9.7
44.8	.....	55.2	0	5300	4660	45	15.1	10.5
42.4	1.22	56.38	3.0	6380	5100	36.2	14.8	9.6
40.4	2.24	57.36	5.5	6780	5100	33.8	11.7	9.2
40.0	2.45	57.55	6.0	7100	5280	33.8	10.4	9.1
39.72	2.61	57.65	6.5	7020	5300	32.5	11.3	9.2
33.7	.....	66.3	0	5740	4260	25	12.2	9.6
31.8	0.94	67.26	3.0	6120	4500	20	11.8	8.9
30.2	1.65	68.15	5.5	6460	4520	21	10.1	8.6
30.1	1.75	68.15	6.0	6580	4380	21	11.3	8.6
29.8	1.92	68.28	6.5	6600	4440	18.8	9.2	8.6
		100	0	1780	1792	39	5.6	7.91

*S. J. Nightingale and O. F. Hudson, Tin Solders, *Br. Non-ferrous Met. Res. Assoc. Res. Monogr.* 1, 1942.

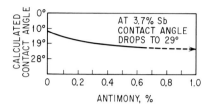

Fig. 3-32 Effect of antimony additions on the wetting of eutectic tin-lead. Test simulates only self-spreading operations.

as absolute chemical composition of the alloy. Each result in the table is the average of values for five specimens. The impurities in the alloys were unfortunately not quoted.

In several sections of this chapter, we have mentioned the loss of spreading ability of the tin-lead solder with the addition of antimony to the alloy. Figure 3-32 shows the actual loss of wetting power (similar to the test described in Fig. 3-11), and the graph clearly shows the increasing dihedral angle, or the loss of wetting, with the additions of small quantities of antimony to the solder. The loss appears to be rather rapid until approximately 0.5 percent of antimony has been added (the top limit of the QQ-S-571 specification) and levels off until, at 3.7 percent antimony, the spread is 86.4 percent. This decrease in wetting power of the antimonial tin-lead is quite marked in comparison with alloys of the same tin-lead composition. Reputable solder manufacturers therefore try to maintain the lower limit of the QQ-S-571 specification (0.2 to 0.25 percent antimony) rather than the higher percentages, although there would be a slight saving in price, as shown earlier.

The shear strength of antimonial lead for the optimum antimony content as used in British solders is presented in Fig. 3-33. Here the antimony addition was 6 percent, as specified. The material joint was copper, the flux was zinc chloride, the soldering temperature was 280°C (536°F), and the thickness of the joint was 0.003 in. The effect of the addition of antimony on the solder joint was discussed in Sec. 3-18

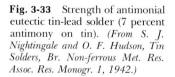

Fig. 3-33 Strength of antimonial eutectic tin-lead solder (7 percent antimony on tin). (*From S. J. Nightingale and O. F. Hudson, Tin Solders, Br. Non-ferrous Met. Res. Assoc. Res. Monogr. 1, 1942.*)

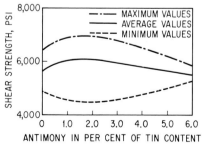

together with the effect of tin-lead solders on solder joints and the various parameters.

3-21 Functional Alloys Containing Antimony   The concept of a functional alloy was developed by the author while searching for a better solder. The only "scientific" alloys in the tin-lead system are the eutectic (63/37), some high-lead alloys above the solid solubility range (from 81 percent lead and up), and similar high-tin alloys (from 98 percent tin and up). The rest of the common solders, like 60/40, 50/50, 40/60, and 30/70, are historical mixtures of convenient rounded-off percentage composition that date back to the days of the guilds in the Middle Ages.

The price of tin, however, being much higher than lead, has always dictated frugality in the selection of alloys. In addition, there is no need to use only the eutectic compositions, since other tin-lead ratios are quite useful for specific applications. The 60/40 tin-lead solder should be considered a "near eutectic" and can be universally used unless we have very close spacings, thick multilayer printed-circuit boards, or similar critical applications.

The approach taken to a better solder started with a list of desired properties. The trend toward single-sided printed-circuit boards with straight-through leads coincided with this project. Industry's biggest worry therefore was to make sure it had a solder fillet that was strong enough. This need was even more urgent in the light of the ever-increasing trend toward larger hole-to-wire ratios necessitated by automatic insertion, etc.

A better solder could resolve this problem if it provided a stronger solder alloy. This was a natural application for antimony-containing alloys, which, as we have indicated, have better strength. A second major fallout from this approach was an obvious conclusion: in many cases, the true needs of the solder joint can be fulfilled by lower-cost high-lead alloys far removed from the eutectic composition. In other words, if such needs as strength, conductivity, and ease of manufacturing are satisfied, the solder alloy is adequate. Industry's acceptance then hinged on the ease of application, repair, and inspection. Let us review those in detail.

Ease of manufacturing and repair seemed a problem in light of decreased wetting power, both with lower tin content and/or the addition of antimony. But a close analysis of manufacturing methods indicated less trouble for hand applications with cored solder and liquid dipping in wave or still pots. The alloys may not be as suitable for automatic preform applications or the use of paste and cream; this requires further study.

Why isn't hand soldering affected by the more sluggish alloys? In the hand-soldering sequence (Sec. 7-18) we melt the solder into the joint and draw it in the direction of flow. This process is much more temperature- and technique-sensitive than material-sensitive. The telephone industry is living proof, since they have been using 30/70 tin-lead solder for most of their hand applications for years. Many appliance, television, and radio manufacturers have also been saving by using such alloys as 40/60 tin-lead in core solder.

The liquid (molten) solder application in still pots has also used low-tin alloys in substantial quantities for many operations. Here the reduced attack on copper, the lower drossing at elevated temperatures, and the need for piggyback joints was answered by these cheaper alloys. In some applications, like wave soldering of printed-circuit boards, the solidification characteristics and higher liquidus temperature prevented their use. Here the use of solders with narrow or no pasty range is mandatory. More will be said about a unique solution to this problem later.

Finally, we ought to consider the ease of inspection. Low-tin alloys are inherently dull and tend to get worse with increasing lead content. Historically, the electronics industry has associated quality with shiny solder. This cosmetic attitude must be changed if functional alloys are to be used. In some cases, it is actually easier to discern the fillet contour visually when the solder is not so reflective. The rule of thumb here is that *a solder joint need only be as shiny as the parent alloy.* In all other respects the inspection of functional alloys is no different from that of conventional alloys.

Let us then define the term *functional alloy.* It refers to an alloy that meets or exceeds the true needs (function) of a joint and at the same time has the lowest metal cost possible. In that respect it differs from the historical alloys that have been used blindly without consideration for the function of the solder in the joint.

In 1976 the author was granted U.S. Patent 3,945,556, entitled Functional Alloy for Use in Automated Soldering Processes. This solder, which is commercially available,[1] is an extension of this concept for an alloy useful in wave soldering and similar applications of liquid solder from a molten reservoir. Functional alloys have a higher tensile and shear strength than equivalent historical alloys and a similar elongation but are lower in cost[2]. These alloys, which contain 2.5 to 3.0 percent antimony in a tin-lead matrix can be used in wave soldering because they have a pasty range (between solidus and liquidus) of the same order of magnitude as the eutectic and near-eutectic tin-lead solders. Industrial experience to date, with millions of pounds successfully used, indicates that they fulfill all the technical requirements.

3-22 Fusible Alloys as Solders Although most people associate the word *solder* with tin-lead alloys, many other families of alloys can be used for soldering. These are sometimes referred to as *fusible alloys.* If they wet

[1]Patent assignee, Alpha Metals, Inc., Jersey City, N.J.

[2]J. P. Langan and L. Souzis, The Functional Alloy Approach to Soldering, *Weld. J.,* January 1977.

the base metals and thus effect a joint, however, they are by definition soldering alloys. The wetting characteristics of fusible alloys are not necessarily such that they automatically can serve as a solder. Table 4-3 gives a list of materials which have a known history of successful use. This list does not include many proprietary alloys whose detailed analysis and properties cannot be published.

Figure 3-34 shows a temperature scale in the form of a thermometer with both soldering and brazing superimposed on the scale. Disregarding brazing in this discussion, we see that the tin-lead alloys cover only a small portion of the temperature range allotted to soft soldering. However, many additional alloys cover the soldering-temperature range all the way from 100 to 800°F (37 to 426°C). In the following chapter we shall discuss the more prominent groups, touching on bismuth-base alloys, covering the lower part of the soldering range, the indium-base alloys, covering the medium-temperature range, and the silver-bearing alloys, covering the top range of the temperature scale.

The suitability of an alloy with a melting point under 800°F for a specific soldering operation is a subject for individual study. No scientific rules are available today to predetermine wetting characteristics (remember that the wetting characteristics are an integral part of the flux–base-

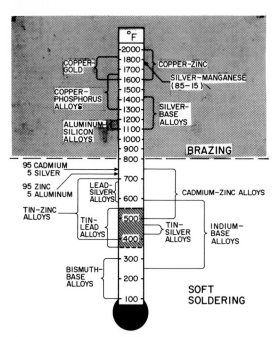

Fig. 3-34 Common alloy families for soldering and brazing.

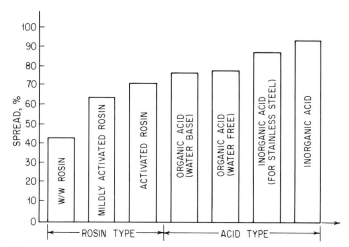

Fig. 3-35   Percentage spread as a function of flux type; average spread of Alpha Metals alloy 111 on copper, measured on spread-rate analyzer under identical conditions.

metal–solder system used and as such should be considered together). To demonstrate this point, let us look at Fig. 3-35. In an investigation of a proprietary alloy (Alpha Metals alloy 111) the author checked the activity of a series of proprietary fluxes as shown in the diagram. The spread-rate analyzer was used to evaluate the wetting characteristics of this alloy over copper surfaces. The results clearly indicate that the activity of the flux greatly influences the degree of wetting of this particular solder over copper and that the alloy cannot be used satisfactorily with rosin-base fluxes or organic-acid fluxes but must be used with inorganic-acid fluxes in order to obtain a high degree of reliability for the particular assembly.

Thus, if an unusual alloy is required for a particular joining operation where no history of application can be found, it is relatively simple to check the wetting characteristics using a simple spread-factor test. This does not assure trouble-free use, however, because of other metallurgical, physical, and chemical considerations covered in Chap. 4.

3-23 Low-Temperature Solders Containing Bismuth and Cadmium   Bismuth additions to tin, lead, cadmium, and other alloying agents make a uniquely attractive combination for low-melting-point alloys which are also suitable for soldering. Table 3-16 shows a typical series of tin, bismuth, lead, and cadmium alloys with some small additional alloying constituents covering a solder-temperature range of 133 to 289°F (56 to 143°C).

Table 3-16 is by no means a complete list of all the possible combinations of the materials which can make alloys that melt at various tempera-

TABLE 3-16 Tin-Bismuth with the Addition of Other Elements (Fusible Alloys)*

Hardness, tensile properties, solidus, and liquidus

| Composition, % | | | | | Freezing range | | | | | Brinell hardness 10/500/60 | | Tensile properties | | |
Sn	Bi	Pb	Cd	Others	Solidus °C	°F	Liquidus °C	°F	Ref.	Value	Ref.	Ultimate tensile strength, tons/in²	Elongation %	Ref.
43	14	43	…	…	143	289	163	325	2	…	…	3.06	…	7
34	46	20	…	…	…	…	…	…	…	10†	5	3.55	100	5
33	34	33	…	…	95	203	143	289	2	…	…	…	…	…
26	54	20	…	…	102	216	…	…	2					
25	50	25	…	…	93	199	…	…	1					
20	35.5	35	9.5	…	67	153	90	194	1	18		2.66	15	1
17	67	16	…	…	95	203	149	300	2	…	…	2.73	…	7
16	52	32	…	…	94	201	…	…	1					
15	39	31	15	…	68	154	85	185	1	1		2.90	9	1
13	50	27	10	…	70	158	…	…	5	…	…	2.03	50	6
13	48.5	26.5	10	2 Tl	68	154	…	…	3					
13	49	27	10	1 In	68	154	69	156	4					
12.5	50	25	12.5	…	68	154	73	163	1	25		2.02	3	1
11.5	45	24	9.5	10 Tl	67	153	81	178	3					
11.5	45	24	9.5	10 In	52	126	55	131	4					
11.5	57	23	8.5	…	67	153	73	163	1	12		2.18	10	1
11	42.5	38	8.5	…	70	158	90	194	2					
10.5	40	21.5	8	20 In	48	118	50	122	4					
10	28	27.5	34.5	…	71	160	120	248	1	23		2.82	15	1
9.5	50	34.5	6	…	70	158	80	176	2					
9	33	18	7	33 In	56	133	59	138	4					

*Reproduced by permission of the Tin Research Institute from L. T. Greenfield and F. G. Forrester, "The Properties of Tin Alloys," 1961.
†Testing conditions unspecified.

References: 1 = N. F. Budgen, *J. Soc. Chem. Ind.*, vol. 43, p. 200T, 1924; 2 = W. C. Smith, *Metals Alloys*, vol. 22, p. 397, 1945; 3 = S. J. French and D. Saunders, *Metals Alloys*, vol. 7, p. 22, 1936; 4 = S. J. French, *Ind. Eng. Chem.*, vol. 27, p. 1464, 1935; 5 = Cerro de Pasco Copper Corp., Matrix Alloy, 1934; 6 = E. J. Daniels, *Int. Tin Res. Dev. Coun. Tech. Pub. 5*, B; 7 = A. J. Philips, *Metal Ind. (Lond.)*, vol. 62, p. 150, 1943.

tures. Table 4-3 gives a select list of those alloys for which the author had ample evidence of successful use and should be consulted whenever solders of low temperatures are to be used.

Low-melting-point alloys consist of eutectic and noneutectic combinations of metallic elements which in themselves melt at a much higher temperature; they have been known for many years and have been called by many names. Many of these alloys have more than one name, or the same name is attributed to various compositions. It is therefore well today not to refer to these alloys by such names as Wood's, Lipowitz's, Onion's, Newton's, or Rose's metal. On the other hand, many of these alloys are sold under either trade names or proprietary numbers. This obviously is done in order to protect the manufacturer who develops a specific composition of material. With today's chemical facilities, it is easy to break down the composition of such materials if the information is really essential. With a reputable manufacturer, however, this system has an advantage inasmuch as the manufacturer sells a product rather than an alloy composition with specific properties and specific characteristics which he advertises and which he promises to supply. This practice, which is widespread in the steel, aluminum, and copper industry, can easily be carried into the region of low-melting-point alloys.

We should remember that some low-melting-point alloys are molten at temperatures below that of boiling water. This is unique, especially as the parent metals in the pure state have the melting points presented in Table 3-17.

This can be explained by the same considerations used in explaining the phase diagram. Many metals form eutectic compositions with a melting point much lower than that of the parent metals. Therefore, if we have a binary alloy system, we can achieve relatively low melting temperatures which can be lowered even farther by the use of ternary and quaternary eutectic compositions with melting points as low as shown in Table 3-16.

TABLE 3-17   Melting Temperatures of Solder Elements in Order of Increasing Temperature

Metal	Melting point		Price per lb*
	°F	°C	
Indium	313.5	155	170.00
Tin	449.4	231.9	6.50
Bismuth	520.3	271	8.00
Cadmium	609.6	321	3.75
Lead	621.3	327.4	0.32
Antimony	1166.9	631	2.38
Silver	1706.9	960.5	78.00

*The prices quoted are the approximate commodity prices as of November 1977.

A survey of Table 3-17 will reveal that the cheapest additions to tin-lead are antimony, cadmium, and bismuth. Antimony is not desirable in large quantities because of its relatively high melting point. This is why most fusible alloys with low melting temperatures contain bismuth, cadmium, lead, and tin in large quantities. Only when the addition of indium is absolutely necessary is this more expensive material added to the composition. As shown earlier, the addition of cadmium is not very desirable, whereas additions of bismuth improved the wetting of the solder. Additions of indium are discussed in Sec. 3-25.

3-24 Contraction and Expansion after Freezing   A temperature change in materials is always accompanied by a dimensional change. In metals, however, in addition to the regular expansion and contraction due to the temperature change, there is an additional change due to some lattice rearrangement of the atoms. This effect is especially prominent in bismuth and bismuth-containing alloys and also in some antimony alloys. If the amounts of bismuth and antimony are large enough, the resulting alloy will expand rather than contract after freezing because of a rearrangement of the lattice structure of the alloy itself. This physical change in the crystal structure is not to be overlooked when using bismuth-base solders. Although these forces may seem small, they are really of such magnitude that they can hold heavy loads, as is evident from some of the uses of these materials. It is common practice in the stamping industry to hold and position the guides of large sheet-metal presses with bismuth-base alloys.

In a simple system such as a bismuth-tin binary, it is simple to predict whether the alloy will expand or contract upon cooling. Alloys containing more than about 47 percent bismuth are reported to expand, whereas those which contain less bismuth are reported to contract, just like any other metal. An alloy that will have little expansion or contraction during cooling can therefore be designed by properly adjusting the addition of bismuth.

Let us consider the various factors which affect the volume change of a frozen solder alloy from the cooling period to room temperature and their time dependence. First we have a linear cooling-contraction function which is common to all materials. Modern physics teaches us that the atoms are not stationary inside the metal but, according to the amount of free energy they contain (mostly in the form of heat), vibrate around a location in the lattice which is their statistical position. With a rise in temperature, the thermal agitation increases, and the end result is an increase in the lattice spacing between the atoms and an overall growth in the dimensions of the material. This thermal agitation decreases proportionately with the temperature. The material, upon cooling, will therefore shrink. As indicated earlier, this is a linear function and can be considered instantaneous.

Nine materials have definite changes in their crystallographic structure during solidification which involve large changes of the outside dimensions. These changes are usually a function of the material itself and of the rate and method of cooling. The changes are always instantaneous and when arrested by a unique cooling process will not occur at room temperature.

Finally, there is the type of volume change that occurs with aging. This time-dependent change is the result of the readjustment of the lattice structure at a low temperature to the conditions it would have assumed if it had been cooled extremely slowly. The major difference between this and the previous type of dimensional change is that the first group of materials does not anneal or recrystallize at room temperature whereas the second group of materials can lower its internal free energy by crystallographic readjustment of the atoms even at room temperatures. The dimensional changes in bismuth-type alloys fall within the second group.

Figure 3-36 shows the dimensional changes of high-bismuth alloys as a function of time. Alloys A and B grow rapidly and start growing while

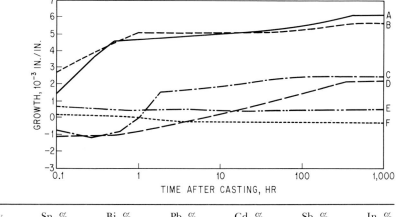

Alloy	Sn, %	Bi, %	Pb, %	Cd, %	Sb, %	In, %
A	14.50	48.00	28.50	. . . .	9.00	
B	13.30	50.00	26.70	10.00		
C	11.30	42.50	37.70	8.50		
D	. . . .	55.50	44.50			
E	42.00	58.00				
F	12.00	49.00	18.00	. . . .	. . . .	21.00

Fig. 3-36 Dimensional changes due to shrinkage and subsequent growth of some fusible alloys. (*From "Fusible Alloys Containing Tin," rev. ed. Tin Research Institute, Columbus, Ohio, 1956.*)

they are still warm. Alloys C and D show a normal contraction upon freezing, which is counteracted by slow growth, and after several hours they reach an overall larger dimension than the metal as initially cast. Alloy E shows a slight growth immediately upon casting, which is then maintained throughout the aging period, whereas alloy F grows slightly after freezing and then slowly contracts until it ends up slightly smaller than when it first froze. Note that zero in the chart indicates the dimension of the specimen upon freezing and that the scale starts from it, showing either growth or contraction.

The time dependency of the dimensional changes is very clearly brought out in Fig. 3-36. Some alloys start growing immediately after casting, whereas others seem to shrink and then start growing eventually, and finally there is a group that does not seem to be affected at all.

Hand in hand with the dimensional changes are definite changes in the physical properties of these alloys. However, from the standpoint of general design and use of solder, it is generally safe to use the characteristics of the joint after 1000 h, when there appears to be no more change in the properties of the material. This is the closest to what would occur with a normal solder joint which is not used immediately after it is made, and this time of aging is usually possible before use. When a joint must be used before it has aged completely, the safety factor will usually cover the situation adequately.

These unique expansion characteristics of bismuth-type solders can be utilized in some more delicate soldering operations such as intricate printed-circuit designs, where mechanical strength and reliability are important. For further details, see Sec. 6-20.

The mechanical properties of many low-melting alloys change gradually because of the minute changes in structure which take place during aging. This aging at room temperature is sometimes also referred to as *room-temperature recrystallization* or *annealing*. This effect increases the hardness and strength of the alloy and is shown in Table 3-18, where the

TABLE 3-18   Change in Mechanical Properties of Chill-cast Quaternary Eutectic on Aging*

Condition of test piece	Tensile strength		Elongation, %
	lb/in²	kg/mm²	
As cast	4547	3.20	50
Aged, 4 weeks	5734	4.03	43
8 weeks	5645	3.97	35
12 weeks	6451	4.54	30
16 weeks	6877	4.84	24

*S. F. Barclay, laboratories of Messrs. Mather & Platt Ltd., private communication, May 5, 1936.

results of some tests on chill-cast bars of the quaternary eutectic (tin-bismuth-cadmium-lead) are shown.

3-25 Intermediate-Temperature Indium-Base Solders   As stated earlier, the melting point of indium is 313.5°F (155°C) which in itself is the lowest melting point of any of the elements used normally for soldering alloys in extensive percentages. Indium can be described as a noble metal with good oxidation resistance, high luster, and good electrical and heat conductivity. In its pure state, the metal has low tensile strength (515 lb/in²), high ductility (elongation of 61 percent), and good electrical conductivity (22.1 percent IACS). Because of its softness and low tensile strength, the material tends to cold-weld easily. The term *cold weld* is generally used to describe the joining of two metallic surfaces mechanically by pressure so that the surfaces at the interface of the bond have been deformed and have flowed into one another, forming metallic continuity. The ease with which indium cold-welds can be attributed to (1) the relatively small amounts of surface tarnish due to the nobility of the metal and (2) the extreme softness and ductility of the metal, making it easy to reflow it with relatively small forces. The ease with which indium surfaces cold-weld has prevented them from being used extensively in mechanical electrical connections, where noble-metal surfaces are used because of their greater tarnish resistance. Mechanical electrical connections and make-and-break contact, as in relays and switches, are normally made of silver-, gold-, and platinum-plated surfaces according to the price the components can support. If these surfaces were indium-coated, they would give extremely good mechanical connections the first time, but several openings and closings of the connections would cause a complete rupture of the indium surfaces and leave a highly unreliable surface behind. In addition, the make-and-break forces would have to overcome the cold-welding operation physically.

Indium by itself is seldom used for soldered connections, although its extreme softness and pliability make it an ideal soldering material between surfaces which have different coefficients of thermal expansion, for instance.

As seen at the beginning of the chapter, solution hardening is the process whereby the strength of a pure element is greatly increased by the presence of small alloying additions. This also holds true for indium-base solders. However, because of the high price of indium, as seen in Table 3-17, it is not economically feasible to use high-indium alloys with their low strength, and indium is therefore considered to be an alloying addition to many materials. In small quantities, indium will improve wetting and electrical conductivity and will lower the melting point of the alloy. In larger quantities, it will predominate in the metallurgical structure of the alloy and impart some of the ductility and high conductivity of the parent

TABLE 3-19 Physical Properties of Indium Solders[a]

Indalloy solder number	Composition Indium, %	Composition Containing	Hardness[b]	Tensile strength lb/in²	Tensile strength kg/cm²	Elongation in 2 in, %	Melting point, °C Solidus	Melting point, °C Liquidus	Bonding-holding strength lb/in²	Bonding-holding strength kg/cm²	Electrical conductivity vs. copper[c]	Approx. weight 0.030 wire,[d] lb/1000 ft
1	50	Sn	4.94	1720	120.9	83	117	125	1630	114.6	11.7	2.37
2	80	Ag, Pb	5.24	2550	179.3	58	...	149	2150	151.2	13.0	2.51
3	90	Ag	2.68	1650	116.0	61	141	237.5	1600	112.5	22.1	2.48
4	100	......	e	515	36.2	41	...	156.7	890	62.6	24.0	2.28
5	25	Pb, Sn	10.24	5260	369.8	101	274	358	4300	302.3	7.8	2.75
6	5	Ag, Pb	8.98	4560	320.6	36.5	280	285	2830	199.0	5.5	3.50
7	50	Pb	9.60	4670	328.3	55	180	208.9	2680	188.4	6.0	2.80
8	18.3	f	4.80	g	g	g	...	38	100	7.0	g	g
9	12	Pb, Sn	11.96	5320	374.0	135.5	150	174	4190	294.6	12.2	2.44
10	25	Pb	10.24	5450	382.2	47.5	226.8	264.3	3520	247.5	4.6	3.00
11	5	Pb	5.98	4330	304.4	52	292.8	314.3	3220	226.4	5.1	3.48
12	5	Ag, Pb	9.00	5730	402.9	23	290	310	3180	223.6	5.6	3.47

[a]Data from the Indium Corporation of America.
[b]Hardness is modified Brinell; 100-kg load for 30 s. This is one-fifth of the standard 500-kg load.
[c]The conductivity is expressed in terms of the International Annealed Copper Standard, i.e. copper having a resistance of 0.15328 aΩ (meter, gram) at 20°C, a density of 8.89 g/cm³ at 20°C, and a temperature coefficient of resistance at 20°C of 0.00393.
[d]To obtain weights of other wires multiply figures as follows:

Weight	0.015	0.020	0.025	0.035	0.040	0.045	0.050
%	25	44.4	69.4	136.1	177.7	225	277.7

[e]Too soft to measure.
[f]Modified Lipowitz.
[g]This Indalloy cannot be extruded; it is shipped in cast shapes, bars, etc.

metal. Table 3-19 gives a list of 12 alloys (including pure indium) which have found uses in industry. The table lists their physical properties and even their bonding-holding strength. Since this list was published, additional alloys have been developed using indium for such special application as cryogenic solders and solders for thin films of gold.

Cryogenic solder,[1] which is covered by a patent, is composed of 19 percent lead, 30 percent tin, and 50 percent indium with 1 percent silver added to slow down the solution of silver from the tab to the solder. Although the standard composition proved mechanically stronger, the ingredients may be varied within the following limits: lead 16 to 40 percent, indium 35 to 70 percent, tin 10 to 33 percent, silver 1 to 3 percent. These materials have been applied to silver-fired glass surfaces, have been used at temperatures as low as 4.2 K, and seem to behave like superconductors.

A high-indium solder has been developed by the author which when applied to thin coatings of gold (several angstroms thick, as in vapor depositions) will not scavenge the gold off the surfaces rapidly. The alloy is the eutectic indium-zinc (96 percent indium, 4 percent zinc) with a melting point of 143.5°C (290°F). Tin-rich solders with good wetting characteristics are inadequate for this application because they dissolve the gold immediately upon contact. Although indium will dissolve gold eventually, if the temperature of the alloy is low enough and the alloying additions are chosen properly, users can apply the indium solder in the molten state for several minutes over the gold without dissolving the thin layer. Indium has a large appeal for this and similar thin-film solder applications.

Belser[2] claims that it is possible to make excellent bonds between indium and metals, nonmetals, and thin metal films. For specific application details of recommended indium alloys, see Table 4-3.

[1]A. W. Grobin, Jr., Special Solder for Use in Cryogenic Circuits, *Rev. Sci. Instrum.*, vol 30, no 11, p. 1057, November 1959.

[2]R. B. Belser, *Sci. Instrum.*, vol. 30, November 1959.

TABLE 3-20   Effect of Cleaning Technique on Shear Strength of Indium-Glass Bond

Alloy: 90 % In, 5 % Ag, 5 % Pb

Cleaning agent	Average shearing force at failure*	Shear strength, lb/in²
Tide	6.0	390
Tide and chromic acid	9.5	610
Tide and acetone	9.7	620
Tide, KMnO₄, and NaOH	11.5	730

*Area of contact: 1/8 by 1/8 in.

TABLE 3-21 Materials Successfully Soldered by Indium

Metals and alloys	Thin metal films	Nonmetals
Aluminum	Aluminum	Glass
Antimony	Cadmium	Quartz
Bismuth	Chromium	Mica
Brass	Copper	Porcelain
Cadmium	Germanium	Tile
Cobalt	Gold	Concrete
Copper	Iron	Brick
Germanium	Lead	Marble
Gold	Magnesium	Granite
Iron	Nickel	Seashells
Lead	Palladium	Aluminum oxide
Magnesium	Platinum	Copper oxide
Manganese	Rhodium	Germanium oxide
Molybedenum	Silver	Iron oxide
Monel	Tin	Magnesium oxide
Nickel	Titanium	Nickel oxide
Palladium	Zinc	Titanium oxide
Platinum	Zirconium	Zirconium oxide
Silicon	Brass	
Silver	Monel	
Stainless steel	Stainless steel	
Tin		
Titanium		
Tungsten		
Zinc		
Zirconium		

*Experiments performed by R. B. Belser, Georgia Institute of Technology. Wetting was considered to occur if a solder glob placed on a substance could not be scraped off with a scalpel as a single piece. Where tensile shear tests were conducted, values above about 350 lb/in² were obtained for nonmetals or thin metal films.

In his paper, Belser reports that, because of its position in the electrochemical series, indium solders wet to any of the metals below zinc in the series, using fluxes commonly used for tin-lead solders. However, as will be shown later, the oxidized surface of the same metals can be adhered to without the use of flux, which is the method of adhering to a nonmetallic form. A tool had to be used which heated the base metal to a temperature slightly above the melting point of indium (155°C). Either a torch or soldering iron may be used with acceptable procedures. Belser further states:[1]

> For soldering to metals above zinc in the electrochemical series, no flux appears desirable—even the application tool must be cleaned. Temperature

[1] R. B. Belser, "Intermediate Indaloy Solders," The Indium Corporation of America.

control is more critical and the temperature should be only slightly above the melting point of indium for the best results. The indium does not flow but must be smeared or rubbed over the surface to be wetted. Good joints are less certain than where a flux may be used.

In soldering to nonmetals such as glass and quartz maximum cleanliness is essential. Fluxes are not desirable and appear to inhibit wetting. Cleaning by scrubbing with detergents and subjecting to chromic acid, by flaming or by positive ion bombardment, is excellent. Table [3-20] shows some representative shear tests obtained with different cleaning methods.

A soldering tool of small capacity such as the Ungar soldering pencil (20 W, $1/8$-in tip) is a good choice for most purposes. A method of voltage control such as a Variac is desirable. In general, a minimum temperature to accomplish the joint should be used. A voltage of 75 to 80 V supplied to the tool is satisfactory for most purposes.

The soldering tool bit must be kept uncontaminated by flux for good results, and it is better not to use a soldering bit which has been used with flux unless thoroughly cleaned.

The solder is applied by a rubbing motion. It does not flow over nonmetals.

In general, soldering to thin metal films is conducted as described above for nonmetals. Temperature control is critical. A flux is not used. For soldering to certain metal films such as silver, the solder must be enriched with silver 5 to 10 percent to prevent absorption of the film. Excellent mechanical and electrical bonding are obtained. Metal films to which solder joints have been made are listed in Table [3-21]. It should also be pointed out that indium adhered to frictionally applied metal films on nonmetallic substrates as well as those deposited by evaporation, sputtering, or electroplating. Evaporated films of aluminum are also readily wet. In fact, all metal films tested thus far have been readily wet. Some metal films such as aluminum require somewhat higher temperatures of the soldering tool (105 V) than for gold or silver.*

*See R. B. Belser, "The Frictional Application of Metal to Glass Quartz and Ceramic Surfaces," Indium Corp. of America.

3-26 Silver-bearing Solders   Soldering of such items as silver-fired ceramic parts or silver-plated lugs and contacts is frequently a delicate operation because of the solubility of silver in tin (*silver scavenging*). The soldering operation must be accomplished before the thin silver coating becomes dissolved in the solder. One means of overcoming this difficulty is to use an alloy containing very little tin, since it is the tin which readily dissolves the silver. However, this generally requires higher temperatures and involves alloys with relatively poor wetting qualities. Since the solubility of silver in tin increases with an increase in temperature, a suggested solution might be to use a lower-melting alloy. Unfortunately, most low-melting soft solders, exclusive of bismuth-bearing alloys, are generally rich in tin, which is undesirable from the standpoint of silver solubility.

The most satisfactory solution to the problem of soldering to silver-

TABLE 3-22   Physical Properties of Liquid Solder Elements*

| Element | Temp., °C | | Latent heat, cal/g | | Vapor pressure | | Density | | Heat capacity | |
	mp	bp	Of fusion	Of vapor	mmHg	°C	g/cm³	°C	Cal/per g/per (cm³)	°C
Sb	630.5	1440	38.3	383.0	1	886	......	......	0.0656	650–950
					10	1033	6.49	640	......	....
					100	1223	6.45	700	......	....
					200	1288	6.38	800	......	....
					400	1364	6.29	970	......	....
Bi	271	1477	12.0	204.3	1	917	10.03	300	0.0340	271
					10	1067	9.91	400	0.0354	400
					100	1257	9.66	600	0.0376	600
					200	1325	9.40	802	0.0397	800
					400	1400	9.20	962	0.0419	1000
Cd	321	765	13.2	286.4	1	394	8.01	330	0.0632	321–700
					10	484	7.99	350	......	....
					100	611	7.93	400	......	....
					200	658	7.82	500	......	....
					400	711	7.72	600	......	....
In	156.4	1450	6.8	468	1	1249	7.026	164	0.0652	156.4
					10	1466	7.001	194	......	....
					100	1756	6.974	228	......	....
					200	1863	6.939	271	......	....
					400	1982	6.916	300	......	....
Pb	327.4	1737	5.89	204.6	1	987	10.51	400	......	....
					10	1167	10.39	500	0.039	327
					100	1417	10.27	600	0.037	400
					200	1508	10.04	800	0.037	500
					400	1611	9.81	1000	......	....
Ag	960.5	2212	24.9	556	1	1357	9.3	960.5	0.0692	960.5–1300
					10	1575	9.26	1000	......	....
					100	1865	9.20	1092	......	....
					200	1971	9.10	1195	......	....
					400	2090	9.00	1300	......	....
Sn	231.9	2270	14.5	573	1	1492	6.834	409	......	....
					10	1703	6.761	523	......	....
					100	1968	6.729	574	0.0580	250
					200	2063	6.671	648	0.0758	1100
					400	2169	6.640	704	......	....
Zn	419.5	906	24.4	419.5	1	487	......	......	0.1199	419.5
					10	593	6.92	419.5	0.1173	600
					100	736	6.81	600	0.1076	800
					200	788	6.57	800	0.1044	900
					400	844	......	......	0.1012	1000

*From Charles L. Mantell, "Engineering, Materials Handbook," copyright 1958, McGraw-Hill, Inc.

Viscosity		Thermal conductivity		Resistivity		Surface tension		Volume change on fusion % of solid
cP	°C	Cal/(s)(cm)(°C)	°C	μΩ·cm	°C	dyn/cm	°C	solid
......	....	0.052–0.05	630–730	117.0	627	...	.....	−0.94
1.296	702	....	....	117.65	700	383	635	
1.113	801	....	....	120.31	800	384	675	
0.994	900	....	....	123.5	850	383	725	
0.905	1002	....	....	131.00	900	380	800	
......	....	....	....	....	....	376	300	−3.32
1.662	304	0.041	300	128.9	300	373	350	
1.28	451	0.037	400–700	134.2	400	370	400	
0.996	600	....	....	145.3	600	367	450	
......	....	....	....	153.5	750	363	500	
......	....	....	....	33.7	325	564	330	4.74
2.37	350	0.106	355	33.7	400	608	370	
2.16	400	0.105	358	34.12	500	598	420	
1.84	500	0.105	380	34.82	600	611	450	
1.54	600	0.119	435	35.78	700	600	500	
......	....	0.09–0.12	mp	....	....	...	.....	2.5
......	....	....	....	29.10	154	...	170–250	
......	....	....	....	30.11	181.5	340		
......	....	....	....	31.87	222			
......	....	....	....	34.84	280.2			
2.116	441	0.039	330	94.6	327	...	.....	3.6
2.059	456	0.038	400	98.0	400	442	350	
1.700	551	0.037	500	107.2	600	438	400	
1.349	703	0.036	600	116.4	800	438	450	
1.185	844	0.036	700	125.7	1000	431	500	
2.98	1200	....	....	17.0	1000	923	995	4.99
......	....	....	....	18.2	1100			
......	....	....	....	19.4	1200			
......	....	....	....	20.5	1300			
......	....	....	....	21.0	1340			
1.91	240	....	....	47.6	231.9	526	300	2.6
1.67	300	0.080	240	51.4	400	522	350	
1.38	400	0.081	292	56.8	600	518	400	
1.18	500	0.079	417	62.7	800	514	450	
1.05	600	0.078	498	68.6	1000	510	500	
......	.....	....	....	35.3	419.5	...	.....	6.9
3.17	450	0.138	500	35.4	500	785	510	
2.78	500	0.136	600	35.0	600	778	550	
2.24	600	0.135	700	35.65	700	768	600	
1.88	700	....	....	35.70	800	761	640	

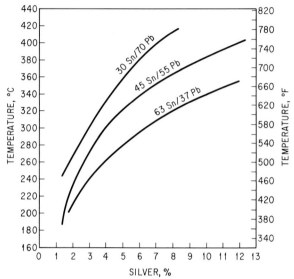

Fig. 3-37 Solubility of silver in tin-lead solder at various temperatures.

plated parts appears to be the use of silver-bearing solder. The solubility of silver in tin is greatly decreased by using a solder alloy already partially saturated with silver without adversely affecting the inherent solderability of the tin-rich alloys. Excessive amounts of silver should not be used, however, because of the bad effect of excess silver, as discussed earlier, and because silver is an expensive ingredient.

In determining the proper alloy for use in a particular application, the graph in Fig. 3-37 will prove valuable. For example, if an alloy with the solderability of the tin-lead eutectic (63/37) is desired for use in a pot operating at a temperature of 460°F (238°C), a 2 percent silver alloy is recommended, since the solubility of silver under these conditions is just under 3 percent. Such an alloy might be 62 percent tin, 36 percent lead, 2 percent silver.

Aside from the important use of silver in soft solders to prevent scavenging of silver from plated parts, it is widely used in applications requiring a high-melting alloy. Silver solders generally include silver alloys melting above 1100 or 1200°F (593 to 649°C); in the field of soft solders, however, lead-silver alloys provide a series of solders in the 500 to 600°F (260 to 315°C) melting range. The lead-silver eutectic, for example (97.5 percent lead, 2.5 percent silver), melts at 581°F (305°C). In general, these alloys are not so readily workable as the tin-lead solders and require more heat and active fluxing.

The 97.5/1/1.5 tin-lead-silver eutectic, which melts at 305°C (581°F), is an important solder for component and hybrid assembly.

Alloys that have been found useful in the soldering industry in order of melting point are as follows

Tin, %	Lead, %	Silver, %
98	....	2
...	97.5	2.5
1	97.5	1.5
...	95	5
61	36	3
56	39	5
62	36	2

3-27 Other Alloys Used for Soldering   A variety of fusible alloys which have limited use in industry today are discussed in Chap. 4, and technical data concerning them are presented in Table 4-3. In this chapter, we have discussed only the properties and metallurgy of the major alloying families without going into the details of the various minor additions other than the effect of contamination on the tin-lead solder system. In addition to this chapter and the chapters on design and selection of the solder alloy, there is one more approach to finding the correct solder alloy for an application.

In cases where no specific technical data can be obtained in the literature, Table 3-22 can be consulted in order to estimate some of the properties of the liquid elements as they can be used for soldering. In many cases it will be necessary to extrapolate the values from those given in the table and to estimate the information required. The only other alternative is a thorough testing program to obtain the information required. Table 3-22 is given for the convenience of the readers so that they can make up some theoretical values and see which alloys and which combinations may be useful for their applications.

FOUR

Designing the Solder Joint

4-1 The Need for Solder-Joint Design A solder joint must be designed to bring out its major advantages, as listed in the Preface.
1. Electrical conductivity
2. Mechanical durability (strength)
3. Heat dissipation
4. Ease of manufacture
5. Simplicity of repair
6. Visual inspectability

The first three properties depend on both the material and the geometrical design. Mechanical durability also depends to a large extent on the process used.

We have already seen that solder interacts with the base metals it wets, thus changing the metallurgical system. Here time and temperature of soldering are critical factors depending on the process. Durability of the solder alloy, on the other hand, depends on its ability to act as a *stress coupler*. Excessive alloying and intermetallic crystal formation detract in most cases from the ductility of the solder and may shorten its long-term reliability. Thus, in item 2 above we refer to mechanical durability rather than mechanical strength.

The first part of this chapter will be devoted to the material-selection aspects of the solder joint, including the base metals, the flux, and the solder alloys. The material selection is then tied together in a simple flowsheet leading to geometrical design.

The geometrical configuration of the solder joint influences every one of the six characteristics listed above. The task of laying out a solder connection is greatly simplified, however, in this chapter. We design a solder joint around the desired geometry taken from points 4 to 6 and then calculate its dimensions for electrical conductivity. We check these dimensions for mechanical durability and heat dissipation. In the author's experience *a joint designed to match electrical conductivity is also mechanically durable and a good heat dissipator.*

The chapter contains two geometry-oriented sections, one to cover the mechanical design for structural joints and one to cover electrical conductivity. These two sections provide more than just design information and can often help make decisions on quality and workmanship standards, acceptance criteria of marginal joints, etc.

Ease of manufacture, repair, and inspection require no special discussion. Many good assemblies, however, have failed because one of these important factors was not properly considered in the design stages. There is a right way and a wrong way to make a solder connection, and the responsibility lies entirely with the designer to tailor each joint to the specific requirements of each application. It is poor practice simply to note on the blueprint "to be joined by solder". Since this leaves many important parameters open to the discretion of less qualified personnel, the designer may find his end product far from the desired quality and reliability.

A good designer must also consider production problems involved in soldering and should give reasonably explicit instruction on related procedures, like methods of cleaning, methods of heating, etc.

MATERIAL SELECTION

4-2 A Selection Outline   As shown in Chap. 1, the solder system consists of three major components, namely, the base metal, the soldering flux, and the soldering alloy. These three materials must be in perfect balance to give the soldered connection the characteristics the designer seeks. We shall discuss these one by one, show how they can be correlated and matched in the various systems, and give the designer a way to evaluate the material selection and a flow chart summarizing this phase of the design.

4-3 Choosing the Base Metal   When a base metal is being considered for a part in a soldered assembly, the designer normally weighs such properties of the metal as shear, elongation, and corrosion but may neglect factors important in soldering; these can be critical omissions. When a soldered assembly is designed, base-metal characteristics pertinent to the soldering operation should be considered. Table 4-1 lists common metals in order of their electromotive potential and indicates properties pertinent to soldering. These properties are discussed in detail below.

Electromotive potential is important in terms of possible corrosion. Corrosion will occur where there is a great potential difference between the metals joined or between the metals and the solder joining them if an ionic liquid is present. Table 4-1 shows the metals commonly used in solder alloys in italics.

To demonstrate this point, let us view the electrochemical potential between aluminum (line 3) and tin-lead (lines 11 and 12). The potential between the aluminum and the tin-lead solder is over 1½ V. This explains why tin-lead solder is rarely used for exposed aluminum, even disregarding the inherent difficulty of soldering aluminum. For further details, see Sec. 6-2.

In many new assemblies with very small and delicate parts, the thermal coefficient of expansion has become an important factor. All metals expand in all three directions upon heating. For engineering considerations, however, the linear thermal coefficient of expansion gives a better idea of the thermal behavior of the metal. In structural assemblies, the ductility of the solder itself plus the rigidity of the assembly are such that we need not stress the proper use of the thermal coefficient of expansion. However, parts in small and delicate assemblies can be buckled or pushed out of the required tolerances. It is important here to be sure that the coefficients of thermal expansion of the base metals and the solders lie in a range that will not cause stress or dimensional change in the assembly upon use at changing temperatures. This is especially important when soldering to fired surfaces on glass and ceramic materials. Here a high coefficient of expansion on one side and a low coefficient of expansion on the other side may well cause a failure of the solder joint upon heating and cooling, e.g., as a result of electrical-component warmup or frictional temperatures in running motors.

Failure due to heating and cooling is often referred to as *thermal fatigue*. When the design of assemblies which failed under these thermal-fatigue stresses is considered, however, it is usually found that the configuration is such that stresses are set up during thermal cycling. These stresses may be due to the relative expansion and contraction of the base metal and solder during the heating and cooling. Regardless of where these stresses originate, they put the solder in the fillet under continuous movement. If the solder is ductile and good wetting occurred, there will be no failure of the

TABLE 4-1 Design Data for Metals (in Order of the Electromotive Series)

No.	Metal	Electromotive potential, V	Solder-ability*	Electrical resistivity, μΩ·cm	mp, °F	Physical data Density, g/cm³	Linear thermal coefficient of expansion, in/°F	Modulus of elasticity in tension lb/in² ×10⁶	Modulus of elasticity in tension kg/cm² ×10⁵	Brinell hardness
1	Magnesium	2.34	2	4.46†	1204	1.74	0.0000143	6.5	45.7	30
2	Beryllium	1.70	3	106.8§	2343	1.82	0.0000069	4.6	32.3	60–125
3	Aluminum	1.67	2	2.655†	1215	2.70	0.0000133	10	70.3	15
4	Zinc	0.76	1	5.916†	787	7.14	0.000018	12	84.4	31
5	Chromium	0.71	3	6.24†	2822	7.148	0.0000045	36	253.1	108
6	Iron	0.44	1	9.71†	2795	7.87	0.0000066	28.5	200.4	60
7	*Cadmium*	0.40	1	3.43§	610	8.65	0.0000166	3	21.1	35
8	*Indium*	0.38	1	8.37§	313.5	7.31	0.000018	1.57	11.0	0.9
9	Cobalt	0.28	3	6.24†	2714	8.9	0.0000067	30	211.0	124
10	Nickel	0.25	1	6.84†	2645	8.9	0.0000076	30	211.0	110
11	*Tin*	0.14	1	11.5†	450	7.30	0.000013	6	42.8	5.2
12	*Lead*	0.13	1	20.65†	621	11.36	0.000016	2.6	18.3	3.9
13	Hydrogen	0.00								
14	Stainless steel	−0.09	2	74.0†	2550	8.02	0.000092	29.0	203.9	200
15	*Antimony*	−0.10	1	39.0§	1166	6.62	0.0000063	11.3	79.5	42
16	*Bismuth*	−0.20	1	119§	520	9.80	0.00000747	4.6	32.3	4.7
17	Silicon	−0.26	3	1 × 10⁵§	2588	2.33	0.0000041	16	112.5	240
18	Copper	−0.34	1	1.726¶	1981	8.94	0.0000091	16	112.5	42
19	Steel	−0.58	1	18.0†	2760	7.7	0.0000067	...	...	175
20	*Silver*	−0.80	1	1.59†	1761	10.5	0.0000105	11	77.3	95
21	Palladium	−0.82	1	10.8†	2829	12.0	0.0000066	17	119.5	35
22	Platinum	−0.86	1	9.83§	3224	21.45	0.0000043	21	147.7	37
23	Gold	−1.68	1	2.19§	1945	19.3	0.000008	12	84.4	28

*1=solders under normal conditions, 2=solders under special conditions, 3=not normally soldered.
†20°C.   §0°C.   ¶23°C.

127

joint although sometimes the surface may appear frosty (this is concentrated in areas of stress). The ductility of the solder, of course, is a function of its purity, which can be related to the amount of alloying and intermetallic formation during soldering and/or time and temperature of the process. It is also the function of the composition of the alloy, high-lead alloys being more soft and ductile and many higher-temperature solder alloys becoming harder and less of a stress coupler. In thermal fatigue the temperature cycling range is also important. While a cursory review might indicate that the size of the thermal excursion is important, this is not the case. Since solder is very ductile at elevated temperatures, it is the low-temperature extreme which is detrimental. Remember that the lower the temperature the harder and the less ductile the alloy becomes.

In Table 4-1 the solderability of the various metals is listed in three major categories: those which are normally soldered, those which are soldered under special conditions, and those which are not normally soldered. For further details on this problem of solderability and how it affects the choice, see Sec. 4-4.

Solderability refers to the bare metal and does not include details like surface treatment for improving solderability. There are various methods of improvement, the most common of which is the pretinning of the surfaces by *hot tinning* (coating with a fusible alloy that is going to be used as solder) or by plating with the solder alloy or with an alloy that is easily wetted by the solder alloy, e.g., silver, cadmium, or tin-nickel. This surface preparation may take place not only on the areas to be soldered but on the whole part. However, here again a close look should be taken at the corrosion resistance of the assembly. When severe galvanic corrosion can occur, other means of joining may be preferred.

4-4 Choosing the Right Flux   In Chap. 2 we discussed the important characteristics, both theoretical and practical, which make a material a good industrial flux. However, we have not yet indicated how to go about the actual selection of fluxes for reliable soldering.

To determine proper soldering fluxes for specific application, two steps are required:

1. Consult Table 4-2. The metals normally used in conjunction with soldering are represented in the table in three major groups.
   a. *Soldering easy.* These are the materials that can be soldered with rosin-base organic fluxes as well as the stronger fluxes. Note that under favorable conditions water-white rosin can be applied. This means that unless heavy tarnishes or similar contamination of the surfaces is present, these metals can be soldered with water-white rosin. Normally, they can be soldered with activated rosin or the nonrosin organic materials. The inorganic fluxes here are marked as not normally used because they have too

TABLE 4-2 Flux Guide for Engineering Metals*

Metal surfaces	Inorganic	Nonrosin organic	Rosin, organic w/w	Activated
Soldering easy:				
Brass	1	1	2	1
Bronze	1	1	..	2
Cadmium	3	1	2	1
Copper	3	1	2	1
Gold	3	1	2	1
Lead	1	1	2	1
Nickel	1	1	..	2
Palladium	3	1	2	1
Platinum	3	1	2	1
Silver	3	1	2	1
Tin	3	1	2	1
Tin-lead, hot dip	3	1	2	1
Plate	3	1	2	1
Tin-nickel, plate	3	1	2	1
Tin-zinc, plate	3	1	2	1
Soldering less easy:				
Aluminum	..	..	Special solder and flux	
Beryllium copper	1	2	Special flux only	
Cast iron	1			
Copper alloys	1	2		
Germanium	1	2		
Inconel	2	..	Special flux only	
Kovar	1	2		
Magnesium	..	..	Special flux only	
Monel	2	..	Special flux only	
Nichrome	..	..	Special flux only	
Rhodium	1	2		
Rodar	1	2		
Steel	1	2		
Galvanized	1	2		
Stainless	1	..	Special flux only	
Zinc	1	2		
Die-cast	..	..	Special flux only	

Normally not soldered: beryllium, chromium, molybdenum, niobium, silicon, tantaum, titanium, tungsten, zirconium

*1 = normally used; 2 = used only under favorable conditions; 3 = not normally used.

much chemical activity and too much corrosion danger is involved. However, they can be used if desired.

b. *Soldering less easy.* No materials in this group can be soldered with regular rosin-base fluxes, and even the nonrosin organic fluxes can be used only under favorable conditions. In many cases, special fluxes and solders are required to solder these materials.

c. The group of materials mentioned at the bottom of the table is not normally soldered without preparation of the surfaces. Unless special techniques are used, these materials should not be considered for soldering.

The first step in the selection of a flux is to find out which group of fluxes is suitable for the base material and then to select the one which is most suitable and at the same time least corrosive for the assembly at hand. The reader is referred to Chap. 2 and Table 4-2 for a comparison of the flux materials and their tarnish-removal and corrosive properties as well as cleaning procedures and other data.

2. Step 2 involves careful consideration of the fluxes suitable for the specific application and selection of the one which is compatible with the rest of the assembly according to the outline given later in this chapter. If, for reasons described later, the flux-selection chart reveals that the only fluxes suitable for a particular base metal are not suitable for the assembly, various surface-preparation methods are still available to the designer to eliminate the need for strong fluxes. Precoating surfaces by either hot dipping or plating is a simple method which will ensure good solderability with milder fluxes than those required originally. The reader is reminded that when the surfaces are prepared in such a manner, strong chemicals are used (plating solutions or strong fluxes for hot tinning) and then the corrosive residues are removed before the assembly is put together. Thus we do not eliminate the use of a strong flux but change the time and operation sequence.

This brings to light another important consideration in the flux selection, namely, the post- and precleaning operations involved. The precleaning operation makes the task of the flux easier but does not eliminate the need for fluxing. For further details on precleaning assemblies, the reader is referred to Chap. 5. The presence of foreign materials such as oils, waxes, and paints interposes a layer between the flux and the metal and makes the action of the flux impossible.

Postcleaning operations, however, have great bearing on the selection of fluxes, as will be shown later. Unless the flux materials, corrosive or noncorrosive, can be adequately removed from the surfaces according to the specification of the particular assembly, the selection of the flux is unfortunate and should be changed. Here sometimes a slightly more corrosive flux calling for easy removal techniques is more suitable than the less corrosive flux which is tenacious and expensive to remove.

4-5 Choosing the Right Solder   Though solders are normally thought of as tin-lead alloys, a large variety of other fusible alloys are suitable for soldering. Table 4-3 contains a series of such fusible alloys covering the melting range from 117 to 600°F (47 to 315°C) and consisting mostly of tin-lead, bismuth, indium, antimony, cadmium, and silver combinations. They are by no means the only alloys suitable for solder in this range, but they are among the most characteristic and readily available. For more details on the metallurgy of solders, see Chap. 3.

In Table 4-3 we have tried to list eutectic alloys which have very desirable freezing characteristics and alloys which are either very common or have a very narrow plastic range. The alloys in Table 4-3 are listed in order of increasing melting temperature, but one should keep in mind that melting points are by no means recommended soldering temperatures; these should be 70 to 150°F (21 to 65°C) above the melting point in order to give the molten alloy fluidity and good wetting characteristics. Actually, the 70 to 150°F above the liquidus temperature is an arbitrary figure and is also a function of the time of soldering. If long soldering times can be tolerated, a much lower soldering temperature can be used. The value 70°F above liquidus is generally appropriate for the lower melting points, while 150°F is appropriate for the higher points. For specific tin-lead recommended wetting temperatures see Fig. 3-8.

The parameters most frequently used to determine which fusible alloy is the most suitable solder for a given application follow.

Temperature. The effect of temperature on the selection of solder falls into two categories. There is the upper limit, determined mainly by the heat deterioration of the assembly, and there is the lower limit, which is mostly determined by the operating temperature of the solder joints.

THE UPPER TEMPERATURE LIMIT. Heat may distort or buckle thin sections in a given assembly; it may change the characteristics of materials, especially organic and nonmetallic; it may discolor the materials. The choice of the solder is therefore usually affected by the temperature which can be tolerated by the assembly. In this respect, the alloys listed can be chosen easily. Sometimes it is necessary to use a solder with temperature characteristics which are detrimental to the assembly due to some other consideration which outweighs thermal distortion. There are several methods of eliminating the hazards of unwanted heat. First, the use of a heat sink can reduce the amount of heat flowing away from the soldering area. Second, thermal shocks caused by large heat gradients can be avoided by slowly preheating the entire assembly. Third, the use of a heat barrier can prevent flames from reaching sections of the assembly which should not be heated.

The use of a heat sink is a relatively simple operation. If the configuration of a part allows us to attach a heavy metallic object of good heat conductivity, such as copper, we can clamp it immediately behind the soldering area, where it will lower the overall temperature. This is not always the solution. but it is helpful in many cases; a good example of this is in the prevention of heat from reaching delicate electronic components when leads are soldered. Here, a copper clip between the component and

TABLE 4-3   Properties of Recommended Soft-Solder Alloys (in Order of Increasing Liquidus Temperature)

| | Alloy composition, % | | | | | | | Melting temp., °F | | Mechanical properties | | | | | |
| | | | | | | | | | | Tensile strength | | Shear strength[a] | | Elonga- | Brinell |
No.	Sn	Pb	Bi	In	Sb	Cd	Ag	Solidus	Liquidus	lb/in²	kg/cm²	lb/in²	kg/cm²	tion, %	hardness
1	8.3	22.6	44.7	19.1	...	5.3	....	...	117	5400	38.0	NA	NA	1.5	12.0
2	12.0	18.0	49.0	21.0	...	....	....	...	136	6300	44.3	NA	NA	50.0	14.0
3	13.3	26.7	50.0	....	...	10.0	....	...	158	5990	41.5	300[b]	2.1[b]	200.0	9.2
4	12.5	25.0	50.0	....	...	12.5	....	158	165	4550	32.0	NA	NA	30.0	25.0
5	11.3	37.7	42.5	....	...	8.5	....	160	190	5400	38.0	300[b]	2.1[b]	220.0	9.0
6	50.0	....	....	50.0	...	....	....	...	243	1720	12.1	1630[c]	11.5[c]	83.0	4.9[d]
7	37.5	37.5	....	25.0	...	....	....	...	280	5260	36.9	4300[c]	30.2[c]	101.0	10.2[d]
8	42.0	....	58.0	....	...	....	....	...	281	8000	56.2	500[b]	3.51[b]	200.0	22.0
9	....	15.0	....	80.0	...	....	5.0	...	314	2550	17.9	2150[c]	15.1[c]	58.0	5.2
10	....	....	....	100.0	...	....	....	...	314	515	3.6	890[c]	6.3[c]	41.0	Too soft
11	70.0	18.0	....	12.0	...	....	....	302	345	5320	37.4	4190[c]	29.5[c]	135.5	12.0[d]
12	63.0	37.0	....	....	...	....	....	...	361	7700	54.1	5400	38.0	28–30[e]	17.0
13	70.0	30.0	....	....	...	....	....	361	367	7800	54.8	5000	35.2	20.0	17.0
14	60.0	40.0	....	....	...	....	....	361	370	7600	53.4	5600	39.4	27–40[e]	16.0
15	50.0	47.0	....	....	3.0	....	....	365	399	8400	59.1	6850	48.2	29.0	15.6
16	50.0	50.0	....	....	...	....	....	361	417	6200	43.6	5200	36.6	38–98[e]	14.0
17	....	50.0	....	50.0	...	....	....	...	419	4670	32.8	2680[c]	18.8	55.0	9.6[d]
18	96.5	....	....	....	...	....	3.5	...	430	8900	62.7	4600	32.3	73.0	40.0
19	....	....	....	90.0	...	....	10.0	...	448	1650	11.6	1600[c]	11.2[c]	61.0	2.7[d]
20	....	75.0	....	25.0	...	....	....	...	448	5450	38.3	3520[c]	24.7[c]	47.5	10.2[d]
21	40.0	60.0	....	....	...	....	....	361	460	5400	38.0	4800	33.7	39–115[e]	12.0
22	95.0	....	....	....	5.0	....	....	452	464	5900	41.5	6000	42.2	38.0	13.3
23	95.0	....	....	....	...	....	5.0	430	473	8000	56.2	NA	NA	30.0	13.7[f]
24	20.0	80.0	....	....	...	....	....	361	531	4800	33.7	4200	29.5	22.0	11.0
25	....	90.0	....	....	...	5.0	5.0	...	554	5730	40.3	3180[c]	22.4[c]	23.0	9.0[d]
26	....	97.5	....	....	...	....	2.5	...	579	4400	31.0	2900	20.4	42.0	Too soft
27	1.0	97.5	....	....	...	....	1.5	...	588	4420	31.1	NA	NA	23.0	9.5
28	....	95.0	....	5.0	...	....	....	...	599	4330	30.4	3220[c]	22.6[c]	52.0	6.0[d]

[a]Ultimate stress alloy alone, except as noted.
[b]Recommended working stress in joint.
[c]Ultimate stress in joint.
[d]Modified Brinell hardness, using 100-kg load, ½ min.

[e]Depends on specimen preparation.
[f]Vickers pyramid diamond.
NA = Not available.

the soldered area keeps the component free from heat damage (Fig. 4-1). For further information see Secs. 5-36 and 7-16.

THE LOWER TEMPERATURE LIMIT. As an alloy approaches its melting point, the tensile strength and shear strength fall off. This drop in strength increases as we approach the melting point in an asymptotic manner. Thus a solder alloy cannot be used to hold components together unless some mechanism gives mechanical security. With fusible alloys used as solder, this limitation is quite serious. The following empirical rule has been established:

$$\text{Upper useful temperature} = \frac{T_{\text{sol}} - T_{\text{room}}}{1.5} + T_{\text{room}}$$

For 63/37 eutectic tin-lead solder we get

$$T_{63/37} = \frac{183 - 18}{1.5} + 18 = 128°C = 262°F$$

| Physical properties | | | | |
Electrical conductivity, % IACS	Electrical resistivity, $\mu\Omega\cdot cm$	Expansion coefficient, $\mu in/(in)(°F)$	Specific gravity	General notes
3.3	51.62	13.9	8.86	Expands, then shrinks to zero in 30 min
3.0	57.47	12.8	8.58	Expands, then shrinks to zero in 60 min
4.0	43.10	12.2	9.38	Expands to 0.0057 in/in permanently
3.1	55.61	NA	9.50	Nonelectrical solder for low-ambient temp.
4.3	40.38	NA	9.43	Shrinks to 0.0025 in/in, then expands to zero in 60 min
11.7	14.74	NA	7.74	Low vapor pressure; good for glass
7.8	22.10	NA	8.97	Very good resistance to alkaline corrosion
5.0	34.48	8.3	8.72	Expands to 0.0007 in/in, then shrinks to 0.0005 in/in
13.0	13.26	NA	8.20	Good for thin precoat on ceramics
24.0	7.18	18.3	7.44	Expensive, bonds to nonmetallics
12.2	14.13	NA	7.96	Good strength, low-cost indium alloy
11.5	14.99	13.7	8.46	Used where pasty range is intolerable
12.5	13.79	12.0	8.17	Good pretinning alloy
11.5	14.99	13.3	8.52	Good electrical-grade solder
9.6	17.96	NA	8.75	Similar to 50/50 Pb–Sn, resists creep well
10.9	15.82	13.0	8.90	General-purpose solder
6.0	28.74	NA	9.14	Very good resistance to alkaline corrosion
14.0	12.31	NA	10.38	High-temp. electrical solder for instruments
22.1	7.80	NA	8.10	Solders silver, fired glass, and ceramics
4.6	37.48	NA	9.80	Very good resistance to alkaline corrosion
10.1	17.07	13.9	9.28	Inexpensive utility solder
NA	NA	15.0	7.20	Lead-free, used in food equipment
12.6	13.70	NA	NA	High-temp, electrical solder
8.7	20.50	14.7	10.04	Cheap solder for body work and plumbing
5.6	30.79	NA	11.30	Tin-free indium solder
8.8	19.50	NA	NA	Torch solder, poor corrosion resistance
NA	NA	NA	11.28	Slightly better corrosion resistance than no. 26
5.1	33.80	NA	11.35	Zinc-free indium solder

Notice that the formula uses the solidus temperature rather than the melting point, which for the 63/37 solder is the same and equal to 183°C (361°F). Room temperature was taken as 18°C (65°F).

Thus if the equipment containing the solder joints has to operate at known elevated ambient temperatures the selection of the solder must be

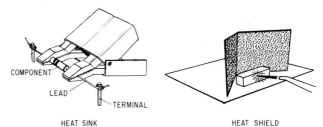

COMPONENT
LEAD
TERMINAL
HEAT SINK   HEAT SHIELD

Fig. 4-1   Reducing heat shock (also avoiding thermal gradient by uniform heating).

carefully matched to make sure that the solder joints still have enough strength to hold the connection together.

The Mechanical Strength. While we must differentiate between structural joints, which are made for their mechanical properties, and electrical joints, where the strength is not that important, both types of connections must be checked for mechanical properties. Here the reader is reminded that the solder is often used as a stress coupler and that assemblies often defy good design practices, relying on the solder to absorb the strain. Each case must be judged on its own merits.

Since solder is normally the weakest link in the assembly, the strength of the solder joint is very important. A certain amount of flexibility is available here to the designer; if the use of a weak soldering alloy is dictated by other properties, a stronger type of joint can be used. See the list of joints for more information (Figs. 4-3 and 4-4). Table 4-3 lists the tensile strength,[1] the elongation, and the hardness of the soldering alloys. Since these values are greatly dependent on the type of specimen used and the way it was prepared, we may find some variation between the data presented and those found in the literature or in the vendor's listings. The variations are not crucial and will not affect the value of the chart in solving design problems.

Electrical Characteristics. The electrical conductivity or resistivity is an important factor whenever the solder joint is part of an electric circuit. Here, it is important to ensure that enough current-carrying capacity is available in the solder joint to avoid hot spots. Sections 4-14 to 4-18 contain full instructions on how to apply electrical characteristics.

Density. When a solder joint is part of a moving assembly, it is important to be sure that the density of the solder in the joint is similar to that of the material used. Otherwise, particularly in rotating parts, centrifugal force may cause imbalance. The density of the solder is also useful information when the overall weight of the assembly is a consideration, as when a floating device is designed.

The Linear Coefficient of Thermal Expansion. This is especially important when glass-to-metal or ceramic-to-metal joints are soldered. If the coefficient of thermal expansion of the solder is very different from that of the glass or ceramic material and the part is subjected to thermal shocks during its service, the stresses and strains which are built up may cause breakdown of an assembly. This is especially true of delicate electronic parts (see also Sec. 4-3).

Other Properties. The designer often needs a solder which has special properties. Some of the more common ones are listed in Table 4-3,

---

[1]Do not confuse with the solder bond strength.

namely, vapor pressure, expansion characteristics after solidification (not connected with the thermal coefficient of expansion), corrosion characteristics, and suitability for various jobs.

Once a solder is considered for a certain assembly and the base metals have been chosen, the problems of corrosion discussed in Sec. 4-3 hold true also for the solder alloys themselves.

Piggyback soldering is used to solder two consecutive joints in the same area. When it is imperative that the first joint should not be disturbed, two soldering alloys are used. The first alloy has a solidus temperature above the soldering temperature of the second one.

Another imporant parameter affecting the selection of the solder alloy in conjunction with the base metal (already selected) is the formation of intermetallic compounds between the base metal and the solder alloy. Intermetallics have been briefly discussed in Sec. 3-1. For the purpose of this book, we shall limit ourselves to those intermetallic compounds which form crystalline structures in the solder matrix upon alloying. These crystals are not always compatible with the alloy and therefore are regarded as an undesirable constituent in the solder-joint fillet. Their presence, as in the case of the copper-tin intermetallic, not only seems to weaken the joint (failures usually will occur along lines of intermetallic concentrations, as in thermal fatigue) but also seems to cause grittiness in the solder. This hampers the fluidity of the alloy as it is attempting to spread and wet the base metal. For further details on specific intermetallic compounds and their effect on the tin-lead matrix, see Sec. 3-12.

Table 4-4 gives a list of the intermetallic compounds of the 7 common solder elements and the 14 basic engineering metals normally used in conjunction with solder. The table lists intermetallics which are stable at room temperature. For our purposes the intermetallics reported were defined as the single-phase field which was not bounded by either of the terminal solid solutions in a phase diagram. The shape and the similarity to standard configurations are reported.

To avoid the formation of intermetallic compounds when they have proved detrimental to the function of the solder connections, it is possible to plate the surfaces of the base metal with a diffusion barrier. This will eliminate the formation of the intermetallic compounds by its physical presence. The fact that the materials cannot mutually diffuse into each other prevents the formation of these intermetallic layers. As an example, the author used nickel plating over copper contacts which were resoldered several hundreds of times using tin-lead solders. No detrimental effect was recorded. However, the same operation on unplated surfaces formed large amounts of intermetallic compounds which finally made the resoldering of the same connection impractical. The complete removal of

TABLE 4-4 Intermetallic Compounds of the Common Solder Elements and Base Metals (for Low Temperatures Only)[a]

hcp = hexagonal close-packed, bc = body-centered, bcc = body-centered cubic, fc = face-centered, fcc = face-centered cubic; parentheses with formulas mean that the formula is assumed from atomic proportions; intermetallics defined as the single-phase fields which are not bounded by either of the terminal solid solutions; number of atoms = atoms per unit cell; O = orthorhombic, M = monoclinic, H = hexagonal, T = tetragonal

Metal	Common elements in solder alloys and their intermetallic compounds						
	Antimony (Sb)	Bismuth (Bi)	Cadmium (Cd)	Indium (In)	Lead (Pb)	Silver (Ag)	Tin (S)
Aluminum (Al)	AlSb, cubic ZnS type	None	None	None	None	Ag_3Al, β-Mn cubic (Ag_2Al), hcp	None
Antimony (Sb)		None	Cd_3Sb_2, O, 16 atoms CdSb, M, 20 atoms	InSb, zinblende (B-3) type	None	Ag_xSb_y, hcp (8% Sb) Ag_3Sb,O	(SnSb), NaCl type[b]
Bismuth (Bi)	None		None	In_2Bi, H, AlB_2 type InBi, T, Pb0 type	Pb_3Bi, hcp	None	None
Cadmium (Cd)	Cd_3Sb_2, O, 16 atoms CdSb, M, 20 atoms	None		None	None	(AgCd), bcc Ag_5Cd_8, γ brass complex cubic, 52 atoms (AgCd_3), hcp	Intermediate phase β, simple H (β is 95 at % Sn; decomp. under 130°C
Copper (Cu)	$(Cu_{13}Sb_3)$, hcp Cu_2Sb, T, 6 atoms	None	Cu_2Cd, $MgZn_2$ type $Cu_4Cd_3{}^c$ Cu_5Cd_8, γ brass $CuCd_3{}^c$	Cu_9In_4, γ brass[d] ($CuIn_2$), NiAs type (Cu_4In_3)[c]	None	None	Cu_3Sn, O, 64 atoms Cu_6Sn_5, O, 530 atoms

Gold (Au)	$AuSb_2$, cubic FeS_2 type	Au_2Bi, fcc, reported to decomp. at low temp.	Au_3Cd, *T* $AuCd$, bc *T*[b] or *O*[e] $AuCd_3$ complex *O*, ~90 atoms	Au_4In, hcp Au_9In_4, γ brass $AuIn$, triclinic $AuIn_2$, cubic	Au_2Pb, $MgCu_2$ type, decomp. on cold working $AuPb_2$, bc*T*	None	(Au_6Sn), hcp $AuSn$, NiAs type $AuSn_2$, *O* primitive $AuSn_4$, complex *O*, $PtSn_4$ type
Indium (In)	$InSb$, zincblende (B-3) type	In_2Bi, *H*, AlB_2 type $InBi$, *T*, PbO type	None		(In_3Pb), fc, *T*	(In_3Pb), fc, *T*	Ag_3In, hcp Ag_2In, γ brass $AgIn$ $CuAl_2$ type	$(InSn)$, *T*[b] $(InSn_4)$, simple *H*
Iron (Fe)	Fe_3Sb_2, NiAs type $FeSb_2$, FeS_2 type	None[f]	None[f]	No information available	None[f]	None[f]	FeSn, CoSn type $FeSn_2$, *T*, $CuAl_2$ type
Lead (Pb)	None	(Pb_3Bi), hcp	None	(In_3Pb), fc *T*			None	None
Nickel (Ni)	Ni_5Sb, cubic, 32 atoms Ni_3Sb, *H*, 64 atoms Ni_7Sb_3, *T*, 52 atoms $NiSb$, NiAs type $NiSb_2$, marcasite (C-18) type	$NiBi$, NiAs type $NiBi_3$, *O*[r]	"	Ni_3In, *H*, Ni_3Sn type $NiIn$, CoSn type $NiIn_3$, *H* Ni_2In, NiAs type Ni_3In_7, γ brass	None	None	Ni_3Sn, hcp Ni_3Sn_2, *H*, NiAs type Ni_3Sn_4, *M*, 14 atoms
Platinum (Pt)	Pt_4Sb[r]	"	"	"	Pt_3Pb, cubic, Cu_3Au type $PtPb$, *H*, NiAs type $PtPb_2$, *T*, $CuAl_2$ type	(Ag_3Pt), fcc $(AgPt)$, cubic[d] $(AgPt_3)$, fcc	Pt_3Sn, Cu_3Au type $PtSn$, NiAs type Pt_2Sn_3, *H*, 10 atoms $PtSn_2$, CaF_2 type $PtSn_4$, *O*, 20 atoms

137

TABLE 4-4 Intermetallic Compounds of the Common Solder Elements and Base Metals (for Low Temperatures Only)^a (Continued)

Metal	Common elements in solder alloys and their intermetallic compounds						
	Antimony (Sb)	Bismuth (Bi)	Cadmium (Cd)	Indium (In)	Lead (Pb)	Silver (Ag)	Tin (S)
Silver (Ag)	Ag_6Sb_p, hcp (8% Sb) Ag_3Sb, O	None	(AgCd), bcc Ag_5Cd_8, γ brass complex cubic, 52 atoms (AgCd₃), hcp	Ag_3In, hcp Ag_2In, γ brass $AgIn_2$, $CuAl_2$ type	None		(Ag_6Sn), hcp Ag_3Sn, hcp^d
Tin (Sn)	(SnSb), NaCl type^b	None	Intermediate phase β = simple H (β is 95 at % Sn); decomp. below 130°C	(In_2Sn), T $(InSn_4)$, simple H	None	(Ag_6Sn), hcp Ag_3Sn, hcp^d	
Zinc (Zn)	$ZnSb^f$ $Zn_4Sb_3^c$ $Zn_3Sb_2^c$	None	None	None	None	(AgZn), hcp^b (Ag_2Zn_3), γ brass $(AgZn_3)$, hcp Mg type	None

^{a} Adapted from M. Hansen, "Constitution of Binary Alloys," copyright 1958 by McGraw-Hill, Inc. Used by permission of McGraw-Hill Book Company.
^{b} Some disagreement among investigators as to the exact structure.
^{c} Not firmly established.
^{d} Slightly distorted.
^{e} Structure depends on exact proportion of metals.
^{f} Materials insoluble.
^{g} Intermetallics exist, but only limited investigations have been reported.

the old solder and replenishment with new solder (thus removing the intermetallic crystals) was necessary before a new cycle of resoldering was possible.

Another way to avoid intermetallics in the joint, of course, is to select an alloy which has no appreciable intermetallic-compound formation with the base metal. This involves a change of materials to meet these requirements, however.

The real effect of intermetallic compounds on a solder joint is a subject which should be studied carefully for each individual system.

4-6 Evaluating the Material Combination   On the basis of the information already discussed, it is possible to make a tentative selection of materials. We now wish to evaluate the selection and proceed with the design. Figure 4-2 outlines a simple method. Let us start on the top left side with the base metal. It is enough to summarize this step by saying that the base metal is usually selected for its mechanical and electrical properties. These are an integral part of the function of the assembly. Using Table 4-2, it is now easy to determine the type of flux required for the base metal. On the basis of general considerations, we must now determine whether the flux is suitable for the specific assembly and, if not, do one of two things: (1) change the base metal so that a more suitable flux can be used (provided the new base metal still fulfills the requirements of mechanical and electrical function in the assembly) or (2) change the surfaces through plating, pretinning, or a chemical etch, coupled with the preservation of the prepared surfaces.

On the top right-hand side, we have solder-alloy selection. The special consideration required for this step was discussed earlier. The reader is reminded here that under special characteristics not only must the physical characteristics of the alloys be checked but also the possibility of intermetallic formation in conjunction with the base metal already chosen. Once these points have been cleared, we are ready to go to the next major compatibility check.

Corrosion danger to the assembly is one of the most important parts in the consideration of materials for an assembly. Corrosion, from a true metallurgical viewpoint, has been covered, and Table 4-1 gives the electromotive series so that the actual potential between two metal systems can be checked. However, chemical corrosion is just as important. If the flux selection is, by necessity, one that introduces corrosive materials to the assembly, the designer's duty is to make sure that the corrosive remains of the flux, the flux fumes, and the flux residues can be easily and totally removed from the assembly. This eliminates any corrosion hazards. The importance of the adequate removal of chlorides for tin-lead soldered joints is illustrated in Sec. 1-13.

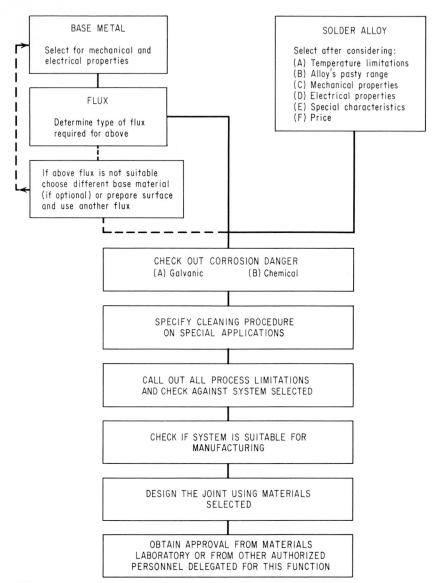

Fig. 4-2   Flowchart for solder-joint design.

Once the corrosion problems are well understood, it is important for the designer to be sure to specify cleaning procedures known to be adequate for that assembly because the designer knows the materials and functions of the assembly better than the production personnel who actually produce it. It is not advisable to recommend specific cleaning equipment, however, because here the experience of the production area and the economy involved are aspects which the designer normally is not familiar with. For further details on proper cleaning procedures and equipment, see Chap. 5.

The next step in the design evaluation involves listing all the process limitations from the standpoint of the design and checking them against the system selected. Such items as temperature limitations to prevent component damage and overcuring or distortion and discoloration of plastic materials should be clearly specified. The type of solvent which might be detrimental to the assembly should be indicated. In general, as much care should be taken with these materials considerations as with tolerances and dimensions in mechanical design.

The designer is often lured into unusual design approaches which may not lend themselves to easy manufacturing. It is well at this stage to consult with production personnel who are used to manufacturing similar assemblies and to check some of the basic ideas in the assembly. Thus it is possible to make sure that there are no serious reasons why the assembly should not be made in the manner called for by the design. It is sometimes possible to achieve large savings in time and equipment and possibly eliminate failure by such discussions between designer and production experts. Simple basic changes in design can often lead to substantial economical benefits and lower the unit cost.

We are now ready for the actual design of the solder joint using the material combination selected.

JOINT GEOMETRY

4-7 Structural Joints and Electrical Connections   We have considered the materials that go into the assembly and have tried to match them to each other in order to eliminate problems stemming from improper combination. It is now time to consider the actual physical design of solder joints. Here two major categories must be considered: (1) the structural joint, which is designed for its mechanical strength, and (2) the electrical joint, which is designed for its current-carrying capacity.

The Structural Joint

4-8 Soldering for Mechanical Strength   In nearly all cases where solder is used in conjunction with base metals, the solder has the lower structural strength, and unless special precautions are taken, the fillet will be the weakest link in the assembly.

For mechanical strengths, the preferred design is a lap joint subject to shear forces only. This is by far the most widely used configuration for mechanical solder joints. A lap joint under pure tensile stresses is also a reliable configuration but is difficult to obtain under everyday conditions. Therefore, unless special conditions exist, it is recommended that all mechanical solder joints be made as lap joints with pure shear stresses. The butt joint has serious limitations in the design of solder joints for mechanical applications. A major disadvantage is in the limited area available for joining, which is always dictated by the parts themselves; whereas in a lap joint the magnitude of the overlap controls the strength of the overall joint, which can easily be made to match the strength of the assembly. A butt joint is limited to the wettable area available, which determines the cross section of the fillet.

4-9 Basic Structural Joints   In order to help the reader visualize these configurations, the most common structural joints available to the designer are listed below.

Butt Joint (Fig. 4-3, type 1). This is the simplest and weakest type. The amount of soldered interface is limited by the cross section of the part. This process is usually used for sealing. Great uniformity in the joint is required because poor interfaces or slight imperfections in the solder affect the strength and the sealing property of the joint. Under direct pull, the joint is exposed to tensile forces which are transmitted to the bulk solder, assuming that the base metal is stronger than the solder. Side loading causes shearing forces in the solder joint under the same conditions.

Scarved Joint (Fig. 4-3, type 2). This butt joint has a larger cross-sectional area with greater strength. This type of joint has locating properties which are sometimes useful when assemblies are complicated. Both direct pull and side loading cause tensile and shear forces in the joint. This type of joint is not used very often.

Lap Joint (Fig. 4-3, type 3). This is the most common solder joint. Two metals are overlapped and a solder film between them holds them together. The strength of the joint varies with the size of the overlap. Under direct pull, this joint is subjected to shearing forces. Under bend-

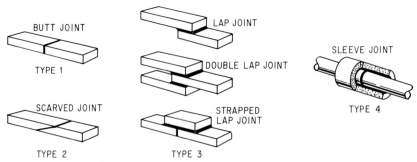

Fig. 4-3   Basic mechanical-joint configuration. No mechanical security before soldering in types 1 to 3. Full mechanical security before soldering in type 4.

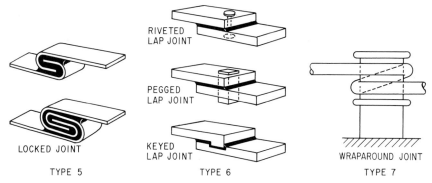

Fig. 4-4 Basic mechanical-joint configurations with mechanical security before joining.

ing forces this joint is subjected to either tension or compression. This is a strong joint. Various configurations this joint can take are the double-lap joint and the strapped lap joint.

Sleeve Joint (Fig. 4-3, type 4). This is for pipes or shafts. When two pipes are to be connected, a sleeve is placed over them. A butt joint is made where the pipes meet and a lap joint where the sleeves come in contact with the pipes. Under direct pull this joint is exposed mostly to shear stresses. Under side-loading stresses the solder itself is under compression and tension, and added strength is given to the joint by the presence of the stronger metal sleeve. This is a superior and common type of joint (make sure the clearance falls in the recommended limits).

Mechanically Reinforced Joint. There are several ways of mechanically reinforcing the solder joint.

LOCKED JOINT (FIG. 4-4, TYPE 5). The metal is folded over and locked mechanically before soldering. This uses not only the strength of the solder joint itself but also the mechanical rigidity of the folded metal to hold itself in place. This is the usual joint in the can industry.

RIVETED (PEGGED) LAP JOINT (FIG. 4-4, TYPE 6). As implied by the name, the joint consists of two overlapping metals that are held together not only by solder but also by a metal peg through both of them. Sometimes this peg is a rivet; other times plain metal pegs are used.

KEYED LAP JOINTS (FIG. 4-4, TYPE 6). This joint is strengthened by keying the base metals, but the joint is limited by the type of assembly in which it is used and the amount of machining time available.

WRAPAROUND (FIG. 4-4, TYPE 7). The strength of this joint lies in the mechanical wrapping, which is held in place only by the solder. This type is used mostly for electrical connections and will be discussed in Secs. 4-14 to 4-18.

4-10 Some General Structural Considerations Whatever the joint configuration and the type of stresses used, it is important to make sure that the solder is subjected to evenly distributed stresses across the wetted surfaces. If poor design causes stress concentration, the joint is subject to tearing failure, which exposes only part of the total wetted area to the stress at a time and thus exceeds the ultimate strength of the fillet. The failure occurs by propagation from one stress zone to another without utilizing the overall strength of the fillet.

If an experimental part or existing design shows progressive cracking of the solder joint from a specific location, it is well to reappraise the stress configuration carefully and see whether the stresses at that specific place can be alleviated either by a change in design or by tapering and building flexibility into the stressed area in order to distribute the stresses more evenly along the solder joint.

With stress concentration it is erroneous to assume that strengthening the structural members or increasing the solder fillet of the joint will solve the problem. Such measures generally cause the same type of failure to continue. They only change the location of the origin of the failure. It is recommended that no bending moments on a lap joint should be allowed. This will cause localized stresses and tearing in the solder, which is a progressive way of affecting each individual cross section of the solder joint and does not utilize the strength of the overall joint. These joints should be reinforced mechanically or designed in such a way that no direct bending moment is exerted on the solder joint.

Another important parameter for good solder joints is the spacing (solder thickness of the joint), which contributes largely to the joint strength. Proper spacing is discussed in Sec. 3-19, but generally speaking, it is controlled on the one hand by good access for flux and solder to the joint area (this eliminates too close a spacing) and on the other hand by the capillary and surface energies holding the solder in the joint (which would eliminate too big a fillet). Remember also the solution-hardening process which takes place during the joint formation (this increases the strength of the solder alloy by a large percentage).

For the calculations of the solder joint in this section, it is difficult to estimate the amount of solution hardening, embrittlement, and other side effects which occur during the wetting of the surface. For conservative design, therefore, it is feasible to use only the strength of the pure bulk solder.

The ductility of the base metal is also important. The more ductile a base metal, the more even the distribution of the stresses across the critical cross-sectional area of the joint. It is difficult to change the base metal in many applications, however, and the lack of ductility then has to be compensated for with a greater safety factor.

The proper design of the joint covers also the provisions for complete displacement of both gas and flux from the wetted surfaces by the solder

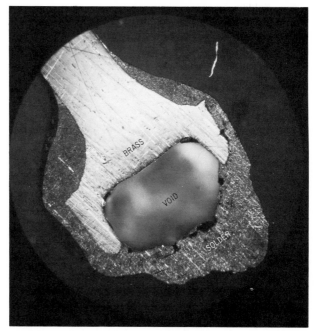

Fig. 4-5   Blind hole in a brass cup showing a large air pocket with solder around it. This will occur when the cup is dipped face down into molten solder. The cup will easily fill when turned up. Note the good wetting on the brass terminal.

alloy to form a completely solid and uniform fillet. Thus blind holes, cavities, and similar air pockets must be eliminated. If this is not the case, the expanding air or flux vapor will cause gas pockets which greatly reduce the effectiveness of the cross-sectional area in the joint and cannot be counted on to obtain the required strength. These inaccessible joint areas usually result in large gas pockets, as indicated in Fig. 4-5.

When dissimilar base metals are soldered together, the differences in their coefficient of expansion, ductility, and other vital properties are accounted for. Just as these parameters are considered for the base metal and the solder, they should also be considered for the fixturing device which holds the parts immobile relative to each other during the soldering process.

4-11 Analysis of Solder-Joint Strength   The design usually is based around the weakest cross-sectional area of the weaker member in the base metal. When high-strength materials are joined by soldering, the solder alloy is usually the weakest link in the assembly unless special precautions are taken. For good design, it is possible either to make the joint to the

TABLE 4-5 Joint Strength for Common Configurations

		Lap joint	Butt joint
When used	Shear	Preferred design	Under favorable conditions
	Tension	Special conditions only	Under favorable conditions
When strength will match weakest link of assembly		Flat to flat: $L_j = \beta H$ Round to round: $L_j = \dfrac{\pi}{2}\beta D_b$ Round to flat: $L_j = \dfrac{\pi}{4}\beta D_b$ Flat to round: $L_j = \dfrac{\beta W_b H}{D_b}$	Possible only if $\beta \leq 1.0$ (only when solder alloy is stronger than base metal)
When strength will be specific		Flat to flat: $L_j = \dfrac{S_j}{\sigma_s W}$ Round to round: $L_j = \dfrac{2S_j}{\sigma_s D_b}$ Round to flat: $L_j = \dfrac{S_j}{\sigma_s D_b}$	Only if strength required is: Flat to flat: $S_j = \sigma_s W H$ Round to round: $S_j = \sigma_s \dfrac{\pi}{4} D_s^2$
Strength ratio β consists of	For shear	$\beta = \dfrac{\sigma_b \text{ tension}}{\sigma_s \text{ shear}}$	$\beta = \dfrac{\sigma_b \text{ shear}}{\sigma_s \text{ shear}}$
	For tension	$\beta = \dfrac{\sigma_b \text{ shear}}{\sigma_s \text{ tension}}$	$\beta = \dfrac{\sigma_b \text{ tension}}{\sigma_s \text{ tension}}$

S = strength (shear or tension) β = strength ratio Subscripts:
A = critical area W = width b = base metal
L = length H = height s = solder alloy
σ = yield strength j = solder joint

specific strength of the work anticipated or to make it equal to the weakest cross section of the assembly. Let us analyze the situation for the common structural configurations of the solder joint for both tension and shear (see Table 4-5).

In consideration of what was stated earlier, the following basic assumptions must be made in order to calculate the solder-joint strength:

1. Proper design of joint assures uniform stresses of pure tension or shear.

2. Totally wetted areas and void-free fillets assure full utilization of cross-sectional area.

3. No solution strengthening or other changes in solder material affect fillet strength (conservative assumption).

4. All deviations from the ideal conditions described in assumptions 1 to 3 will be compensated for by an engineering safety factor.

In both tension and shear the assembly strength is a function of the critical areas involved. If we know the weakest link, we can express the strength of the assembly as

$$\text{Strength} = \sigma_b \times \text{critical area} \quad \text{or} \quad S_b = \sigma_b A_b \quad (4\text{-}1)$$

TABLE 4-6 Strength of Engineering Alloys

Minimum values independent of shape

Metal or alloy	Tensile strength, lb/in²	β (eutectic Sn–Pb)	Metal or alloy	Tensile strength, lb/in²	β (eutectic Sn–Pb)
Aluminum*			Phosphor bronze:†		
EC-0	12,000	1.56	Alloy A-soft	43,000	5.6
EC-H14	16,000	2.08	Alloy A-¼H	60,000	7.8
EC-H19	27,000	3.51	Alloy A-½H	80,000	10.4
1100-0	13,000	1.69	Alloy A-¾H	96,000	12.5
1100-H14	18,000	2.35	Alloy A-H	108,000	14.0
1100-H18	24,000	3.10	Alloy C-soft	53,000	6.9
1060-0	10,000	1.30	Alloy C-¼H	74,000	9.6
1060-H14	14,000	1.82	Alloy C-½H	95,000	12.3
1060-H18	19,000	2.47	Alloy C-¾H	113,000	14.6
Beryllium					
copper:†			Alloy C-H	125,000	16.2
A	60,000	7.8	Alloy D-soft	60,000	7.8
¼H cold	75,000	9.8	Alloy D-¼H	83,000	10.8
½H rolled	85,000	11.0	Alloy D-½H	108,000	13.0
H	100,000	13.0	Alloy D¾H	125,000	16.2
AT	160,000	20.8	Alloy D-H	135,000	17.5
¼ HT	170,000	22.0			
½ HT	180,000	23.5	Stainless steel		
HT	185,000	24.0	annealed:‡		
Copper:†	45,000	5.85	302	73,500	9.55
Oxygen-free	47,000	6.10	316	83,000	10.80
H.O. wire	49,000	6.40	410	114,000	14.80
Medium H.D.			430	69,000	8.95
wire	42,000	5.45			
Annealed wire	36,000	4.70			
Nickel silver:†			Steel:§		
A-¼H	68,000	8.8	C1010 HR	47,000	6.1
A-H	99,000	12.8	C1010 CD	53,000	6.9
B & B-1¼H	74,000	9.6	C1018 HR	58,000	7.5
B & B-1 H	112,000	14.5	C1018 CD	64,000	8.3
B & B-1 spring	120,000	15.6	C1095 HR	120,000	15.6
C-½H	75,000	9.8	C1117 HR	62,000	8.1
D & E-¼H	73,000	9.5	C1117 CD	69,000	8.95
D & E-H	108,000	14.0	Music wire		
D & E spring	112,000	14.5	(ASTM A228)	240,000	31.2

*"Alcoa Aluminum Handbook," 1956.　　　　　　†ASTM Standards, Part 2.
‡"Stainless Steel Handbook," Allegheny Ludlum Corp.　§"ASM Metals Handbook," 1961.

where σ is in either shear or tension according to the configuration. By itself the solder fillet strength can be expressed in the same way as

$$S_s = \sigma_s A_s = S_j \qquad (4\text{-}2)$$

If we require that the joint strength equal the weakest link in the rest of the assembly, we get, from Eqs. (4-1) and (4-2),

$$S_b = S_j \qquad \sigma_b A_b = \sigma_s A_s \qquad (4\text{-}3)$$

If we define σ_b/σ_s as β, the *strength ratio*, we get

$$A_s = \beta A_b \qquad (4\text{-}4)$$

Thus we can see that for all base metals whose strength is equal to or smaller than that of the solder, β is equal to or smaller than unity. In this case,

$$S_b \leq S_s \qquad \beta \leq 1 \qquad A_s \leq A_b \qquad (4\text{-}5)$$

where the cross-sectional area of the solder can be equal to or smaller than that of the base metal. A short survey of Table 4-7 reveals, however, that not many engineering alloys fall in this category for eutectic tin-lead alloys. In most cases,

$$\beta > 1 \qquad A_s > A_b \qquad (4\text{-}6)$$

thus requiring a solder fillet several orders of magnitude larger in order to equal the strength (all the above without safety-factor compensation). σ could be for both tension and stress, but let us analyze these separately for the two basic configurations, the butt and the lap joint.

The Lap Joint. This is the preferred configuration because the overlap can easily be adjusted to the required strength. It is suitable for both a specific required strength and to match the strength of the weakest link (see Table 4-6). Here σ_b is in tension, while σ_s is in shear.

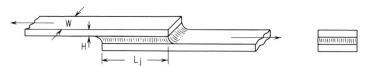

Fig. 4-6

In shear, this configuration will look like Fig. 4-6 for flat stock, and the overlap L_j will be calculated from Eq. (4-4). From Fig. 4-6

$$WL_j = \beta WH \qquad L_j = \beta H \qquad (4\text{-}7)$$

For round stock, where inspection requirements give the width of the fillet as equal to or larger than the smaller radius, the configuration is depicted in Fig. 4-7.

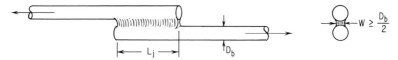

Fig. 4-7

Taking L_j from Eq. (4-4) gives

$$\frac{D_b}{2} L_j = \beta \frac{\pi}{4} D_b^2 \qquad L_j = \frac{\pi}{2} \beta D_b \qquad (4\text{-}8)$$

For round to flat stock, where inspection requires the fillet width to be equal to, or larger than, the diameter of the round member (assuming it to be the weaker) the configuration is shown in Fig. 4-8.

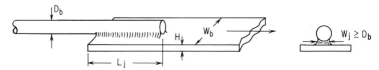

Fig. 4-8

From Eq. (4-4) we get

$$D_b L_j = \beta \frac{\pi}{4} D_b^2 \qquad L_j = \frac{\pi}{4} \beta D_b \qquad (4\text{-}9)$$

However, should the flat stock be the weaker member, then from Eq. (4-4) we get

$$D_b L_j = \beta W_b H \qquad L_j = \beta \frac{W_b H}{D_b} \qquad (4\text{-}10)$$

If a specific joint strength is required, we can use Eq. (4-2) and get the strength of the solder fillet for the basic three configurations as follows:

$$S_j = \sigma_s A_s \qquad (4\text{-}2)$$

Flat to flat (Fig. 4-6): $\qquad S_j = \sigma_s W L_j \qquad L_j = \dfrac{S_j}{\sigma_s W} \qquad (4\text{-}11)$

Round to round (Fig. 4-7): $\qquad S_j = \sigma_s \dfrac{D_b}{2} L_j \qquad L_j = \dfrac{2 S_j}{\sigma_s D_b} \qquad (4\text{-}12)$

Round to flat (Fig. 4-8): $\qquad S_j = \sigma_s D_b L_j \qquad L_j = \dfrac{S_j}{\sigma_s D_b} \qquad (4\text{-}13)$

Any other configurations can be calculated the same way, using minimum inspection and quality criteria for the calculations.

The Lap Joint in Tension. This is a very unlikely configuration because in most cases it ends up as a butt joint. However, if it should occur, Eqs. (4-7) through (4-10) apply, except that σ_b in this case is for shear, while σ_s is in tension.

The Butt Joint in Tension. Although butt joints are inherently weak, their application in tension or shear involves the simple calculation of maximum joint strength which either fulfills the requirements or not. Here both base metal and solder are subjected to the same shear or tensile forces (not like lap joints).

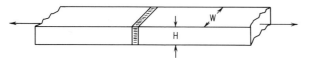

Fig. 4-9

The butt in tension is stronger than the butt in shear. For a rectangular-to-rectangular butt joint (Fig. 4-9) we get (assuming the solder to be the weakest link)

$$S_j = \sigma_s WH \qquad (4\text{-}14)$$

and in the same way we get for round members

$$S_j = \sigma_s \frac{\pi}{4} D_b^2 \qquad (4\text{-}15)$$

If we want to establish whether it is possible to match the weakest link, we follow the same reasoning as before, keeping in mind that

$$\beta = \frac{\sigma \text{ in tension of base metal}}{\sigma \text{ in tension of solder alloy}}$$

from Eq. (4-4) and in a butt joint $A_s = A_b$, and they are equated only if $\beta = 1$. However, only if $\beta < 1$ is the solder joint stronger than the weakest link.

The Butt Joint in Shear. This is a similar configuration, but here

$$\beta = \frac{\sigma \text{ in shear of base metal}}{\sigma \text{ in shear of solder alloy}}$$

otherwise the same reasoning applies. For summary, see Table 4-5.

4-12 Evaluating Proposed Design

1. Make sure where the weakest link in the joint is located. Normally the metallic interface is much stronger than the solder bulk in the fillet. In this case, the weakest link is the solder fillet itself. However, suppose an intermetallic compound is formed between the base metal (or one of its components) and the metals in the fusible alloys. Then these intermetallic compounds, which are congregated at the interface, may lower the strength at this location. This is especially true if they are brittle in nature

because they do not blend in with the more ductile matrix of the surroundings. The interface, in this case, is the weakest link in this assembly.

2. Check whether the solder in the assembly would be subjected to large stresses for prolonged periods. Even if these stresses are below the normal yield strength of the solder, they will cause creep (creep is defined as the plastic deformation which occurs when metals are subjected to stresses below their yield strength for prolonged periods). Creep occurs more readily near recrystallization temperatures. These are very low for most fusible alloys, some of which recrystallize at room temperature. When creep must be avoided, mechanical reinforcement is useful; otherwise brazing should be considered. Useful creep information on fusible alloys can be found in vendors' literature.

3. Make sure that the assembly can withstand the volume changes occurring in the solder during cooling. Most fusible alloys, like other metals, shrink upon cooling; the bismuth alloys expand. It is important that these volume changes between the solidification temperatures and room temperature do not distort the joint.

4. In most design work, a safety factor is included with each material in the design calculations. In solder joints, no established rule is available. Such factors as solderability, flux, solder alloy, temperature-time, and method of application provide too many variables. The designer must therefore base his safety margin both on the degree of uniformity and reliability of the soldered connections and on the type of abuse they are liable to get, putting special emphasis on fatigue.

5. Finally, the designer is reminded that the general considerations of stress distribution must be carried over to solder joints. Localized stresses should be avoided. Flexibility should be provided where the tearing of a solder joint might start.

4-13 Example Problems in Structural Design

Example 1 (Fig. 4-10).   It is required to solder a $\frac{1}{8}$-in diameter copper ball to a $\frac{1}{32}$-in steel shaft. The assembly is part of a hydraulic system for silicone oils (specific gravity 1.1 g/cm^3) at room temperature. The ball serves as a safety plug (well under the oil level). Total weight of ball and solder is not to exceed 0.1 oz in oil. The design does not allow any increase in diameter at the solder joint. However, the ball has a small flat ground on top for soldered connection and filleting. The ball is connected to a spring, which contributes 3.75 lb load to the stresses in the joint. Recommend the type of solder suitable for this joint.

solution: The design dictates the use of a butt joint. The stresses in the solder joint are due to tensile forces originating from (1) the spring loading of 3.75 lb and (2) the copper ball and solder, which can be neglected for design purposes. Assuming a total of 3.75 lb and adding a 50 percent safety factor, we get the desired strength of the joint $S = 3.75 \times 1.5 = 5.625$ lb.

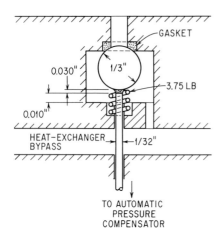

PRESSURE LINE TO HEAT EXCHANGER

Fig. 4-10   Safety relief valve.

A standard solder in this industry is the ASTM 40A alloy. From line 21, Table 4-3, we get a tensile strength of 5400 lb/in², which gives a joint strength $S_j = 5400 \times 0.0007 = 3.78$ lb (where 0.0007 in² is the area of a circle with $\frac{1}{32}$ in diameter). This is not sufficient.

Therefore, we need a stronger solder. From Table 4-3 we see that alloy 15, which is a universal antimonial solder, is the strongest in this range of temperature and cost.

Using this solder, we get $8400 \times 0.0007 = 5.88$ lb, which is over the 5.625 lb required.

Let us check whether this solder fulfills the other requirements of the joint.

1. Weight of ball and solder is less than 0.1 oz. The weight of the copper ball, neglecting the ground section, is

$$\text{Weight copper} = (0.02)(8.94 - 1.1)(16.39)(0.0353) = 0.09 \text{ oz}$$
$$\text{Weight solder} = (0.003)(0.087)(2)(8.75 - 1.1)(16.39)(0.0353)$$
$$= 0.0023 \text{ oz}$$
$$\text{Total weight} = 0.0923 \text{ oz} < 0.1 \text{ oz}$$

where 0.2 in³ = volume of copper ball
   0.003 in = spacing of solder fillet
   0.087 in = area of $\frac{1}{32}$-in diameter joint
   2 = factor to compensate for excess solder in fillet
   8.94 − 1.1 = density of copper in oil
   8.75 − 1.1 = density of solder in oil
   16.39 = conversion factor, in³ to cm³
   0.0353 = conversion factor, g to oz

2. Antimonial solders have excellent creep resistance. In spite of a steady load, there is little danger from creep because the joint operates at room temperature.

3. A well-wetted joint will not have stress concentrations that could cause localized tearing. Regular fluxes and soldering techniques can be used for this solder to obtain a good joint. Danger from initial shrinkage and thermal expansion is not anticipated.

4. Corrosion danger is not anticipated because of the operation in silicone oil, which does not react with the assembly metals.

conclusion: A butt joint using 50/47/3 tin-lead-antimony soldering alloy is suitable for this design.

Example 2 (Fig. 4-11).   A vapor-damper plate is to be soldered to an $\frac{1}{8}$-in crossbeam hanging perpendicular to the gas flow. The design permits attaching the beam only at the top of the plate. The gas flow is intermittent, and the plate will swing slightly with the draft pulses. The plate is 316 stainless steel (18 by 18 by $\frac{3}{16}$ in). The crossbeam is plastic-coated steel, stripped at the joint area. Soldering is chosen as a low-temperature joining technique which would not affect the plastic coating. Bolting or riveting to the $\frac{1}{8}$-in crossbeam is not feasible. The vapor tunnel is fixed to the floor, and soldering on location is required.

solution: Let us figure out the weight of the damper plate. From Table 4-1, we get the density of 316 stainless steel as 8.02 g/cm³

$$(18)(18)(0.188)\frac{8.02}{27.68} = 17.637 \text{ lb}$$

where 1/27.68 is the conversion factor from grams per cubic centimeter to pounds per cubic inch. Because of the nature of the work, a low degree of uniformity and reliability is expected, and a safety factor of 100 percent seems in order. Thus the required joint strength is

$$S = (2)(17.637) \approx 35 \text{ lb}$$

The lap joint is the simplest configuration suitable for this purpose. Find the width of the lap joint necessary to hold up the plate. Using ASTM 20A

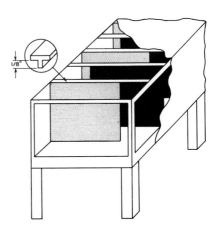

Fig. 4-11 Damper plates in vapor tunnel.

solder, which is the cheapest tin-lead used for this application, we get, from line 24, Table 4-3, the shear strength of 4800 lb/in². Therefore

$$\frac{35}{18W} = 4800$$

$$W = \frac{35}{(18)(4800)} = 0.0004 \text{ in}$$

So we see that a lap joint 18 in long and 0.0004 in wide would be sufficient to hold up the plate. However, we have nearly ⅛ in to use. This additional width would contribute more strength than required.

The use of the plate involves a slight swinging with the vapor pulses. This calls for tapering the plate immediately adjacent to the joint area and soldering the plate to the beam on the side facing the flow direction in order to avoid stress concentration. (This is a general consideration, and if we have an idea of the magnitude of the vapor draft, we can determine whether this taper is really necessary.) Although the galvanic potential between the solder and the stainless steel is small, the presence of vapor dictates the use of a protective coating. Possibly the same material used for the crossbeam should be applied over the joint and overlap some of the plate.

conclusion: A lap joint using ASTM 20A solder is suitable for the application. A taper of the plate at the joint area is desirable. The joint should be totally covered with a protective coating.

The Electrical Connection

4-14 The Electrical Characteristics of the Soldered Connection   In the beginning of the nineteenth century, Georg Simon Ohm showed that the resistance of a simple conductor such as a length of uniform wire is given by

$$R = \rho \frac{L}{A} \qquad (4\text{-}16)$$

where R = resistance of conductor
L = length of conductor
A = cross-sectional area of conductor
ρ = constant characteristic of material of conductor, defined as resistivity

The resistivity of the material is measured as the resistance of a piece 1 cm long having a uniform cross-sectional area of 1 cm². The dimensions of resistivity are resistance times length, expressed in microhm-centimeters. A quick survey of Table 4-1 will reveal that the resistivity of silver and copper is the lowest and that tin, lead, and the other fusible-alloy elements have a much higher resistivity, making them poorer conductors. Non-metals such as carbon, ceramics, and mica have a considerably higher resistivity, varying from 0.04 to 0.07 $\Omega \cdot$cm for carbon to 9×10^{15} $\Omega \cdot$cm for mica.

We see, therefore, that the resistivity shows the dependency of the resistance of a conductor upon its material and dimensions. The resistivity is also a function of temperature. The relationship between the resistivity and temperature T, compared with the resistivity of the material at 0°C, is given by

$$\rho_t = \rho_0[1 + \alpha(T - T_0)] \tag{4-17}$$

In many cases ρ at 20°C is used rather than at 0°C, in which case the 20 can be substituted for 0 in the above formula (α can be obtained from a physics textbook).

The reciprocal value of the resistivity is the conductivity K. Its dimensions are reciprocal ohm-centimeters. The value of K is numerically equal to the length of the conductor measured in centimeters which has a resistance of 1 Ω when its cross-sectional area is 1 cm²

$$K = \frac{1}{\rho} \tag{4-18}$$

The definition of conductivity is important when we consider the flow of current through two dissimilar metals joined together. A conductor with a larger conductivity value would be more suitable for application as a conductor, whereas a small conductivity number under the same conditions would give a higher total resistance. This is why copper is universally adopted for conductors. This point can be actively illustrated in the following manner. When a piece of copper wire and a piece of tin wire, both 1 cm long and 1 cm² in diameter, are joined in some manner, their resistance can be presented by a graph as in Fig. 4-12. If we disregard the joint area completely for the time being, we see that the total resistance of this combination of copper and tin is equal to the resistance of the copper wire by itself and the tin wire by itself. If, however, it is desirable to keep the resistance uniform, we must increase the cross-sectional area of the tin correspondingly until its resistance is equal to that of the copper. The major objection to the increased resistance over a short distance due to the

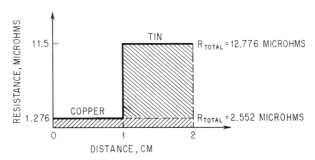

Fig. 4-12   Resistance gradient across copper-tin interface.

tin wire comes from the energy loss, which is directly related to the amount of heat formed. The amount of heat given off by the wire is equal to

$$Q = I^2 R \tag{4-19}$$

where Q = heat, cal
$\quad I$ = current, A
$\quad R$ = resistance of wires, Ω

In order to keep energy losses down to a minimum and stop the assembly from heating up unnecessarily, the current-carrying capacity of the conductor employed is usually computed. Table 4-7 gives these computed values of the current-carrying capacity of wires for copper as a function of their diameter and temperature rise. If the current flowing through the assembly is known, the correct wire is chosen from the table and any other conducting members in the circuit are matched by varying their diameter to bring their conductivity up to that of the conductors.

Assuming that the conductors going into the solder joint are adequate for current-carrying capacity and will not heat over the temperature allowed by the general consideration of the joint, let us consider and analyze the solder joint made with this kind of a conductor. The solder joint is represented by Fig. 4-13.

In order to do this, we consider an infinitesimally thin cross section of the conductor perpendicular to the current flow. The resistance of the conductor is

$$R_c = \rho_c \frac{\Delta L}{A_c} \tag{4-20}$$

The same will be true for a similar section in the solder

$$R_s = \rho_s \frac{\Delta L}{A_s} \tag{4-21}$$

where R = resistance, Ω
$\quad \rho$ = resistivity, $\mu\Omega \cdot$ cm
$\quad A$ = cross-sectional area perpendicular to current flow, cm^2
$\quad \Delta L$ = infinitesimally small width of section in direction of current flow

We can now establish the equivalent cross-sectional area of solder necessary to accommodate a current flowing through a specific cross-sectional area of conductor. As stated earlier, the resistance of the conductor should be equivalent to the resistance of the solder over the same length; therefore

$$R_s = R_c \qquad \rho_s \frac{\Delta L}{A_s} = \rho_c \frac{\Delta L}{A_c} \qquad \frac{\rho_s}{A_s} = \frac{\rho_c}{A_c} \qquad \frac{\rho_s}{\rho_c} = \frac{A_s}{A_c} \tag{4-22}$$

TABLE 4-7 Current-Carrying Capacity of Insulated Annealed Copper Wire*

Insulation emissivity 0.9; Ambient temp. 68°F

AWG (wire size)	Solid wire diam., in	Bare stranded wire		Insulated wire diam., in	Current to give these wire temperatures, A						Ultimate tensile strength, lb		AWG (wire size)
		Overall diam., in	Strands per gage (AWG)		122°F (50°C)		158°F (70°C)		212°F (100°C)				
					Solid	Strand	Solid	Strand	Solid	Strand	Solid	Strand	
6	0.1620	0.210	266/30	0.296	75.7	79.4	98.1	102.8	122.9	128.7	762	794	6
8	0.1258	0.167	168/30	0.224	55.8	58.6	72.5	76.1	90.9	95.4	480	499	8
10	0.1019	0.120	104/30	0.166	40.9	41.7	53.1	54.3	66.8	68.2	310	314	10
12	0.0808	0.096	65/30	0.146	30.7	31.1	39.8	40.4	50.0	50.7	197	198	12
14	0.0641	0.077	41/30	0.129	23.2	23.5	30.1	30.5	37.8	38.3	124	124	14
16	0.0508	0.060	26/30	0.115	17.6	17.7	22.9	22.9	28.6	28.7	78	79	16
18	0.0403	0.046	16/30	0.104	13.6	13.3	17.5	17.2	21.9	21.5	49	50	18
20	0.0320	0.038	10/30	0.055	9.8	10.0	12.6	13.1	15.8	16.4	31	31	20
22	0.0253	0.031	7/28	0.048	7.5	7.8	9.7	10.0	12.2	12.6	19	19	22
24	0.0201	0.024	7/32	0.044	5.9	6.1	7.6	7.8	9.5	9.8	13	13	24

*Contains data calculated on a solid-state IBM 7090 data-processing system. Data are applicable to copper conductors at various operating temperatures, with conventional sleeve insulation. Normal ambient temperatures are assumed, with heat dissipated by natural convection and radiation. Taken from H. H. Manko, How to Design the Soldered Electrical Connection, *Prod. Eng.*, June 12, 1961. For more data see H. H. Manko, The Right Conductor Size, *Prod. Eng.*, June 11, 1962.

The ratio between the resistivity of the solder and the resistivity of the conductor is important because it is used in all the calculations of the electrical solder joints and is termed the *resistivity-ratio delta:*

$$\delta = \frac{\rho_s}{\rho_c} = \frac{A_s}{A_c} \tag{4-23}$$

Values for δ are plotted in Fig. 4-14 for the particular conductors. Then, using the above formula, we get

$$\delta = \frac{A_s}{A_c} = \frac{L_j D_{c_1}}{(\pi/4)D_{c_1}^2} = \frac{4L_j}{\pi D_{c_1}} \tag{4-24}$$

and
$$L_j = \frac{\pi}{4}\delta D_{c_1}$$

So we see that, for round wires to a flat lapped connection (Fig. 4-13), the overlap L_j is equal to a geometrical constant $(\pi/4)$ times the resistivity ratio δ times the wire diameter. Then, by following the same procedure, we can work out the formulas for any required configuration of solder joints. For further details, see Table 4-8.

If the conductors going in and out of the joint are unequal, the calculations should be based on the key conductor, which is the one with the highest resistance per unit length. All the calculations are based on the assumption that the full current is carried by the solder fillet itself. This is not always the case. In actual connections, the metallic conductors may be in contact. In many cases it is impossible to predetermine the extent to which these conductors will be in contact and to design a reliable connection around these figures. The solder joint is designed around the worst case, namely, that of the conductors not touching each other, giving a built-in safety factor that yields a conservative design.

In the next section, we shall discuss the major electrical connections, stating the specific assumption used to arrive at these results. This is done

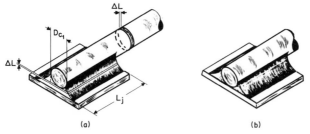

(a)   (b)

Fig. 4-13 Round wire to flat lapped connection. (*a*) Note the solid solder fillet, which carries all the current. (*b*) Only part of the current is carried by the solder; the rest flows through the contact between the round wire and the flat.

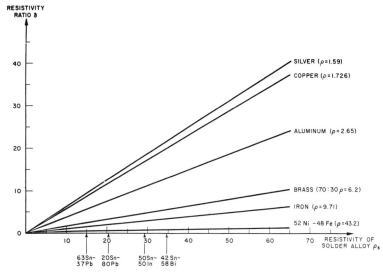

Fig. 4-14   Resistivity ratio for six base metals as a function of solder resistivity.

in an effort to help designers and engineers to establish whether the results fit their specific case, rather than to categorize the results.

Although many formulas based on the worst possible case have an inherent safety factor built in, no effort has been made to include such safety factors. These should be added later after the situation has been analyzed.

The addition of safety factors in electrical solder joints follows the same general procedure as in other design fields. No clearly established rules can be made because too many variables interact: solderability of base metals, flux, solder alloy, solder method, time-temperature cycle, etc. However, a clear understanding of the specific application nearly always helps in estimating the magnitude of the safety factor required.

4-15 Calculating the Basic Solder-Joint Configurations   The considerations applied previously to finding the parameters of a solder fillet in an effort to maintain a uniform voltage gradient throughout the circuit have been applied to all other configurations of solder joints. The results of these computations are listed in Table 4-8. Solder joints are arranged in three major groups according to their mechanical security before soldering. The first group includes those which are not mechanically attached (butt and lap connections). The second group includes those which are partially attached mechanically (hook type). The third group has mechanical security (wrapped-around joints). In all these cases, we shall try to analyze the configuration of the joint for the worst case that will occur when the solder joint itself carries all the current. This is very important

TABLE 4-8 Data for Electrical-Connections Design*

GROUP I—NO MECHANICAL SECURITY PRIOR TO SOLDERING						
BUTT CONNECTIONS						
NO.	TYPE	DIAGRAM	CONTROLLING FORMULA	CONDITIONS	FIXTURES	CURRENT
1	ROUND TO ROUND		$D_s = \sqrt{\delta}\, D_{c_1}$	$\rho_{c_1} \geq \rho_{c_2}$ $D_{c_1} \leq D_{c_2}$	YES	SMALL
2	SQUARE TO SQUARE		$D_s = \sqrt{\dfrac{4}{\pi}\delta}\, T_{c_1}$	$\rho_{c_1} \geq \rho_{c_2}$ $T_{c_1} \leq T_{c_2}$	YES	SMALL
3	RECTANGLE TO RECTANGLE		$T_s = \delta T_{c_1}$	$\rho_{c_1} \geq \rho_{c_2}$ $W_1 = W_2 = W_s$ $T_{c_1} \leq T_{c_2} \neq T_s$	YES	SMALL
LAP CONNECTIONS						
1	ROUND* TO ROUND		$L_j = \dfrac{\pi}{2}\delta D_{c_1}$	$\rho_{c_1} \geq \rho_{c_2}$ $D_{c_1} \leq D_{c_2}$ $W_s \geq \dfrac{D_{c_1}}{2}$	YES	LARGE
2	ROUND TO FLAT		$L_j = \dfrac{\pi}{4}\delta D_{c_1}$	$\rho_{c_1} \geq \rho_{c_2}$ $A_{c_1} \leq A_{c_2}$	OPTIONAL	LARGE
3	FLAT TO FLAT		$L_j = \delta T_{c_1}$	$\rho_{c_1} \geq \rho_{c_2}$ $W_1 = W_2 = W_s$ $T_{c_1} \leq T_{c_2}$	OPTIONAL	LARGE
4	WIRE TO POST		$L_j = \dfrac{1}{2}\delta D_{c_1}$	$\rho_{c_1} \geq \rho_{c_2}$ SOLDER FILLET $\geq \dfrac{D_{c_1}}{2}$	NO	MEDIUM
5	WIRE TO CUP		$L_j = \dfrac{1}{4}(\delta - 1)D_{c_1}$	$\rho_{c_1} \geq \rho_{c_2}$	NO	LARGE
6	WIRE TO HOLE		$L_j = \dfrac{1}{4}\delta D_{c_1}$	$\rho_{c_1} \geq \rho_{c_2}$	OPTIONAL	MEDIUM

*H. H. Manko, How to Design the Soldered Electrical Connection, *Prod. Eng.*, June 12, 1961, p. 57.

GROUP II—PARTIAL MECHANICAL SECURITY PRIOR TO SOLDERING						
HOOK CONNECTIONS						
NO.	TYPE	DIAGRAM	CONTROLLING FORMULA	CONDITIONS	FIXTURES	CURRENT
1	ROUND TO ROUND		$D_{c_1} = \dfrac{2}{\delta} D_{c_2}$	$\rho_{c_1} \geq \rho_{c_2}$ $D_{c_1} \leq D_{c_2}$ HOOK $\geq 180°$	NO	LARGE
2	ROUND TO FLAT		$D_{c_1} = \dfrac{1}{\pi\delta}(8L_j + 4T_{c_2})$	$\rho_{c_1} \geq \rho_{c_2}$ $A_{c_1} \leq A_{c_2}$ HOOK $\geq 180°$	NO	MEDIUM
GROUP III—FULL MECHANICAL SECURITY PRIOR TO SOLDERING†						
WRAP CONNECTIONS						
NO.	TYPE	DIAGRAM	CONTROLLING FORMULA	CONDITIONS	FIXTURES	CURRENT
1	ROUND TO ROUND		$L_j = \dfrac{\pi}{2}\delta D_{c_1}$	$\rho_{c_1} \geq \rho_{c_2}$ $D_{c_1} \leq D_{c_2}$ $n > 1$	NO	LARGE
2	ROUND TO FLAT		$D_{c_1} = \dfrac{8}{\pi\delta}(L_j + T_{c_2})$	$\rho_{c_1} \geq \rho_{c_2}$ $A_{c_1} \leq A_{c_2}$ $n = 1$	NO	MEDIUM
3	ROUND TO POST		$D_{c_1} = \dfrac{4n}{\delta} D_{c_2}$	$\rho_{c_1} \geq \rho_{c_2}$ $D_{c_1} < D_{c_2}$ $n \geq 1$	NO	LARGE

D_{c_1} – DIAMETER OF SMALLER CONDUCTOR
A_{c_1} – AREA OF SMALLER CONDUCTOR
S – SOLDER
W – WIDTH
L_j – LENGTH OF JOINT

T – THICKNESS
N – NUMBER OF TURNS
δ – RESISTIVITY RATIO $\dfrac{\rho_s}{\rho_{c_1}}$
ρ – RESISTIVITY (MICROHM–CM)

*USE ONLY WHEN LARGE CONDUCTOR DIAMETER IS 3 TO 4 TIMES LARGER THAN SMALL DIAMETER; OTHERWISE USE ROUND–TO–FLAT LAP–JOINT FORMULA.

†IN CASES WHERE LOOSENING OR BREAKING OF THE JOINT WOULD RESULT IN A HAZARDOUS CONDITION, MECHANICAL SECURITY SHOULD BE SPECIFIED.

since under certain circumstances the solder may not be carrying the full load of the current because of direct physical contact between conductors when part or most of the current is transferred from one conductor to the other through this interface. We shall also see that in each case there is a controlling factor which can be varied and can thus change the current-carrying capacity in the joint. We shall also analyze unequal conductors and use the lower-conductivity member for our calculations and controlling factor (key conductor). In addition, we shall analyze the actual length of the solder joint that figures in the computation of the resistance of the solder joint and its contribution to the overall resistance of the assembly.

GENERAL NOTE. In all the above considerations, we have to take δ, the resistivity ratio, as the ratio of the resistivity of the solder used to the resistivity of conductor 1, which is always the conductor having the smaller conductivity either because of larger resistivity or smaller cross-sectional area.

4-16 Analysis of the Resistance of the Solder Connections   The types of solder connection discussed in Sec. 4-15 make varying contributions to the overall resistance of the solder joint. Our calculations are based on a uniform voltage drop across the soldered connection. In order to achieve this, we established the controlling factors, which we can vary so that the solder joint will have the same current-carrying capacity as the conductors coming in or going out of the solder joint. Once this is achieved, we can estimate the additional resistance of the solder joint to the overall assembly and we shall look at it for the same groups and types discussed in Sec. 4-15. in group I, types 1, 2, and 3 (the lap connections), the situation is very simple. Unless the conductors are in physical contact, the distance or spacing between the two conductors is the actual distance the current has to flow through the solder joint in the metallic continuity. This distance can simply be added to the length of the conductors themselves, making the overall length of the conductor, regardless of the solder in between, the factor to be considered in analyzing the resistance of this type of assembly. In the same group, the lap connections have a tendency for this spacing to approach zero, but because of the configuration we are not always able to measure this distance. In addition, it is possible for part of the conductors to touch and the other part to have some type of spacing where the overall effect is close to zero. In this type of configuration some addition to the overall resistance could be contributed by the solder joints. This depends on the individual configuration and cannot be generalized.

Group II, the hook connections, types 1 and 2, also have a spacing which approaches zero and possibly is in most cases zero, and here the same considerations apply as for the lap connections.

Group III, however, assures definite connection between the conductors, and the spacing here is actually equal to zero unless the wrap is not tight, in which case the situation is similar to that in group II.

In general, it can be said that the solder joints in the average electronic assembly contribute little to the overall resistance of the assembly if the connections are properly designed according to Table 4-8. When some spacing is known to exist, the number of connections in this spacing will probably take up only a small fraction of the total length of conductors in the assembly and therefore in most cases can be neglected.

4-17 Designing the Electrical Connection Electrical soldered-joint design involves many factors which require special engineering. These include flux chemistry, galvanic action, heat transfer, oxidation, mechanical stress, conductivity, and process control. But a straightforward design procedure like the following can meet these requirements.

1. *Sketch the joint.* Usually the electrical and physical requirements dictate approximate configuration, and a few simple sketches will pinpoint one of the 14 types in Table 4-8.

2. *Check the process limitations.* Are the components or terminals made of special materials? Will pretinning or plating of components be necessary for protection or manufacturing reasons? Is there a temperature or thermal-shock limitation on any of them? Is a different solder or flux needed for any individual terminal? If two terminals are close together, must they be separated with thermal barriers or protected by heat sinks during soldering? Must one terminal be made with higher-temperature solder so that it will not melt when the second terminal is soldered? Does the solder have a pasty range, or does it go directly from solid to liquid? Will the solder flux or the cleaning solution react chemically with the conductors, insulators, or supports? Will intermetallics be formed? What about ease of automatic manufacture? Relative cost? All these questions are discussed in detail elsewhere in this book, and the electrical example in this section accounts for many such limitations.

3. *Pick conductor size.* Current-carrying capacity and mechanical strength are important. To choose the right lead wire quickly, see the current-carrying capacity (Table 4-7).

4. *Select solder.* Table 4-3 applies to all ordinary electrical applications. Where a pair of terminals requires two separate soldering techniques, work out the overall sequence.

5. *Look up the resistivity ratio* of each solder vs. conductor in Fig. 4-14.

6. *Calculate important joint dimensions* with the equations in Table 4-8.

7. *Add a 50 to 100 percent safety factor.* Round off numbers to be consistent with a reasonable manufacturing tolerance.

8. *Design a trial joint* complete with dimensions of lead wire and terminals, based on electrical requirements.

9. *Check the trial joint for mechanical strength* by calculation or by building and testing a prototype. Redesign if necessary.

10. *Make final engineering drawings* with specifications.

4-18 Example Problem in Electrical-Joint Design   A thin fired-silver terminal of a glass component is to be connected to a cup terminal ¼ in away using insulated copper lead wire (see Fig. 4-15). The component is rated for 3½ A at 70°C (158°F). A maximum of 6 lb pull on the assembly was established separately. Design the silver-fired terminal, select the proper soldering alloys, select the proper cup size, determine the copper wire diameter and length, and suggest suitable soldering procedures.

Step 1: Establishing the Process Limitations. The short distance between the two solder joints calls for the selection of two solders with a large melting-point difference similar to that used in piggyback soldering. If we are not careful, it is conceivable that enough heat will be conducted from one end of the wire to the other during the second soldering operation to melt the solder from the first operation. There are two ways to avoid this.

1. A heat sink can be used. The sink would be clipped on between the component and the area to be soldered. However, with the size of wire and the short lead we are dealing with, this does not seem an advisable solution. The AWG no. 26 wire selected below might break, and/or the heat sink would be in the way and could possibly rob heat from the soldering area and thus slow down the soldering operation. In addition, the use of insulated wire as required by the design conditions necessitates stripping the wire as close as possible to the solder joint to avoid short circuiting.

2. Use two solder alloys, the first having a solidus temperature higher than the soldering temperature of the second alloy. This is by far the better method for our application and, as discussed earlier, would solve the problem of the process limitations involved.

The material of the insulation selected should match the general design requirements and in addition should withstand the soldering tempera-

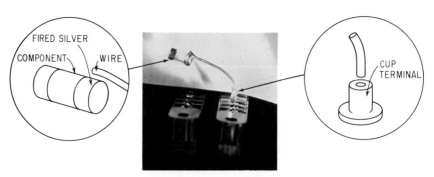

Fig. 4-15

tures without distortion or charring. The use of a specific insulation material will also greatly affect the choice of the flux and the postsoldering cleaning methods.

When we are dealing with a glass component, thermal shock may become an important factor. For added reliability, a uniform preheating of the glass components before soldering should be considered in order to minimize the heat effect. With this added operation, the use of higher-temperature soldering alloys becomes feasible.

Step 2: Select the Wire Size for the Application. It it always desirable to use the smallest-diameter wire possible because this reduces stresses in the joint when the wire stiffens. Table 4-7 gives both the current-carrying capacity as a function of permissible temperature rise and the wire strength. For our application, we use AWG no. 26 solid wire with a current rating of 3.68 A for a 50°C (122°F) temperature rise and a tensile strength of 8 lb.

Step 3: Select the Soft Solders. In step 1 we established the need for two different-temperature solder alloys. Let us consider the one for the ceramic first.

A thin fired-silver film might have a porous structure and/or many unwetted areas around the glass. It is therefore advisable to use a soldering alloy which in itself is reported to wet glass, namely, an indium-containing alloy. When used, such an alloy would penetrate through the porous silver surface and the nonwetted areas, wetting the glass and thus strengthening the joint.

On line 19 of Table 4-3 we find such a suitable alloy with a high soldering temperature which is also recommended for silver-fired glass and ceramics. The alloy is rather expensive, and a decision should be made between choosing a poorer solder or paying the premium price for the application. This is up to the designer. For our problem, we shall adopt this alloy because it has all the positive properties required. The alloy has a solidus temperature of 448°F (231°C).

The use of a silver-containing alloy is beneficial in other aspects. The tendency of the fired-silver layer to dissolve in the solder alloy is greatly reduced. This alloy can be used with a soldering iron provided it has a compatible bit such as an ironclad tip and has been adjusted for the elevated soldering temperature. These tips should be reserved for this alloy only.

SOLDERING TO THE TOP. On line 14 of Table 4-3 we find a suitable alloy for this second soldering operation. This 60/40 tin-lead is a common electrical solder available in all shapes and sizes. The liquidus temperature for this alloy is 370°F (188°C), and the alloy can therefore be soldered at around 425 to 445°F (218 to 229°C). This soldering temperature will not disturb the alloy used for the fired-silver surface and is therefore more than adequate. The soldering operation can be best achieved by using a pretinned and solder-filled cup terminal. No temperature limitation exists during the pretinning. The lead wire can then be prefluxed and/or pretinned and refluxed. The terminal cup is heated to no higher than 450°F (232°C) and the wire is inserted into the molten solder. The heat source is removed and the joint freezes. A preset resistance iron cycle or a thermally controlled iron can be successfully used for this operation. Great care should be specified in *bottoming the cup* in each soldering operation.

Step 4: Obtain the Resistivity Ratios. Here we use Fig. 4-14 to obtain the numerical values of δ. For alloy 19 (90 indium 10 silver) Table 4-3 gives a resistivity value of 7.80 $\mu\Omega \cdot$ cm. From Fig. 4-14 we get a δ value of 4.5 using copper conductors. For alloy 14 (60 tin 40 lead) we get from Table 4-3 a resistivity value of 14.99, which gives us from Fig. 4-14 a δ value of 8.75, again using copper conductors.

Step 5: Calculate the Controlling Factors. Using Table 4-8, we get the two formulas controlling our design. For the silver surface, we use the round-to-flat lap connection where the controlling formula is

$$L_j = \frac{R}{4} \delta D_{c_1} = \frac{R}{4} (4.5)(0.0159) = 0.056 \text{ in} \qquad (4\text{-}25)$$

Since $D_{c_1}/D_{c_2} < \frac{1}{4}$ we use flat to round rather than round to round.

For the cup terminal, we use the wire-to-cup connection, where Table 4-8 gives us the controlling formula as

$$L_j = \frac{1}{4} (\delta - 1)D_{c_1} = \frac{7.75}{4} 0.0159 = 0.032 \text{ in} \qquad (4\text{-}26)$$

Neither of the above values includes the safety factor.

Step 6: Add the Safety Factors. According to the design, a safety factor should be added at this point to the dimension found from the controlling formula. These safety factors are directly dependent on the overall design conditions and in our case should increase by 50 percent. (Bring them up to the next rounded-off figures.)

$$0.056 \text{ to } 0.080 \text{ in} \qquad 0.032 \text{ to } 0.050 \text{ in}$$

Step 7: Design the Terminal Shape. Let us break this up into two configurations. (1) The land on the fired-silver glass component can easily be made as a band around the whole glass circumference. If the widths of the band are smaller than the value obtained above, the wire can be soldered in a radial position using the circumference of the glass component as the length of the solder joint L_j. (2) The cup size is determined by the length of the soldered connection. Its inside diameter should be 0.003 to 0.005 in larger than the wire size. (Bring it up to the nearest rounded-off value.)

It is now possible to determine the overall lengths of the wire by taking the spacing of $\frac{1}{4}$ in plus the lengths of wire needed to solder to the fired-silver land and to bottom in the cup.

Step 8: Check the Mechanical Strengths. Let us separate the two cases again. First consider the joint to the silver-fired surface. If we assume shear forces, which would be normal in stressing this type of assembly, we get

$$S_s = \sigma A_s \tag{4-27}$$

$$A_s = \frac{3}{1600} = 0.0019 \text{ in}^2 \tag{4-28}$$

$$L = \frac{A_s}{D_{c_1}} = \frac{0.0019}{0.0159} = 0.157 \text{ in} \tag{4-29}$$

which is larger than that length required for the electrical connection. A small safety factor should be added to the figure (0.157 in) obtained above. This would mean the redesign of the silver-fired surface. Now consider the connection to the cup.

$$S_s = \sigma A \tag{4-30}$$

$$A = \frac{3}{5600} \, 0.0005 \text{ in}^2 \tag{4-31}$$

$$A = R\pi L \qquad L = \frac{A}{\pi D_{c_1}} = \frac{0.0005}{(3.14)(0.0159)} = 0.001 \text{ in} \tag{4-32}$$

This is less than the value required for electrical strength; the previous design is therefore adequate.

Conclusion. On the basis of the above calculations, the following recommendations are made:

1. The silver-fired terminal should be 0.08 in long and at least twice the diameter of the wire (or 0.08 in bent around).

2. An AWG no. 26 insulated copper wire 0.380 in long is appropriate.

3. The joint to the silver surface should be made using a 90/10 indium-slver alloy and applying it with an iron. Preheating the component to avoid thermal shock should be considered.

4. The joint to the cup terminal should be made with a 60/40 tin-lead soldered alloy using preset resistance equipment or a controlled iron. The size of the cup should be 0.020 in inside diameter or the next nearest size and at least 0.050 in deep.

FIVE

Equipment and Production Techniques

5-1 Introduction This chapter is devoted to a discussion of the equipment and techniques involved in the various unit processes of soldering, e.g., fluxing, preheating, soldering, and cleaning. The processes of printed-circuit soldering and hand soldering are treated separately in other chapters. The reader is encouraged to study this chapter first and then proceed to the specific process description in Chaps. 6 and 7.

After the proper materials have been selected for a specific application, it is time to determine the proper technique and equipment to obtain a quality joint. Presoldering procedures and postsoldering operations are as important as the soldering itself. This chapter is intended to help readers make the proper selection of equipment to match their entire assembly and to outline the proper sequence of operation for optimum quality.

Such a variety of industrial soldering equipment is available today that a book of this kind cannot catalog it all, but general types have been chosen to cover the field adequately. Soldering equipment is classified by the mode of heat transfer used, and samples of the equipment are shown. It is important to remember that any new process that affords a unique way of heating, provided it fulfills the considerations outlined in earlier chapters, constitutes an acceptable new type of soldering process. By the same token, it is easy to evaluate any system for a specific application adhering to the same principles.

5-2 Managing the Solder Process and Workmanship Standards

Soldering, like any other industrial processes, is fully controllable, and it is incumbent on management to take the necessary steps. Successful industrial organizations have a person, task group, or soldering committee charged with the responsibility of setting up a full complement of soldering specifications, covering materials, design, process, workmanship standards, etc. This important task must be completed before the reliability and economy of soldering is guaranteed. One of the major pitfalls in this area stems from the practice of accumulating documentation from other industries and applying it to one's own process. More often than not this results in undue cost and often unacceptable reliability standards because of the unique nature of each industry. In order to explain this point further, let us go through a step-by-step discussion of the best way of setting up one's own documentation.

Phase 1: The Management Cost vs. Quality Policy. A soldering program has to start at the top. In addition to program approval and recognition of achievement, management must provide clearly understood cost and quality objectives.

Given unlimited funding, most companies can produce products of unquestioned quality, but usually it is the other way around. Production costs have to be kept within specific limits. The design, material, and manufacturing compromises required to meet cost limits have to be carefully defined, taking into consideration such factors as product pricing, expected useful life, warranties, and repair.

Soldered electrical connections should be a part of any cost and quality evaluation project. With a clear-cut mandate from management, the rest of the organization can proceed with the important task of setting up acceptable soldering quality and workmanship standards. Designers and product engineers, cooperating with the quality-control and manufacturing departments, should correlate the economics of the product with such elements as reliability and performance under service conditions.

Phase 2: Defining Quality and Workmanship Standards. Armed with the management statement, the task group must then proceed to examine one solder joint of every type in the total assembly, i.e., wire to terminal, component to cup, etc., . . . Minimum and maximum criteria are then established according to the basic function of a joint:

1. Durability (or mechanical strength)
2. Electrical conductivity
3. Thermal dissipation
4. Ease of manufacture

5. Inspectability by visual means

6. Ease of repair

On the basis of these decisions, samples are to be prepared of these joints to show the desired configuration, two that represent the limits of acceptability (minimum and maximum), and finally two or more representing unacceptable conditions.

These visual aids can be used together with sketches and other graphic material to set up workmanship standards, quality-control manuals, and training material for employees.

Phase 3: Process Selection and Control. Process and equipment selection should be based on the above considerations. The adaptation of existing equipment may sometimes lead to serious problems, and the trade-offs warrant economic study. Most equipment manufacturers have demonstration facilities where actual production samples can be run. After the new equipment has been installed or the old equipment has been refurbished and set up for the process, it is time to concentrate on the manufacturing parameters. In some cases, time-temperature relationships are all that is required, but in more sophisticated processes, conveyor speed, angles, preheating temperatures, and material control are added to the list.

At this point a full documentation is called for to record manufacturing parameters for peak results. In addition to setting up quality-control procedures and in-process control, it is necessary to set up a well-defined policy for changes in any one of the parameters involved in the operation. It is best to establish either one knowledgeable person or the original task group as the only authority allowed to make any changes. Well-meaning but unauthorized changes in the process may lead to a loss of reliability or economy and in addition will degrade the process into an art rather than a technology.

To summarize, soldering is an industrial process that requires the same care and attention that machining, heat treating, or plating would receive. Continuous monitoring of developments in the field, searching for opportunities to improve quality and lower cost will keep the process up to date. But one must remember that each industry and each product within the industry is a unique entity and requires its own set of conditions.

PRESOLDERING CLEANING

5-3 Surface Conditions and Storage   The success of the soldering operation depends largely on the condition of the surfaces to be joined before soldering. We have seen that fluxes have certain chemical and physical properties which help soldering but that they are rendered useless when films of oil, paint, grease, or wax are present on the surfaces. Any foreign materials on the surfaces from previous machining operations such as chips, cutting oils, and dust in storage should be carefully removed in order to get a repeatably reliable connection. In addition, heavy rust or metallic scales make soldering difficult, and most fluxes are too weak to effect quick bonding.

It is normal to refer to the surface condition as *solderability* (more about this in the chapter on quality control). Lack of solderability on components, printed-circuit boards, terminals, wires, and chassis is often not the responsibility of the vendor but falls squarely on the shoulders of the user. Most items are received adequately packaged and very solderable, as indicated by in-coming inspection checks. The treatment in in-coming inspection and in-house storage unfortunately causes the deterioration of these surfaces. It is mandatory for each organization to set up a rational system to protect the surfaces to be soldered from a loss of solderability. This can be achieved by the following steps:

1. Include packaging requirements in purchasing specifications wherever possible.

2. Check packaging materials on in-coming inspection and repackage wherever necessary. Use only approved materials.

3. Reseal packages whenever samples are withdrawn for in-coming quality control.

4. Store parts in a clean location. Prevent industrial fumes, environmental sedimentation, or chemical vapors from attacking the surfaces. Neither moisture nor temperature control in the storage area is essential, but they are desirable.

5. Initiate a first-in–first-out (FIFO) system to minimize natural aging.

6. Allocate sensible production quantities when removing material from store to be used up within a manufacturing cycle of no more than 1 week at a time.

7. Include storage responsibilities as part of the solderability management program (Chap. 8).

A word of clarification is needed about acceptable and unacceptable packaging materials. While fluxes are perfectly capable of removing most metal oxides, they are often rendered useless in the presence of sulfur-base tarnishes. Sulfur has been shown to be especially poisonous to the soldering operation in the case of copper- and tin-lead-coated surfaces. In addition, its adverse affect on silver-coated parts is well established. Sulfur in storage may come from the following sources:

1. Sulfides, sulfur, or organic compounds can be airborne.

2. Corrugated cartons contain sulfur, which is especially detrimental under humid conditions.

3. Brown paper (bags), newspaper, or high-rag-content tissue contain sulfur, and humidity accelerates its release.

Packaging materials should be reviewed with this information in mind. The most economical and best available substitute for paper is plastic (electrostatic or static-free) in the form of films, bags, or zip-lip pouches. The danger in this type of material comes from several sources:

1. Silicone oils, which may cause loss of solderability, are often used as mold-release materials in the manufacture of plastics, and care must be taken to avoid their presence.

2. In heat sealing plastic fumes are often released inside the bag; they might settle on the work and cause deterioration of solderability.

3. If the plastic film is too thin, it tears easily and is also permeable to many airborne pollutants.

It is well to follow some of the recommendations made in MIL-B-81705. This document also refers to some heat-sealing equipment which apparently does not cause the difficulties mentioned above. The use of static-free material is an advantage whenever electronically sensitive components such as FET or CMOS devices are handled. Printed-circuit boards and nonsensitive components are sometimes packaged in electrostatic films which cling to themselves and seal the package well. Heat-shrinkable films on plastic backing are also used for this purpose.

Cardboard boxes can easily be replaced with reusable plastic containers. It is good to use bins with a cover. Transparent or translucent containers offer better visibility in storage and prevent unnecessary opening for examination.

These packaging and storage considerations are important not only for prolonged storage but also for in-plant holding and handling before processing. Good housekeeping and common sense will prevent many manufacturing difficulties.

The realities of life, however, involve many unsolderable surfaces even with a good storage program. We shall treat the restoration of solderability as an additional precleaning operation. (For specific information on solderability restoration see Chap. 8.) Precleaning operations can therefore be divided into four major groups, the removal of foreign matter, heat-treat scales and excessive corrosion, insulation, and tarnish.

5-4 Removal of Foreign Matter   Cleaning processes similar to those used in electroplating are employed. For particles and cutting oils left from previous machining operations, it is well to use either a solvent system such as a degreasing operation (see Sec. 5-42) or water-base detergents with strong agitation to remove the foreign elements. Ultrasonic cavitation is highly successful for this application (see Sec. 5-43). When such materials as paints, lacquers, and the like have been applied purposely to protect the whole assembly from environmental attack, a selective removal operation may be necessary. All these considerations should follow regular common sense rather than inflexible rules. Only when the surfaces to be soldered are free from foreign materials can precleaning be successful.

5-5 The Removal of Heat-Treat Scales and Excessive Layers of Corrosion   Parts which have been heat-treated after severe deformation in order to obtain specific metallurgical properties necessary for the assembly are often covered with *heat-treat scales.* These are excessively deep layers of oxides and the like formed at the high temperatures used for the heat-treating operation and sometimes by the materials used for quenching. Since most materials with heat-treat scales are very difficult to solder, it is recommended that the scale be removed in a presoldering cleaning operation called *descaling.* This operation should not be confused with the bright-dipping operation necessary for appearance only. For details, see Table 5-1.

5-6 Insulation Removal for Electrical Connections   Inasmuch as most electrical conductors are coated with an insulation of some type, it is often necessary to strip this insulation before soldering. Great care should be taken not to nick or damage the conductor in any way because such damage causes severe mechanical problems after the solder connection has been completed. Depending on the nature of the insulating material (see Table 5-2), three major methods of removal are widely used:

1. *Mechanical removal.* In this method, the insulation is physically cut at the point of removal and the insulation itself is slipped off the part by physical force. As mentioned earlier, extreme caution is necessary to avoid reducing the diameter of the conductor in the cutting operation. Several types of automatic equipment designed for wires are available.

2. *Thermal-insulation removal.* In this method, the insulation is removed either by flowing it away under heat or by using a hot tool to cut through the insulation and then stripping it as shown in Fig. 5-1. Special formulations of insulating material (mostly polyurethanes) are sometimes applied on the wires to be soldered, and these materials, together with the flux, make a rapid and easily soldered combination. Another type of thermal strip utilizes a reducing flame to burn off the insulation material.

3. *Chemical strip.* Some types of insulation, usually preferred for their abrasion resistance, cannot be heat-stripped or mechanically stripped adequately. Such materials are stripped with a chemical solution. This solution is usually of a corrosive nature and should not be left on the assembly. Adequate washing and rinsing after chemical stripping are therefore highly recommended.

TABLE 5-1 Descaling Solutions for Presoldering*

Metal and solution composition†	Instructions and special comments
1. Aluminum:	
1% hydrofluoric acid + 1% nitric acid	To remove heat-treat stain not etched by alkaline solution
Duralumin:	
1 wt% sodium fluoride + 6.25% sulfuric acid	
2. Copper and alloys	
12.5% sulfuric acid + 1–3 wt% sodium dichromate	Room to 175°F
10% sulfuric acid + 10% ferric sulfate (anhydrous)	Low-copper alloys 120–140°F, high-copper alloys 140–175°F
15–20 wt% ammonia persulfate	Room temperature; use until pH drops to 1.5–2 and discard (fresh pH 3.5–4) stable 3–5 days
Beryllium copper:	
10% sulfuric acid	125–175°F; good for low-copper alloys only (<85%)
20–30% sulfuric acid	160–180°F; only to loosen scale; follow by nitric acid dip
3. Gold and alloys:	
12.5% sulfuric acid	Use at 150°F to remove heat scale
4. Iron and steel:	
50% hydrochloric acid + pickling inhibitor	Room temperature; rapid; good for polished steel
6.25% sulfuric acid + pickling inhibitor	Room to 175°F; cheaper than above but slower
6.25% sulfuric acid + 8% hydrochloric acid	Room temperature
Cast iron:	
12.5% sulfuric acid + 12.5% hydrofluoric acid	
5. Magnesium and alloys:	
20% chromic acid	190–212°F
6. Nickel and alloys:	
33% hydrochloric acid + 1.5% cupric chloride	180°F. $CuCl_2$ is only an accelerator; follow with hot rinse and dip in 25% H_2SO_4 + 12.5% HNO_3 solution
10% sulfuric acid + 1.5 wt% ferric sulfate (anhydrous)	180°F For light scales only (annealed in reducing atmosphere or hot worked)
10% hydrochloric acid + 0.75 wt% ferric sulfate (anhydrous)	160–180°F
Inconel:	
33% nitric acid + 4% hydrofluoric acid	150–165°F
10% sulfuric acid + 0.75 wt% rochelle salt	160–180°F

TABLE 5-1   Descaling Solutions for Presoldering* (*Continued*)

Metal and solution composition†	Instructions and special comments
Nickel-silver: 25% sulfuric acid + 1.2 wt% sodium dichromate	Room or elevated temperature
10% sulfuric acid + 0.75 wt% ferric sulfate anhydrous	140°F
7. Silver: 66.7% nitric acid	Room or elevated temperature
90–95% sulfuric acid balance nitric acid	Room or elevated temperature; work should be dry
8. Stainless steel:‡ 10% sulfuric acid 10% sulfuric acid + 10% hydrochloric acid	180°F 130–140°F $\Big\}$ For heavy scale loosening only; to be followed by solutions below
20% nitric acid + 30% hydrofluoric acid + inhibitor	125–150°F
6.25% sulfuric acid + 6.25% hydrofluoric acid + 6 wt% chromic acid + inihibitor 25% hydrochloric acid + 5% nitric acid + inhibitor	Room or elevated temperature $\Big\}$ Scale removal

*Data from Nathaniel Hall, "Metal Finishing Guidebook," 28th ed., Metals and Plastics Publications, Inc., Westwood, N.J., 1960.
†% refers to volume ratio, not concentration.
‡Use solutions for scale loosening before solutions for scale removal.

Table 5-2 shows the method normally used to strip 13 common plastic materials.

5-7 Tarnish Removal by Abrasion It is often desirable to remove surface coating, contamination, and tarnish layers by mechanical abrasion. Several methods are described below which achieve this purpose. With these methods, it is usually difficult to control the amount of the base metal being removed, but they afford a high degree of reliability if properly employed.

Abrasive Blasting. Dry abrasive powders or slurries of abrasive materials can be used to remove undesired layers of tarnish before soldering. Normally a fine particle size of a hard material such as silicon dioxide is used. The dry abrasive or the slurry is blasted through nozzles at high speeds against the surfaces to be cleaned. The surfaces are thus physically eroded, and the net result is a fresh metallic layer suitable for soldering. One of the problems of this method of application is the abrasive material

TABLE 5-2   Insulation-stripping Methods for Electrical Wire

Insulation material	Mechanical	Thermal	Chemical
Asbestos	1	3	3
Cloth	1	2x	3
Natural rubber	1	2x	3
Neoprene	1	2x	3
Nylon	1	2x	3
Paper	1	2x	3
Polyurethane	2	1	1
PVC	1	2x	3
Rolan†	1	3	3
Silicone rubber	1	3	3
Solder Eaze†	2	1	2
Teflon† (TFE and FEP)‡	1	1	3
Varnish	2	1	1

*1 = normally used; 2 = used only under special conditions; 3 = not normally used. x = used mostly to separate unwanted sleeve from desired part of insulation; rest of strip is mechanical.
†Trade name.
‡Provide adequate ventilation for thermal strip.

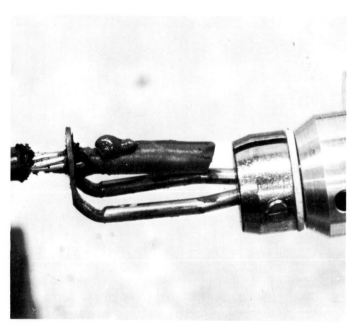

Fig. 5-1   Resistance wire stripper. The heat developed in the top "flows" the plastic, which can then be stripped.

itself, which may settle and lodge in crevices and hard-to-reach places on the work, necessitating a good cleaning operation. If the assembly is fairly simple, additional blasts of air or coolant serve to remove most of the abrasive material together with the particles of metal removed from the surface itself. If this operation is followed immediately by fluxing, the freshly bared metallic surfaces are not given too much of an opportunity to reoxidize. Once the surfaces are fluxed, their solderability is preserved till the soldering operation.

In abrasive blasting as well as in scouring (see below) a basic distinction must be made between hard and soft materials. Abrasive blasting of hard surfaces causes little or no embedment of the abrasive particles in the surfaces treated. Soft surfaces like copper, however, often retain some of the hard abrasive material on the surfaces and thus lose solderability. A general rule of thumb is to use an abrasive material which is of similar or lower hardness than the surfaces being treated. Items like crushed peach pits, coconut shells, etc., can clean a soft metallic surface without the danger of embedment.

Figure 5-2 shows a printed-circuit board being vapor-blasted from underneath to remove tarnish and oxides from the components as well as the printed circuitry. This operation is immediately followed by a dry-air blast to remove all particles and a foam-fluxing operation to protect the surfaces from oxidation. When this method is used, it is possible to solder nickel and Kovar component leads with activated-rosin fluxes with a high degree of reliability. This method makes it possible to dispense with the expensive component plating or pretinning without sacrificing reliability. A similar operation can be carried out with individual parts or specific locations in an assembly by a machine similar to that used to clean spark plugs in the automotive trade. In an operation of this sort it is extremely important to screen out the area where no abrasion can be tolerated. By proper screening techniques, it is possible to avoid the deposition of particles in crevices not normally accessible for cleaning.

Surface Scouring. Pumice and similar scouring materials can be applied in various solutions to the surface, and abrasion is obtained through mechanical rubbing. The liquid media in which the abrasive material is suspended are usually chosen so that they will help in the removal of foreign-material films also present on the surfaces.

There is a certain amount of danger using these materials on soft base metals. If the abrasive particles are embedded in the surface to be soldered, and if they are not solderable, the area of anchorage for the solder is largely decreased and there is a good chance of actual dewetting or incomplete wetting on the surface. The effect of the abrasive material on the surface can be physically examined by dipping a flat piece of treated

Fig. 5-2   Vapor blasting a printed-circuit board.

base metal in the solder alloy to be used for 10 to 15 s and withdrawing it from the solder slowly in a vertical direction, enabling the solder to drain from the surface. Sometimes a fast shake or tapping to remove excess solder helps in this test. A microscopic examination of the drained surfaces upon cooling will reveal pin marks, and dewetted areas will be seen if they are present. The solder should never be allowed to solidify when the surface to be examined is horizontal because hydrostatic pressure will tend to smooth out the solder and will physically cover nonwetted areas without wetting them. The use of water-white rosin flux is recommended for the test.

Wire Brushing. Buffing the areas to be soldered or wire brushing them will achieve the same purpose as the other abrasion methods as long as no greasy buffing compound is used. In most cases, wire brushing is preferable to buffing because it leaves a rougher surface with more surface area, which is better for wetting, as seen in Chap. 1. However, the amount of base metal removed is smaller with careful buffing than with wire brushing, and the specific application will dictate the choice of the method.

Erasing the Surfaces. In recent years, many critical soldered assemblies especially for airborne equipment have been produced by soldering techniques involving erasing tarnish layers and other materials from the surfaces to be soldered with a rubber-type eraser in an effort to increase their reliability. This operation is very expensive and cannot be recommended for everyday solderings. Furthermore many more suitable methods are available to increase the reliability of the solder connection without reverting to such an expensive operation. The rubber eraser used for this application provides a mild degree of abrasion for the base metal that is strong enough to remove any tarnishes or foreign materials. The particles of rubber and tarnish removed in this method have to be carefully removed themselves. Various types of equipment have been built around the eraser, one of which is shown in Fig. 7-17. It looks like a pair of tweezers with rubber erasers on the end.

THE FLUXING OPERATION AND EQUIPMENT

5-8 Ten Methods of Flux Application   Once the flux material has been chosen and the type of flux vehicle has been decided upon, a certain amount of latitude in the flux formulation remains, depending largely on the method of flux application. The true fluxing action occurs at the surface of the base metal. Only those layers of flux immediately adjacent to the tarnish are required for chemical cleaning and retarnish prevention. While the flux also has some thermal characteristics which affect the process, excess flux is wasteful and messy. Thus, we must view the various methods of flux application from the point of view of quantity and uniformity of the deposit. The following methods will indicate in each case what type of material is suitable for the application, and the exact formulation will become obvious. Specific gravity measured with a simple hydrometer is useful for determining, controlling, and maintaining the flux-to-vehicle ratio.

Brushing. This is by far the simplest method of flux application. The equipment usually consists of paint, acid, or rotary brushes. It applies to both liquid and paste application. Although this method lacks uniformity and quantity control, it is possible to flux selective areas in the assembly. The flux vehicle can be any liquid or paste which is not too viscous for brushing. The flux-to-vehicle ratio depends on the amount of flux that must be applied to the area and can be controlled. This method is suitable for automation in the form of a rotary brush immersed on the underside in the fluxing medium. However, this method is generally used in manual applications where liquid or paste flux is desirable. Typical applications are the soldering and sweating of plumbing copper pipes, small-quantity soldering of printed circuits, and flux application on large structural members.

Rolling. This method is suitable for both liquid and paste application. The roller itself is similar to the one used for household painting, but other types of rollers made of special materials, as in the printing industry, are also used. When rolling is done properly, great uniformity and accurate quantity control can be achieved. Because this method is suitable for selective fluxing, it is also suitable for automation. The vehicle-to-flux ratio is relatively low but depends largely on the amount of flux desired on the surfaces. This is a rather common technique used mainly with precision-type soldering as in printed circuits.

Double-sided roller coating is extensively used in infrared fusing of printed-circuit boards. In this application the uniformity of the flux coat is important because it may affect the rate of heat absorption in the process. In fusing, where it is desirable to coat the edge of the printed-circuit-board land, flux must penetrate under the slight overhang which might have been created during etching (the *undercut*).

Spraying. This method is limited to liquids. The equipment is similar to that used in spray painting. A high degree of uniformity and quantity control can be achieved. It is difficult to keep the spray in specific areas, however, and selective fluxing wihout masks is difficult. The vehicle-to-flux ratio is very high, the vehicle usually consisting of a rather volatile material which evaporates on the workpiece itself or in droplets once it leaves the nozzle, thus making a rather viscous and nearly dry droplet of flux hit the surface. Adequate ventilation and other safety precautions may make this method rather expensive. The method is easily adapted to automation and is frequently used with various automatic soldering setups.

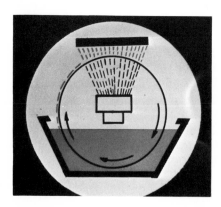

Fig. 5-3 Schematic of rotary-screen spray fluxer. As the screen rotates through the liquid flux, each hole picks up a droplet carried toward the top. When hit by a jet of air the liquid is catapulted toward the work.

Rotary-Screen Spraying (Fig. 5-3). This method is more controllable than the power-spray method. While general considerations are the same, this process permits selective application of the flux to the work. The process is based on a rotary screen (with or without a specific pattern), usually made of stainless steel or other material resistant to the flux. The screen is partially immersed in the flux reservoir, where the openings in the screen

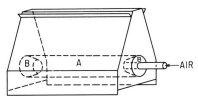

Fig. 5-4   Foam fluxing unit, schematic. All materials to resist specific flux. Tank not shown. Stone, Norton no. P2120 (*A* in figure) stoppers, rubber no. 5 (*B* in figure). Fluid level 1¼ to ½ in above stone. Air pressure ½ to 1 lb/in².

pick up droplets of flux. The screen rotates around its axis, and the rotation speed is variable. Air nozzles are located at the axis of the screen pointing toward the work located above the flux reservoir. The airstream is either continuous or activated by microswitches tripped when the work is located above the spray orifice. The rotation speed and air pressure determine the quantity of flux deposited on the work. This method is often used in printed-circuit soldering when long leads are inserted straight through the board without clinching or cutting before fluxing. While foam fluxing heights are normally limited to ⅝ in (1.5 cm), spray fluxing has been successfully used up to heights of 2 in (5 cm).

Foaming (Fig. 5-4). This method is also limited to liquid fluxes. The equipment consists of a liquid tank with immersed gas nozzles or porous stones designed so that a steady stream of foam is pushed to the surface of the foaming tank (see Fig. 5-5). The work is passed over this foam surface, and the flux adheres to the work in a uniform, controlled layer. This method is only partially suitable for selective fluxing, however. Here the vehicle-to-flux ratio is critical because the formation of the flux foam depends largely on its viscosity, the nature of the vehicle, the gas pressure, the height of liquid over the foaming element, etc. This process is mostly used in automatic printed-circuit soldering lines.

Industrial air produced by standard compressors normally contains small quantities of moisture from the air and oil from the compressor. Water and oil impair the quality of the foam flux and should be removed.

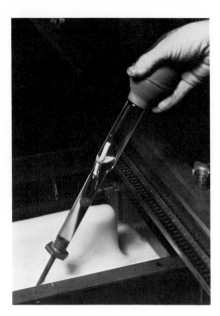

Fig. 5-5 Foam fluxer with wipe-off exit brush. Hydrometer is used to check specific gravity of liquid. (*Electrovert, Inc.*)

An oil trap is normally included with the air-pressure regulator (used to reduce plant air to an intermediate pressure approximately twice that used in the foamer). A clear reservoir on the trap serves as a reminder that it must be drained periodically The fine pressure-regulation valve normally located on the foam fluxer itself has no oil or water trap.

Since the continuous aeration of the foam flux causes an appreciable evaporation of some constituents of the flux formulation, the foam fluxer is normally below room temperature. This becomes important during the measurement of flux density. The specific gravity is a function of temperature, and an adjustment must be made in order to get accurate density readings. Some flux suppliers provide the user with a correction formula, and then it is enough to measure the density and the temperature of the sample. Normally hot water from the tap is used to bring the sample up to the correct measuring temperature by external application to the container.

Using brushes in conjunction with foam fluxing is an additional way of controlling flux quantities on the work (Fig. 5-6); brushes located on both sides of the foam chimney will increase the height of the foam head. A third brush located several inches behind the foam fluxer is used to remove excess flux, which drains to the lower portions of the work. When such brushes are used, they must be kept clean and pliable; otherwise they defeat their purpose.

Dipping. This method is used in conjunction with liquid but can be used for low-viscosity paste. No special equipment is necessary for this type of operation, but uniformity and quantity control cannot be maintained to any great extent. The edges of parts can be fluxed selectively. The vehicle-to-flux ratio is very important and should be maintained to compensate for evaporation. The method is suitable for automation and is frequently used throughout the soldering industry.

Wave Fluxing. This method is suitable only for liquid fluxes. The liquid flux is pumped continuously through a trough, making an exposed wave into which the work is dipped. Work can thus travel above the flux container without having to change its direction, and the flux is pumped to meet the work. The method requires proper selection of flux and vehicle to avoid excessive evaporation, and the flux-to-vehicle ratio should be carefully controlled. This method is frequently used in high-speed fluxing operations on automated lines. The particular wave-fluxing unit shown in Fig. 5-7 is plastic for use with acid fluxes.

Floating Liquid or Molten Salt on Solder Baths. This method is one of the few that can be used with solid fluxes. The solid has a poor contact area

Fig. 5-6 Foam fluxer suitable for both rosin and acid fluxes. Note brushes at sides support the foam head. *(Hollis Engineering, Inc.)*

Fig. 5-7 Wave fluxing mechanism suitable for rosin and acid fluxes. *(Hollis Engineering, Inc.)*

with the surface to be joined and thus is seldom used by itself. In this application, however, the solid flux is heated over its melting point and used as a liquid. The molten salt or a liquid flux is floated on top of the solder baths. Thus the surfaces to be soldered are immersed first in the flux, which is already at soldering temperatures, and immediately following in the liquid solder pool for wetting. A regular solder pot can be utilized if precautions are taken to make sure that the corrosivity of the flux will not attack it. Here the uniformity and quantity of the flux or vehicle are not important, and the only factor controlling the fluxing operation is the dwell time in the flux, which indirectly is a function of the immersion speed and the thickness of the flux layer.

The disadvantages of this method lie in the interaction between the flux and the solder itself. Under certain thermodynamic conditions there is sometimes an ion exchange between the flux and the metallic solder which results in a weakening of the flux and a contamination of the solder pot itself. However, in most systems no such danger is present. On the other hand, there is no need for wiping the solder surface in order to skim the dross and oxides because the metallic surface is kept clean by the flux, which also stops additional oxidation. This method is suitable for automation and is frequently used for the tinning of wires and long strips of material.

Cored Solder. Until now, we have discussed methods of liquid- and paste-flux application, which are completely separated from any solder application. However, the flux can be applied to the work as a core inside the solder wire itself. Here the flux is contained as a solid, powder, or very heavy paste inside the solder wire. When this wire is applied to the heated surfaces (and never to the soldering tools) the flux, which has a much lower melting point than the solder, will flow out of the core and wet and flux the surfaces. Upon further contact, the temperature of the solder itself will reach the melting point and the solder will liquefy, running over the previously fluxed and cleaned surfaces. Thus the flux and the solder are applied at the same time, but because of the melting-point difference the flux has adequate time to prepare the surfaces for a metallurgical bond. The method of introducing the flux into the core of the solder wire is complicated, and a strong possibility exists of small air pockets and the like causing discontinuity in the flux core. When the section of wire containing the discontinuity is used, no flux is available to prepare the surfaces and a poor joint will result. This lack of reliability has brought forward a number of proprietary configurations which are designed to ensure that there will always be a flux to prepare surfaces.

One major advantage of the use of cored solder is that the amount of flux available for each joint is predetermined and is thereby well controlled. The ratio of the flux to the solder itself is indicated by stating either the ratio of the volume of flux to the volume of solder or the ratio of the weight of the flux to the weight of the solder (see Table 5-3 for further details). Cored solder can be made into various preforms and thus becomes very valuable in automatic setups. It is a widely used form of flux application. The correct ratio of flux to solid wire in the core must be determined for each particular application. The general considerations involved are as follows:

1. The amount of flux required for adequate cleaning of the material surfaces to be soldered
2. The temperature and time available for the soldering operation
3. The amount of flux residue which can be tolerated

It is therefore customary to make several trial runs with varying amounts of flux in the solder to establish which core is the most suitable for each specific application.

A danger in the incorrect use of cored solder is the application of extremely high temperatures to organic-cored solders. The high temperature volatilizes the fractions in the flux which have a low melting point, causing a tremendous expansion in volume and giving a spattering effect. This is further aggravated by the fact that extremely-high-temperature

TABLE 5-3   Flux-to-Alloy Ratio for Cored Solder

No.	ASTM B 284–58 T (rosin flux only) weight % Desired	Limit	Core no.	Federal specifications QQ-S-571 vol %	weight %
1	0.55	0.45–0.65	1	10	1.1 ± 0.2
2	1.10	0.90–1.30	2	20	2.2 ± 0.4
3	2.20	1.90–2.50	3	30	3.3 ± 0.6–0.7
4	3.30	2.60–3.90			

Material definitions:
  Natural rosin = dry water-white rosin only
  Plastic rosin = water-white rosin with plasticizers
  Activated rosin = water-white rosin with activators

Material definitions:
  AR = rosin core
  AC = inorganic chloride powder
  Condition D = dry powder
  Condition P = plastic flux

soldering tools, such as soldering irons, may seal the end of the solder tube, making it difficult for the vaporized flux to escape without blowing droplets of solder all over the assembly. Where spattering is encountered, it is suggested that a lower-temperature soldering iron be considered. The spattering is also a function of how the tool is applied. If the soldering iron is brought down on top of the solder in a manner that exposes the flux rapidly to the heat while the wire is sealed at the ends, the amount of spattering will increase. However, if the solder is fed to the hot surfaces, maintaining an open outlet for the flux, the amount of spattering is greatly decreased. Using a lower flux percentage in the core also helps to reduce this problem.

Solder Pastes and Solder Creams. Another method of simultaneous flux and alloy application to the soldered joint is in the form of *soldering paste* or *soldering cream.* Here the solder alloy, in the form of a fine powder (around 350-mesh), is suspended in the fluxing material and its vehicle. This should be the alloy powder and not a powder mixture of the constituents because of the melting-point differences. The flux composition can range from the pure water-white rosin grades to the organic-acid flux and the most aggressive inorganic fluxes. The name *cream* is sometimes applied to rosin formulation while *paste* is normally used for the organic-acid and inorganic-acid fluxes. The ratio of solder to flux by weight is very high, usually between 80 and 90 percent. The amount of flux available is larger than that necessary for the joint formation, but this is the result of the consistency requirements of a solder paste. The paste is applied by brushing, screening, extruding, or rolling onto the surfaces and melts upon heating to give good wetting. The method of paste application determines the uniformity and the ability to flux and solder the work areas selectively. This type of material has one great advantage: it controls the amount of solder available in the joint. It can be used in all applications where no solder is to flow outside the joint area. This method is very adaptable to automation and is used widely in components and hybrid microelectronic soldering.

5-9 Flux Dwell Time and Temperature   As demonstrated earlier, it is not sufficient to flux the work; the flux has to be on the surface at a particular temperature labeled *activation temperature* for a sufficient amount of time. This dwell time and dwell temperature are important to ensure that the flux has adequately prepared the surfaces for soldering. In the sequence of fluxing and before soldering it is often customary to introduce an additional drying station or preheating operation in order to achieve this. The exact time and temperature requirements depend on the system, but several desirable side effects of a preheating operation may make it attractive even if the flux–base-metal system does not require it.

Many assemblies are treated in plating solutions before the fluxing operation. If these assemblies have absorbed (because of porosity) an amount of solution or water, the soldering operation at the elevated temperature will bring these liquids to a boil, causing not only spattering of the solder itself but also the formation of a large amount of vapor which may be trapped in the solder, producing hollow fillets. This is especially true in electronic assemblies and printed circuits. Here a preheating operation either before fluxing or immediately after fluxing is of great benefit. Preflux baking at around 225°F (105°C) for 2 h is very common.

In addition, a heating operation between fluxing and soldering drives off any flux vehicles and diluting agents used in the flux formulations. This tends to minimize spattering and gas formation, which have an effect similar to that of absorbed liquids described above.

When temperature-sensitive materials are to be soldered, a gradual preheating operation tends to minimize the thermal shock imparted to the assembly by the soldering operation. Also, if the assembly is heated gradually and uniformly, the buckling and distortion which accompany such operations as dip soldering are greatly reduced. Figure 5-8a depicts a rugged unit with exposed heating elements, which are self-cleaning if contaminated by dripping flux. An aluminum foil is used as a reflective surface that can be replaced. Figure 5-8b shows a large hot plate attached to a convection heater. The gentle air movement is beneficial for flux drying on horizontal conveyors. Flux drippings should be scraped off periodically.

(a)

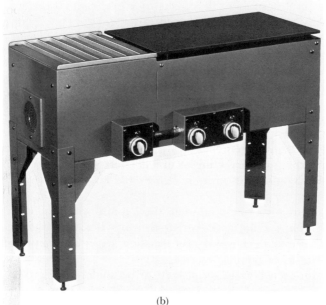

(b)

Fig. 5-8 (a) Radiation type preheater self-cleaning element. Note replaceable aluminum reflector. *(Hollis Engineering, Inc.)* (b) Convection and radiation type preheater. *(Electrovert, Inc.)*

Finally preheating also speeds up the soldering operation by shortening the heating time to wetting temperature during the cycle. The wetting process is instantaneous and will occur at the wetting temperature as soon as the molten solder is introduced to the fluxed surface. Theoretically, preheating to the wetting temperature would enable extremely fast soldering to take place, but in reality this is not the case because of mechanical limitations and the hydraulic effects created when molten solder is placed on a moving surface.

THE SOLDERING OPERATION AND EQUIPMENT

5-10 Classification of Equipment   Soldering equipment is best classified by the mode of heat transfer utilized. Four categories will be discussed: conduction, convection, radiation, and special devices. There is no room in a book of this sort for detailed instructions on the use of every piece of equipment on the market. In that respect, equipment manufacturers' literature is indispensable. However, enough information for the understanding and proper selection of the various types of equipment will be furnished.

The soldering iron is a common tool which has had little or no scientific analysis. In an effort to establish some definite engineering parameters, the author has spent a considerable amount of laboratory time in analyzing and categorizing soldering irons. Chapter 7 is dedicated to the soldering iron itself, its selection, and use.

Conduction

5-11 Conduction Defined   In physics, conduction means the transmission of heat through or by means of a thermal conductor in physical contact with the body, without appreciable displacement of the molecules of the material.

In this group of equipment, the soldering tool itself is heated and the heat transfer to the areas to be soldered is by direct contact, as in the case of a soldering iron. Consequently, it is of primary importance to keep the heat-transferring areas clean and clear of any insulating layers such as oxides which are formed at elevated temperatures. This is an important point, which will be stressed in each of the following soldering methods. We have included liquid-solder applications (soldering pot, solder wave, cascade, and jet) under conduction because the heat is applied to the solder in that fashion.

5-12 The Soldering Iron   Soldering irons are one of the commonest of the conduction soldering tools. Soldering irons with various shapes and types of heating element can be obtained. Electrically heated soldering irons, which are most prominent today, are discussed in detail in Chap. 7. Others are heated by flame or by dipping them into a hot solder pot. Although Chap. 7 deals only with the electrically heated soldering iron, the analysis of tip materials and heat characteristics required in a good soldering iron for a specific use can also be applied to the gas-fired and other units.

Another tool usually included with the soldering irons is the *soldering gun*. Here the temperature control is more difficult, and the amount of heat developed in the iron is a direct function of the time the trigger (electrical contact) is depressed. In most guns, the solder tip is a single turn of a transformed secondary, which therefore carries high currents and heats the tip rapidly. Most of these irons are large, heavy, and cumbersome and suitable only for occasional soldering of diversified joints. Prolonged "on" time will overheat the tip, and the operator can never be sure of the tip temperature. The soldering gun is not recommended for steady production soldering. Figure 5-9 shows a picture of such tools.

5-13 Hot-Plate Soldering   Heat can also be conducted to the work-piece using a high-temperature surface. This is a relatively cheap and simple method but usually requires some special fixtures. It is essential that all heat-transferring surfaces be kept adequately cleaned (surface of hot plate, underside of fixture, top side of fixture, and workpiece itself), which is rather difficult at the elevated temperatures of soldering. This method is quite suitable for automation, and the temperature of the hot plate itself can be used as a safeguard for the maximum temperature permissible for the total assembly (see Fig. 5-10).

The surfaces can be heated by a flame or electric currents. In some cases, the heating elements are built right into the fixtures where the

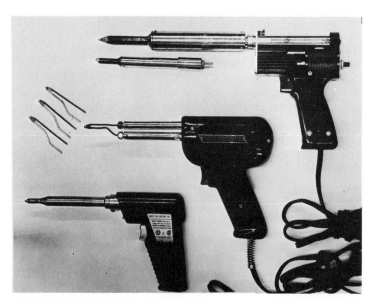

Fig. 5-9 Soldering guns.

Fig. 5-10   Hot-plate soldering.

assemblies are prestacked and ready for soldering. The disadvantage of this method lies in the cooling time required for the solder joint. Once a fixture has reached the soldering temperature and is removed from the heat source, there is danger of movement of the components relative to each other before solder freezing and of deformation or disturbed joints, as discussed elsewhere. Unless quench cooling of the fixtures is part of the setup, this method has some inherent problems.

5-14 The Solder Pot   Another common method of conduction soldering for both manual and automatic techniques is the solder pot. Here an amount of solder is kept molten at the soldering temperature, and the heat content of this mass is large enough to offset any small heat losses due to local cooling during the heating of the work surfaces. In dip soldering, the pot serves as both heat source and solder supply. The prefluxed parts are simply dipped at a slow rate into the solder pot and withdrawn from the solder after a short period (see Fig. 5-11). The advantage of this type of soldering is that gravity and temperature control the amount of extra solder adhering to the assembly. If this is over the allowable limits, a light tapping after withdrawal will get rid of any surplus solder. In properly designed assemblies, a large number of uniformly good soldered connections can be made simultaneously.

The surface of the solder pot must be skimmed continuously to remove all oxides and tarnishes that would serve as a thermal barrier between the solder and the work and to keep contamination away from the soldered surfaces, making direct contact between the solder and the areas to be wetted possible. The skimming is usually done with a nonwetting material that can withstand the elevated temperatures without dissolving in the solder alloy and contaminating it. Such materials as Teflon, heat-resistant glass and ceramic, stainless steel, and blued iron or steel are examples of materials normally used. In addition to skimming the surface, it is possible to maintain an oxide-free solder surface by floating liquids or gases over the surfaces. Because they are mild fluxes, liquids such as palm oil and stearic acid can be used not only to protect the surface from tarnishing but to reduce a certain amount of tarnish already present. Inert liquids such as vegetable oils and greases and various petroleum products can also be floated over the surfaces to protect them from further oxidation. If the protective liquid has a melting point several degrees below the proper solder temperature, it will freeze and coat the assembly, making contact with solder impossible. This is recommended when "freeze-on" (insufficient heat) joints are a problem. It is also possible to maintain a layer or blanket of an inert gas such as nitrogen or carbon dioxide over the solder pot to prevent it from oxidizing. In some instances, solid materials like activated-carbon chunks, cracked nutshells, and granulated asbestos have been floated on the solder surface. These materials serve mostly as a physical shield between the oxygen in the atmosphere and the solder.

The solder pot is usually heated electrically. A large variety of shapes are available on the market (see Fig. 5-12). The choice depends on the individual case. Certain basic considerations apply, however. The pot should be designed so that the bottom is higher in temperature than the

Fig. 5-11   Dip soldering of printed-circuit board.

Fig. 5-12   Solder pots. *(Esico, Inc.)*

sides.This causes natural convection, which keeps the solder agitated. A high degree of solder uniformity is maintained, and no dead spots on the bottom accumulate segregated phases. This design requirement is easily achieved by placing the heating element on the bottom.

In addition, the temperature control is of great importance. Unless the heating element is matched to the solder bulk and temperature required, large temperature drifting results. The ideal case would be a heating element that was on 100 percent of the time, taking care of heat loss and maintaining the soldering temperature steadily. Such an element, however, would take a long time to start the pot from room temperature. An auxiliary fast-heating element is therefore added and used only to start the pot. Unfortunately, it is difficult to design such an ideal element in a commercial all-purpose soldering pot. Consequently, an on-off temperature regulator (mostly a bimetal device) is added to increase the versatility of the equipment. More sophisticated temperature control is available for automatic soldering lines, where small temperature changes cannot be tolerated.

The speed of immersion and withdrawal of surfaces in molten solder has a large effect on the quality of wetting and the uniformity and thickness of coating. Each pair of base metal and flux has an ideal dipping speed which is a function of flux activity and heat transfer in the metal. Plated surfaces do not follow the same rules (see Table 5-4). Careful analysis indicates that the immersion speed of 1 in/s (2.5 cm/s), which is

TABLE 5-4   Maximum Dipping Speeds for Bare Base Metals*

Base metal	Symbol	Sample thickness, mm	Speed, mm/s w/w rosin nonactivated	RMA (mildly activated)
Copper	E.Cu	0–0.3	17.0	>20
(electrolitic)		0.31–1.2	3.5	8.0
Phosphor	CuSn8	0–0.3	6.0	>20
bronze		0.31–1.2	2.5	7.0
Brass	CuZn37	0–0.3	3.0	>20
		0.31–1.2	1.3	4.5
Nickel-silver	CuNi18Zn20	0–0.3	1.0	>20
		0.31–1.2	0.5	2.0

*Data from E. Lenz, Siemens AG. Plated surfaces cannot be categorized this way. Base metals like nickel do not wet with RMA flux. Solder pot, solder alloy and flux are according to *IEC Publ,* 68-2-20, test T, app. C, 1968.

universally specified for solderability testing, is applicable only to thin metals with strong fluxes (RMA). Experience indicates that with activated rosin faster speeds might be possible for the higher heat conductors. Surface uniformity and thickness of deposits depend to a large extent on the rate and uniformity of withdrawal. Standard practice here is to withdraw evenly and smoothly at the rate of insertion. Vibrations which cause rippling on the surface of the solder and jerky motion will cause a rough surface appearance on the parts. The above considerations are especially important for tinning leads and pretinning other surfaces and parts.

The purity of the alloy in the solder pot contributes largely to the quality and uniformity of the solder joints (see also Secs. 3-9 to 3-13). The level of alloy purity should be carefully checked. When the impurities reach undesirable levels, the solder alloy should be changed. Cleaning the solder bath is metallurgically possible, but the materials used are highly corrosive and the fumes given off could be a hazard to both the plant and the ventilation system. Recommended practice is therefore to sell the contaminated solder for scrap and replace it with virgin metals.

The impurities are picked up from the materials dipped in the solder pot. Copper, for example, dissolves readily in tin-lead, and a minimum amount of bare copper should be exposed to the solder pot for as short a time as possible in order to keep copper contamination down. With extremely small cross-sectional parts like fine wires, an additional danger is involved in dip soldering inasmuch as the attack on the base metal may cause embrittlement and even dissolve the part completely, as shown with fine wires of copper[1] and gold immersed for approximately 10 s in a tin-lead solder pot (wires of 0.001 in diameter). The solder pot can also be

[1]H. H. Manko, Soldering Fine Copper Wire, *Electron. Packag. Prod.,* February 1966.

contaminated by the fixtures and clamps which are constantly being inserted into the pot although they are not themselves soldered. One such detrimental impurity is anodized aluminum. Solder pots, clamps, and fixtures are usually made out of iron or other materials (see Secs. 5-33 to 5-35).

Contamination of the bath can result only from contact with the work itself (and fixtures). This means that only a limited number of elements are picked up, depending on the production line. In the dip solder pot, that means copper, cadmium, iron, and zinc, while in wave soldering of electronic assemblies and printed-circuit boards, it means copper, silver, or gold. In other words, a solder bath can only be contaminated by those metals it comes in contact with and which are soluble in it.

To simplify the discussion, consider a single contaminant element of the solder reservoir. A freshly charged bath contains only the elements present in the raw material (Fig. 5-13). For best quality and economy, the purest solder available should be used. In the diagram the finest quality would start nearly at the zero line; whereas the materials meeting either ASTM or QQ-S-571 specifications would start at a higher concentration.

Once the reservoir has been charged with solder, metallic pickup can be considered to proceed at a constant rate for a given amount of work under standard soldering time and temperature conditions. Therefore, the graph would be a straight line within the first regions and would eventually change as the solder is saturated with the soluble element. When the level of solubility where the solution gradient is small is reached, solubility becomes a major factor and tends to flatten out the curve.

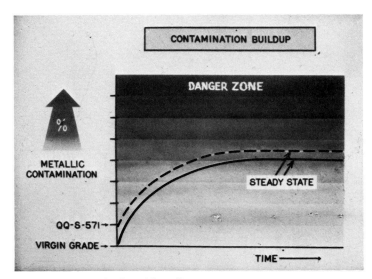

Fig. 5-13   Contamination-pickup diagram.

However, the curve does not show this type of theoretical relationship because of an additional factor. From the outset, as the work passes through the pot, solder is dragged out on each joint. Therefore, the level of the solder in the pot is constantly lowered and must be replenished with fresh solder. The solder which maintains the level in the pot comes in at a low impurity value and thus constantly lowers the impurity level by dilution. This also lowers the curvature of the graph. Experience has shown that after a bath reaches equilibrium, the rate of metallic pickup into the solder is equalized with the dragout and dilution of fresh solder to form a steady-state level at which that particular solder pot operates. This contamination level, then, is typical for a particular reservoir and depends on the production line in which it operates.

As the contamination levels rise, the quality of the solder deteriorates. However, there is no one clear-cut rule for the level of metallic contamination at which the solder becomes unusable for a specific line. This type of situation defies mathematical or theoretical analysis because of the large number of variables involved.

Only rough guidelines have been set up for contamination levels in solder used for wave applications (see Table 5-5). These values were derived from years of practical experience and specific problems encountered in the industry. The variations from line to line and from company to company are such that no rigid rules can be set. One must simply review the difference in manufacturing parameters (time, temperature, heat content, flux, cooling rate, etc.) and in quality-acceptance criteria (appearance of solder fillet, shine, roughness, contour configuration, hole filling, etc.) to understand the reason for these wide variations. These differences in workmanship standards and quality-acceptance criteria are the reason why each organization must set up its own contamination limits

TABLE 5-5   Contamination Guidelines for Wave Soldering with 63/37 or 60/40 Sn-Pb, Weight Percent*

| Element | Typical analysis | | Empirical observations | |
	Virgin grade	ASTM or QQ-S-571, maximum	Troubles noted	Recommended change
Copper	0.002–0.020	0.080	0.200–0.500	0.300–0.350
Gold	0.001	0.080†	0.080–0.300	0.100‡–0.200
Aluminum	0.005	0.005	0.005–0.010	0.005
Cadmium§	0.001	0.001	0.008–0.020	0.005–0.001
Zinc	0.005	0.005	0.005–0.020	0.005

*Although silver might be picked up from the work, it is not considered contamination.
†Included in "all others" category.
‡At economical break-even point.
§Emits poisonous fumes at elevated temperatures.

which can be tolerated in the soldering process according to its quality and cost balance.

Thus, the shading in Fig. 5-13 indicates the approaching danger zone without delineating where the danger zone lies for a specific operation. This is not as serious as it may seem, since it is readily detectable when the solder becomes a problem. At that time two separate actions can be taken to prevent recurrence:

1. A remedial action to limit the amount of contamination per unit work which goes into a solder pot. This is achieved either by shortening the soldering time or lowering the soldering temperature. It is also possible to expose less surface to the solder pot by incorporating a solder resist or masking part of the surface.

2. A preventive step to establish the contamination level that caused the problem. A quality-assurance check can be introduced into the operation to spell out a maximum allowable concentration of an element in the solder; 80 percent of that which has been found to be harmful to the operation is recommended.

Experience has shown that under proper conditions, the same solder (without change) can be used over and over again for many years without reaching detrimental contamination levels. However, if contamination proves to be the reason for production problems, a complete change of solder is mandatory. A word of caution: after changing the solder and before recharging the equipment, a soft bristle brush should be used to clean out any nonmetallic dross adhering to the walls of the equipment. A relatively large amount of such dross is present in the average solder wave or pot. No harsh metal brushes should be used that might expose the metal of the equipment walls to the fresh molten solder. Scouring the walls to a shiny finish might lead to iron contamination of the fresh solder.

A rule of thumb for good housekeeping therefore to obtain the best quality and economy from a solder pot would be:

1. Charge equipment with the purest grade of solder available

2. Operate at the lowest soldering temperature feasible for the operation

3. Make certain that the work touches the solder for the shortest (optimum) time possible

4. Mask off metallic surfaces which do not require soldering

5. Replenish only with the best grade of solder available

6. Prevent clippings and drippings[1] from being reintroduced into the solder

[1]Drippings are defined as solder droplets that fall off the work after the soldering operation and after the parts have left the soldering station. These drippings are high in metallic contamination content, and whereas the average line does not produce more than $\frac{1}{2}$ lb/week, this solder should not be reintroduced into the reservoir. It will increase the contamination level out of proportion to the value of the metal in the drippings.

When trouble starts, the corrective measures should be as follows:

1. Make certain that it is the contaminated solder and not any other parameter within the manufacturing operation that is causing the problem. Replenishing solder is costly in labor as well as materials.

2. If the solder is at fault, change it completely, cleaning out the dross as described earlier.

3. Record the contamination level in the discarded solder and recharge with the best available grade.

4. Take one or more of the preventive measures listed above to minimize future contamination of the pot.

5. Set up a quality-control procedure to monitor the contamination level, and set up a routine change of solder when it reaches 80 percent of the contamination level recorded in item 3.

5-15 Drag Soldering   The use of solder pots for mass production can easily be automated. One of the more sophisticated applications of this type is the drag-soldering system, designed for use with printed-circuit boards and similar planar assemblies. A complete system of this sort has all the necessary ingredients in a regular production machine and incorporates a fluxing station, normally foaming, a preheating and drying station, and a soldering station, consisting of a conveyor mechanism which takes the board through these various preparatory stations and into the solder pot. The sequence in the solder pot is as follows. First a skimmer located immediately in front of the printed-circuit board removes the dross from the surface. This is followed by the printed-circuit board, which is immersed at a slight angle to let flux and flux fumes escape. The board is then dragged through the solder pot to the point of exit, where it is removed at a predetermined angle and allowed to solidify. The end result when the machine is properly adjusted is similar to that obtained with wave soldering (see Fig. 5-14). For increased dwell time, the printed-circuit board can be left stationary on top of the solder for a predetermined amount of time. Some conveyors have an in-and-out motion built into the dipping section to help remove any trapped volatiles from under the board before it is withdrawn.

5-16 The Solder Wave   An automatic method of solder application is the solder wave. Here a continuous stream of solder is pumped up into a spout, forming a head of solder through which the work can be passed. The mechanics of the solder wave, the equipment, and its use for printed-circuit boards are discussed in Chap. 6. It is the most common method of manufacturing printed-circuit boards, but solder waves are also used for other applications, e.g., continuous-wire tinning, soldering armatures for automobiles (Fig. 5-15), attaching leads to hybrids, and pretinning components and flat packs.

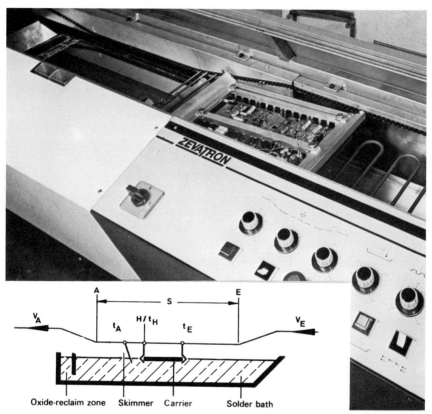

Fig. 5-14 Automatic drag-soldering system with board about to enter the solder pot. Note the direction of the conveyor depicted in the insert sketch. *(Zeva Electric Corp.)*

Fig. 5-15 Wave soldering of armatures. *(Electrovert, Inc.)*

To improve wetting and inhibit the formation of dross, oil can be used in conjunction with wave soldering. The oil is either pumped with the solder (Fig. 5-16) or floated on top as a blanket. For further details see Sec. 6-16.

5-17 Solder Cascade The solder cascade, depicted in Fig. 5-17, is an extension of the solder-wave principle inasmuch as a solder stream is directed over a washboard type of bed in order to produce a number of ripples similar to the solder wave. The work is then passed through these ripples as if through a series of waves. General considerations for the solder wave also apply to the solder cascade.

5-18 The Solder Jet The solder wave and cascade are essentially a solder stream pumped up vertically and then allowed to fall by gravity. The work is introduced at the peak of the solder stream. It is possible, however, to direct the solder stream in any desired direction, horizontal or otherwise, and train it on the area to be wetted. This type of soldering is called *jet soldering*. Whether the stream is continuous or the amount of solder ejected is metered beforehand depends upon the application

Fig. 5-16 Soldering with oil intermix. *(Hollis Engineering, Inc.)*

Fig. 5-17   Cascade soldering of printed-circuit board.

involved. The same considerations that apply to wave or cascade soldering also apply to jet soldering. It is a very simple operation and is used only in special applications.

Convection

5-19 Convection Defined   In physics convection means the transfer of heat by moving masses of matter. The heat transfer occurs by mixing the molecules of one portion of the fluid or gas with another. In convection equipment the heat is applied to the work by a stream of hot gases, which may be reducing or inert in nature. Flame soldering is a good representative of this group. The heat-transfer medium is not limited to gases, and many liquids are used to reflow solder and facilitate joining. In this type of process the heat-transfer medium itself is often used as a flux or in combination with it.

When temperature-stable fluxes are used, protective or reducing atmospheres are unnecessary. If the metallic surfaces have been cleaned of oxides in some previous manufacturing operation, however, a protective atmosphere consisting of argon, helium, nitrogen, etc., can be used for a heat-transfer medium without fluxes, and the thermal stability of the flux is replaced by the chemical inertness of the atmosphere. This prevents formation of surface tarnishes that would hinder the wetting of the base metal.

It is sometimes desirable to solder slightly tarnished surfaces in an atmosphere without the use of external flux, thus introducing no contamination that must be removed later. In that case, the use of a reducing atmosphere as a flux is highly recommended. Pure hydrogen is the most active but the most expensive of the reducing atmospheres. Its use necessitates special safety precautions to avoid explosion when it mixes with the atmosphere. The hydrogen used for soldering semiconductor devices, a major use of this contamination-free fluxing and heat-transfer medium, must have a very low dew point, achieved by drying the hydrogen with activated alumina or silica gel. Another cheaper and less dangerous reducing atmosphere is dissociated ammonia. When gaseous ammonia is heated to 1700°F (927°C), it decomposes to give a reducing atmosphere. A protective atmosphere consisting of 10 to 15 percent hydrogen in nitrogen is sometimes also referred to as *forming gas*. The cheapest and most commonly used atmosphere for industrial soldering is produced by the partial combustion of some hydrocarbon gases. The exact composition of the atmosphere depends on the kind of hydrocarbon used and the degree of combustion under which it was produced. These gases usually contain carbon monoxide, hydrogen, and carbon dioxide, the major portion being nitrogen from the air. This partial combustion of hydrocarbons can be carried out in a nozzle or torch which would result in the reducing flame. Heating the surfaces to be soldered in such a flame would give the same benefits discussed above.

5-20 Flame Soldering   Flame soldering (Fig. 5-18) is mostly a manual operation. It is widely used in plumbing, automotive body work, and for structural joints. It is a very rapid heating method for large masses of metal but usually is accompanied by heavy scale and dross formations.

Flame soldering is not restricted to the use of torches. In many auto-

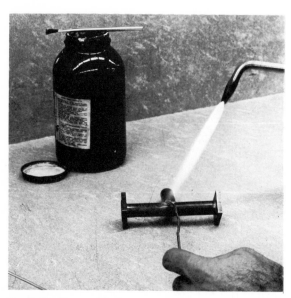

Fig. 5-18   Flame soldering.

Fig. 5-19   Water welder. *(Henes Manufacturing Co.)*

matic applications, the work is passed through a stationary flame. As shown earlier (Sec. 5-19), various types of flames can be used, including reducing flames, which have positive characteristics useful in the joint formation. For instance, insulation can be burned off magnet wires and the like where the reducing flame preserves the unoxidized metallic conductor surface underneath the insulation.

One specific piece of equipment should be mentioned here. Figure 5-19 shows a small generator which converts distilled water into a combustible mixture of hydrogen gas by electrolysis. The gas is emitted from the instrument outlet through a flexible rubber hose into a steel-torch hand-piece and tip, where it is burned. The tip is actually a hypodermic-type needle with a blunt configuration. Changing tip sizes changes the diameter of the gas flame. Regulation of gas pressure is accomplished by increasing or decreasing the input line voltage with a variable voltage transformer.

The length of the flame can be adjusted from one that cannot be seen without magnification to a flame 1½ in (3.75 cm) long. The flame diameter can be adjusted from 0.003 to 0.020 in (0.1 to 0.5 mm). Changing tip sizes and/or flame length does not cause the operating temperature to vary; only the Btu being applied to the part is altered. The normal flame temperature is in excess of 6000°F (3300°C). Normal heat output is approximately 300 Btu/h (75 kcal/h) at this temperature range.

5-21 Furnace Soldering   Furnaces as a source of heat for soldering are another form of convection heating equipment. Furnaces are a well-established industrial tool, and there is no room in a book of this sort to go

Fig. 5-20   Furnace soldering.

into the details of furnace construction. Continuous furnaces, with some sort of conveyer arrangement to carry the parts through the heat zones, are preferable for soldering over batch-type ovens (see Fig. 5-20). The variables to be considered when selecting a continuous oven for soldering are as follows.

In a furnace, the whole assembly will be heated up to soldering temperatures. The rate at which this heat is introduced into the part is rather critical. In addition, the parts should be at flux-activation temperature and solder-bond-formation temperature for a sufficient length of time. Cooling of the joint is an additional factor because no movement between the members to be joined during this cooling period can be tolerated lest disturbed joints be formed. This calls for an extremely smooth conveyance method through the furnace with a sufficient length of travel allowed for cooling in order to reach the solidus temperature of the solder alloy employed.

In all the previous methods of soldering discussed there was no need of introducing the flux and/or solder before heating, but furnace soldering leaves no room for practical methods of flux and/or solder application during the trip through the furnace. The parts must therefore be stacked and fluxed previously with a preform of the soldering alloy in place. It is also possible to use a preform which contains the flux as a core inside or to use a solder paste made of the alloy required suspended in the flux to be used. It is therefore obvious that the solder-joint design should compensate for the special requirements of furnace soldering.

Furnace soldering is widely used in semiconductor-device joining. As mentioned earlier, it is usually carried out in either inert or reducing atmospheres which are closely controlled. Certain alloying operations are carried out simultaneously with the solder-joint formation. The parts are usually stacked in a special tool made of pure carbon or a similar heat-resistant material (stainless steel, aluminum, etc.) which does not contaminate the semiconductor device on one hand and secures the physical location of the parts relative to each other on the other hand.

5-22 Hot-Gas-Blanket Soldering   For small assemblies, especially in electronics, the use of large furnaces or other equipment maintaining specific atmospheres is not always economical. The size of the assemblies and the varieties of operations to be carried out during soldering may make the use of hot-gas-blanket heaters very attractive.

Hot-gas-blanket soldering utilizes rather common equipment and concepts. The atmosphere used is generally contained in a cylinder or other industrial container (compressed air can also be used). It is passed through a normal arrangement of regulators and flowmeters into an air heater, where the atmosphere is heated to the required soldering temperature plus the necessary increment discussed earlier. The assembly to be soldered is then passed under the hot stream (blanket) of atmosphere, where it is heated up and the solder connection is made. See Fig. 5-21 for a typical diagram of such a tube. This can also be used effectively in dip soldering at the point where the work leaves the solder to reduce the thickness of the coating. A strong hot-air blast at the point of departure from the solder pool will help the natural draining and result in thinner coatings. This is not a universal application, however, and is good only where continuous long surfaces are tinned.

Hot-gas-blanket soldering has many advantages. The inert or slightly reducing atmospheres used in conjunction with this method prevent oxidation of the assembly under the elevated soldering temperatures. The system is often coupled with an additional cooling jet using the same

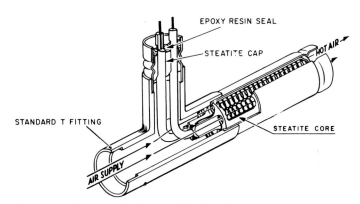

Fig. 5-21   Air heater for blanket soldering. *(Hotwatt, Inc.)*

atmosphere. Under these conditions the importance of the heat stability of the soldering flux is greatly reduced and materials that are not normally adequate for fluxing can be incorporated into the procedure. Tube temperature, gas-flow rate, and exposure time govern the soldering operation and can easily be held within certain limits to give a reproducible and reliable soldered connection. The equipment is relatively simple, and the initial investment is small in comparison with furnace soldering, for instance. Although prestaking of the components and use of preforms or solder paints are recommended, they are not necessary; the solder can be applied during the time of exposure to the hot gases. Figures 5-22 and 5-23 show an effective laboratory setup which was used for many different applications.

5-23 Solder Reflowing   Remelting of a solder deposit on a wetted surface is called *reflow*. Thus, when two pretinned surfaces are reflowed and brought in contact at the same time, a solder joint can result provided that the tarnish layers on top of the molten-solder layers are removed or broken up in some fashion. A relative movement of one molten surface toward the other will break up the tarnish layer, and then these nonmetallic materials will be present in the solder in the form of inclusions. For details on the effect of inclusions on solder, see Chap. 3. A normal solder joint is obtained by using flux on the surfaces before reflowing.

The pretinned surfaces are usually reflowed by immersion in a heat-transfer medium. This is also used to improve electrodeposits of fusible alloys. The surfaces can also be reflowed with a hot gas or any other heating method convenient for the particular configuration.

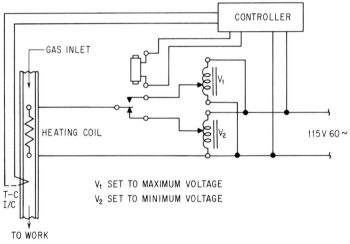

Fig. 5-22   Hot-gas-blanket soldering circuit.

Fig. 5-23   Simple laboratory setup for blanket soldering using circuit in Fig. 5-22.

A large variety of heat-transfer media is available for the reflow solder-ing operation. They are mostly materials with good heat stability com-bined with some chemical activity, which tends to lower the contact angle. A list of these materials is found in Table 5-6. The selection of these materials is usually a function of their temperature range and ease of removal, assuming that their fluxing ability is adequate for the job.

The reflow operation is controlled by the temperature of the medium and the time of exposure. In temperature-sensitive operations where the heat-transfer medium must be kept close to the lower range of the soldering temperature, it is possible to adopt a heat transfer medium of such a nature that the melting point of the material is at the minimum point of soldering. In that case a part submerged in the heat-transfer medium will be enveloped in a solid layer of the medium until proper soldering temperatures are reached. Thus the danger of removing a part too soon is eliminated because its condition can clearly be seen.

The temperature range at which a particular alloy is used in reflowing depends largely on the amount of solder desired, the joint configuration, and the application in mind. It is possible, for instance, to immerse a

TABLE 5-6 Reflow and Fluxing Media in Order of Decreasing Flux Activity

Medium	Density, g/cm³ at 68°F	Melting or softening point, °F	Boiling point, °F	Flash point (Cleveland open cup), °F	Fluxing properties	Corrosivity	Toxicity	Solvents*
Inorganic salts:								
1. Zinc chloride	2.907	567	.....	....	Excellent	High	Slight	1
2. Ammonium chloride	1.527	635	.....	....	Excellent	High	Slight	1
3. Tin chloride	3.393 (469°F)	475	1153	....	Excellent	High	Slight	1
4. Nitrite-nitrate salt (40–50% NaNO₂, 60–50% KNO₃)	1.840	290	.....	....	Poor	....	Dangerous	1
Resins:								
5. Wood rosins	1.070	167†	.....	400	Medium	None	....	2, 3, 4, 5
6. Stabilized rosins (hydrogenated and polymerized)	1.060–1.097	123–323†	.....	435–536	Medium	None	....	2, 3, 4
7. Synthetic resins (hydrogenated and alkyl esters of rosin)	1.020	Liquid	687	361	Poor	None	....	2, 3, 4
Fatty acids:								
8. Oleic acid	1.895 (77°F)	57	680	372	Good	Traces‡	Slight	2, 3, 4
9. Stearic acid	0.847	157	739	385	Good	Traces‡	Slight	2, 3, 4
10. Azelaic acid	1.038 (230°F)	213	673	High	Good	Traces‡	None	1b, 2, 3, 4
11. Palmitic acid	0.853	146	418	421	Good	Traces‡	Slight	2, 3, 4
Petroleum products:								
12. Sulfones	1.266 (86°F)	81	545	356	Poor	None	Slight	1, 2, 3, 4, 5
13. Lubricants§	0.980	−20	450	560	Poor	None	Slight	3, 4, 5
14. Heavy petroleum fractions	0.950	20	600	585	Poor	None	Slight	3, 4, 5
Other organic compounds:								
15. Silicon oil¶	1.100	−8	482	575	Inert	None	None	4, 5, 6
16. Glycol ethers	0.952	Liquid	447	240	Poor	None	None	All
17. Polyalkyl glycols (polyethylene or polypropylene)	1.010–1.204	−68 to 142	.....	385–7475	Poor	None	None	1, 2, 3
18. Polyphenyl ethers (phenoxy phenyls)	1.204	Liquid	982	550	Poor, inert	None	....	2, 3, 4
19. Chlorinated polyphenyls	1.540	Liquid	689	None	Good	High	Medium	2, 3, 4

*Solvents: 1=water; 1b=boiling water; 2=alcohols; 3=ethers; 4=aromatics; 5=naphtha; 6=chlorinated solvents.
†ASTM ring and ball test.
‡The fatty acids give colored compounds with bare metals (a green one with copper, etc.).
§For lubricants other than petroleum fractions, see numbers 15, 18, and 19.
¶Good for reflow of pretinned surfaces only, causes wetting problems on bare base metals.

uniformly pretinned wire vertically in a heat-transfer medium which is kept at 50°F (27.5°C) above its liquidus and as a result obtain a droplet of the fusible alloy on the bottom of the wire with a very thin film at the previously pretinned areas. On the other hand, it is possible to distribute an uneven solder coat over a flat surface by holding it horizontally in the same heat-transfer medium at the same temperature. If migration of the fusible alloy is not desirable, a reflow temperature close to the solidus (or sometimes between the solidus and the liquidus) is used so that the fusible alloy is sluggish and does not move with ease.

In reflow soldering, the exposure time is rather critical because all the alloying considerations discussed in Chap. 3 are rather acute. The use of ionizable fused salts with metallic ions that might contaminate the solder by an ion-exchange reaction should be avoided. Reflow is often used in another context, specifically, when a plated coating is melted onto a surface without effecting a joint.

5-24 Vapor-Phase Soldering   This unusual method offers great temperature control uniformity of heating, and heat-transfer efficiency. It relies on the condensing vapors of the boiling liquid to raise the temperature of the assembly to the liquid boiling point. An expensive fluorinated hydrocarbon having a high boiling point, good heat stability, and oxidation resistance is used. The equipment is designed so that the boiling vapor, which is heavier than air, is retained within the confines of the equipment. When the system reaches equilibrium, the temperature of the vapor is equal to the boiling temperature of the liquid and this temperature is maintained until the cold workpiece is introduced. All the surfaces of the work immediately act as condenser plates, and while the vapor reverts to the liquid state, it gives off its latent heat of vaporization. The work temperature rises to the boiling point of the liquid, when heat transfer stops. Since all surfaces are involved and the heat transfer is intimate, great economies can be achieved in a rather short time. The operation of the unit is similar to that of a vapor degreaser. Once the assembly reaches the soldering temperatures, the work can be slowly withdrawn from the vapor. The flux is washed off the surfaces, dries out, and is filtered from the heat-transfer medium as solid particles. Thus the heat-transfer medium can be used over and over again until it is either dragged out or lost to the environment by poor operating technique or standard diffusion losses.

In order to minimize the loss of the expensive heat-transfer medium (cost approximately $400 per gallon in 1977) it is possible to add a second saturated vapor zone on the top. This second low-boiling liquid helps minimize vapor losses through dryout. Figure 5-24 shows a schematic of such a system.

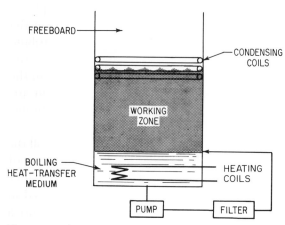

Fig. 5-24 Diagram of vapor-phase soldering system. Note filter system designed to remove dried-up rosin flux from liquid.

Some typical applications of this method of soldering can be found in the manufacturing of computer back planes using preforms for instance. In this method, the printed-circuit board with all the connector blocks in place is fixtured. Solder preforms, normally coated with flux, are automatically located on each pin adjacent to the printed-circuit board where the fillet has to be made. These assemblies are then lowered into the equipment, where all the preforms are soldered at the same time to a high degree of uniformity. When the temperature of the assembly has reached the boiling temperature of the heat-transfer medium, all action stops and the parts are slowly withdrawn from the equipment. Similar applications for preform soldering have been developed since. The method is also suitable for reflowing multilayer printed-circuit boards, where the tin-lead plating on the surfaces is fused by the condensing vapor of the heat-transfer medium and assumes the shape dictated by surface tension and surface energies.

Radiation

5-25 Radiation Defined   In physics radiation includes the total effect of emitting, transmitting, and absorbing energy. Soldering is concerned mainly with the radiation of heat. In practice the three methods of heat-transfer—convection, conduction, and radiation—are difficult to separate in the various types of equipment. Our classification actually divides them by their major features. Thus furnaces at high temperatures have a large amount of radiation, and the process of convection necessitates some conduction before it becomes possible. The use of pure radiation for soldering is rather limited. Most equipment utilizes light or infrared rays

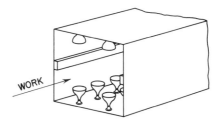

Fig. 5-25 Schematic of a double light bank for soldering.

and is used only in miniature and electronics soldering. However, the advantages gained by the possibility of heating through glass enclosures and the like make radiation soldering an attractive method of heating inside controlled atmospheres and similar hermetically sealed units.[1] An added advantage lies in the fact that only the top layers of the work are heated, minimizing heat damage.

5-26 Unfocused Radiation The most common sources of radiant heat are light waves, covering the spectrum from pure white light to the infrared range. The use of ordinary heat lamps can be highly recommended for a simple type of heating arrangement. Figure 5-25 shows a schematic of a double light bank for soldering. The work is passed between the heat lamps, and the arrangement can be part of a controlled-

[1]Radiation soldering is an extremely clean method of heat application. No contamination of the joint area is introduced with the tool.

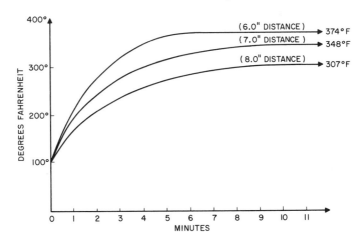

Fig. 5-26 Temperature rise of an XXXP printed-circuit board measured at 0.001 in under the surface using a 500-W infrared lamp at varying distances.

Fig. 5-27 Infrared hot-spot unit: radiation is applied from below for fast heating. *(Argus International.)*

atmosphere apparatus. The distance between the lamp and the work, together with the exposure time, controls the work temperature. Figure 5-26 shows a typical calibration curve giving the temperature of the surface of printed-circuit material as a function of the distance of the lamp and the exposure time. The distance between the centers of the bulbs depends on the uniformity of the heat spot obtained at the distance at which the work is held and is entirely empirical; so is the ratio between work-to-lamp

Fig. 5-28   Variety of assemblies soldered with infrared equipment. *(Argus International.)*

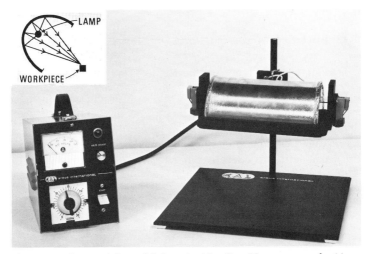

Fig. 5-29   Elongated focused light unit with adjustable power supply. *(Argus International.)*

distance and the exposure time, which must be adjusted for each application. (Shorter distances need shorter times and vice versa.)

5-27 Focused Light   For miniature soldering the energy given out at a source can be condensed and focused at a desired location to generate a high-intensity spot with little heat effect upon the surrounding area. The use of such a concentrated light beam on a particular spot is very advantageous because it can be aimed at hard-to-reach places or through glass capsulating materials. For further details see Figs. 5-27 to 5-29.

5-28 Laser-Beam Soldering   Laser beams offer an additional method of radiation suitable for surface soldering. Like focused and unfocused radiation in the infrared range, lasers can bring the surface temperature up to wetting conditions in a relatively short time. The major difficulty here is to aim the beam at the preferred target area in such a way that the soldering occurs in the correct sequence and both parts to be soldered are heated in the minimum amount of time. The same standard techniques of masking, heat sinking, and heat barriers used in other applications are especially useful in this soldering method.

A typical laser application described in the literature[1] used a beam in the 10-μm range (invisible to the human eye). Power levels were over 1 MW/in². The unit was a 50-W CO_2 laser used for this application. Numerical control was required in order to position the beam accurately above the pad. A rate of up to three joints per second is reported.

[1]J. R. Loeffler, Jr. N/C Laser Soldering, Fast—Low-Cost—No Rejects, *Assem. Eng.*, vol. 20, no. 3, March 1977.

Special Devices

5-29 Special Devices Outlined So far we have classified heat sources by the type of heat transfer used, but this approach is inadequate for resistance and induction soldering.

5-30 Resistance Soldering When a large current is passed through a high-resistance material, a large amount of thermal energy is generated. This heat is instantaneous and can be highly localized, which is suitable for soldering. The heat energy evolved is a direct product of the resistance of the work and the current passed through it ($Q = I^2 R$). Generally the equipment consists of a variable-current source and a set of high-resistance contacts.

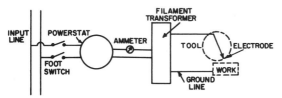

Fig. 5-30 Manual resistance-soldering circuit, single electrode with filament transformer.

Figures 5-30 and 5-31 show sample circuits for resistance soldering with various arrangements for timing (manual or automatic), different transformers (Powerstat with filament transformer or stepdown transformer), and two different tools (single electrode and ground lead, or dual-electrode tool).

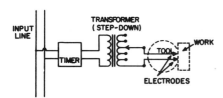

Fig. 5-31 Automatic resistance-soldering circuit, dual electrode with step-down transformer.

Figure 5-30 shows a single-electrode soldering arrangement. Here the work is grounded and the electrode, usually made of carbon or a special high-resistance alloy (stainless steel with copper-clad surface), is placed on the work and closes the circuit. The high current flow will heat up the electrode and the work area, activate the flux, and liquefy the alloy, making the wetting operation possible. This type of application is especially useful when multiple connections must be made to the same piece of base metal, as in a bus bar. Here one ground lead will make the circuits for multiple positions of the soldering electrode. It is also possible to ground a whole set of wires, as in a pin connector, and make individual contacts with each pin.

When this method is used, special care should be taken to ensure that no current-sensitive components are placed in the current passage to avoid overloading. Common sense should also be followed to localize the heat in the soldered area. When soldering multiple-connection items such as connectors, it is possible to plug in a special fixture to make the ground connections on the connector side and solder all the leads into place using a single electrode. It is also possible to obtain the single electrode in the form of a high-resistance wheel capable of heating up strips of material which are grounded and passed over or under the wheel. This is useful in sealing cans, for components, etc. This type of electrode is very easily automated. Here the instantaneous localized heat source is of great benefit, and pretinned surfaces are slightly fluxed before they are pressed together and heated in such a setup.

Figures 5-31 and 5-32 show an arrangement of dual-electrode resistance soldering. Here both electrodes are fixed in the same tool, and the area to be heated is the metal which completes the circuit between the two tips. In this arrangement, a high degree of uniformity can be obtained if the contact time, and thus the amount of current passed through the connection, is predetermined by an electronic timer. The amount of heat depends mainly on the resistance of the electrodes unless the metal

Fig. 5-32 Resistance soldering with dual electrodes.

between the electrodes is of high resistance itself. The electrodes are usually made of an alloy rather than carbon to avoid physical changes in dimensions due to burning of carbon from the electrodes. Figure 5-32 shows both electrodes in the hand tool. It is possible, however, to have the electrodes fixed to a table and press the work in between. For further details, see Fig. 5-33.

A different arrangement is shown on the right-hand side of Fig. 5-33, where a scissor-action dual-electrode soldering tool is used with carbon inserts. Here, there is a double action in the tool where the jaws are clamped over the work first and then the current is passed through. When the current is shut off, the pressure is still applied and the connection can cool down during the solidification with the pressure still on.

Properly controlled resistance soldering is a very reliable method. It is easily automated and is fast and economical. The heat damage can be localized because of the fast supply of heat and the rapid cooling.

Another form of resistance soldering has found application in the microelectronics assembly industry. The *hot-wire bonder* is similar to the schematic in Fig. 5-32, where the workpiece has been replaced by a high-resistance wire connected to the electrodes. When a timed pulse of current passes through this wire, it develops a predetermined amount of energy. This is calculated to be sufficient for the entire bonding operation of the microcircuit. The high-resistance filament can be manufactured into a contoured bar to match the work and is usually made of a nonwetta-

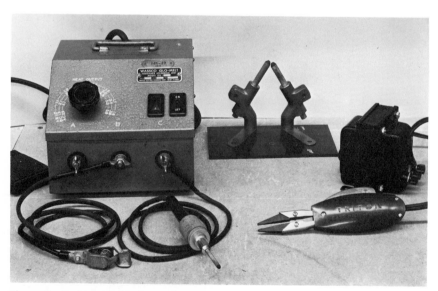

Fig. 5-33 Resistance-soldering equipment.

Fig. 5-34   Induction soldering.

ble material, e.g., tungsten. When the unit is pulsed while the filament is not touching the work, the heating element will glow in the visible frequencies due to the heat being developed. This is used to burn off any flux that might otherwise interfere with the easy operation of the tool. The hot-wire bonder is normally used with prelocated solder and flux in the forms of paste and preforms. It can also be used on pretinned surfaces that have been refluxed. Typical applications are the planar mounting of flat packs onto printed-circuit boards, sealing micropackages, and attaching ribbon to hybrid substrates.

5-31 Induction Soldering   In induction soldering, the workpiece is used as the secondary of a transformer converting electric energy into heat. In this method, no contact with an external heat source is necessary, and the work itself serves as its own heat generator. Furthermore, no physical contact with the energy source (the induction coil) is necessary (see Fig. 5-34). As higher frequencies are used, the skin effect becomes dominant.[1] Thus the depth of heating in the workpiece is easily controlled through the frequency used. It is also possible by proper tool selection to concentrate the power in the region of the solder joint.

Magnetic materials such as iron and steels will heat up more efficiently than nonmagnetic materials with electromagnetic induction. In addition, the heat characteristics of the material becomes of vital importance in conducting the heat away from the skin, where it is formed, thus changing the requirements for the operation.

A guide to the depth of heat penetration as a function of the frequency used is given in Fig. 5-35. It illustrates the depth of penetration as a function of the frequency for a number of metals. Examination of Fig. 5-35 shows that the skin effect is more pronounced for higher frequencies. This heating of the surfaces only is extremely useful for soldering operations because wetting occurs only at the surface and this method of heating minimizes the danger of distortion and oxidation in the areas not

[1]The skin effect is the concentration of current near the surface due to a rapidly alternating current flow in a metallic conductor.

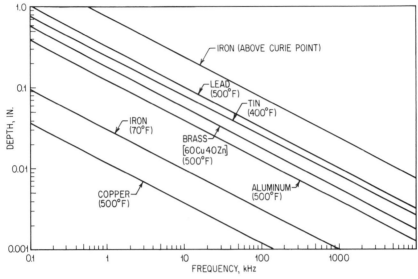

Fig. 5-35   Average depth of penetration vs. frequency for base metals.

being joined. A large cost reduction is often possible when this method is used instead of a conventional heating operation.

As in furnace soldering and other methods, it is necessary to introduce the flux and the soldering alloy to a joint before heating. Figure 5-36 shows a closeup of an AN-connector soldering fixture. Continuous-feed-

Fig. 5-36   Induction-soldering fixture for AN connectors.
(*McDowell Electronics, Inc.*)

ing mechanisms can be coupled to the power unit so that every time a workpiece passes through the coil, the power is sent through the inductor loop. Figures 5-37 and 5-38 show automated arrangements in mass production. Although the time depends entirely on the configuration of the parts (the heat capacity and conductivity characteristics of the work metal), the frequency used, and the soldering alloy employed, induction-soldering time usually ranges from 3 to 15 s.

Most soldering operations are carried out at radio frequencies (400 to 500 kHz) because of the fast rate of power input and the ability to localize the heat to small areas at these frequencies. However, it is also possible to use lower frequencies as generated by machine frequency equipment (up to 10 kHz). In general, smaller parts are normally heated under higher frequencies and larger parts under lower frequencies.

5-32 Ultrasonic Soldering. The term *ultrasonic* does not refer to a mode of heating for soldering but to an additional device used to facilitate soldering without any significant changes in the temperature of the assembly. Once the system has been brought up to soldering temperatures, high-frequency waves are generated into the molten soldering alloy (frequencies higher than the range of sound audible to the human ear, hence the name ultrasonics). These waves travel freely through the liquid alloy, but when the liquid-to-solid interface is reached, *cavitation* takes

Fig. 5-37 Induction soldering of 5-gal cans. *(Lepel High Frequency Laboratories.)*

Fig. 5-38   Automatic induction soldering of jewelry. *(Ther-Monic Induction Heating Corp.)*

place. As the waves reach the solid at the interface, pressure and suction (explosion and implosion) alternate at the frequency of the wave (an oversimplified explanation). The total effect of ultrasonics is an erosion of the surfaces whereby foreign particles and tarnish layers are physically removed and the bare metal underneath is exposed. Consequently, the solder alloy can wet the surface adequately without the use of flux. This is extremely useful wherever fluxing is not feasible.

The equipment usually consists of a heating device, a transducer, a transmission rod, and a high-frequency generator. The transducers can be roughly divided into two major groups:

1. Magnetostrictive transducers, usually based on nickel and its alloys (ferromagnetic materials). These materials change dimensions (the Joule effect) when introduced into a magnetic field. By oscillating the magnetic field backward and forward, a mechanical motion can be generated at ultrasonic frequencies. The magnitude of the dimensional change is a function of the magnetic field applied (independent of its sign), the material, and the temperature used. Magnetostrictive devices for soldering usually range between 18 and 26 kHz. The lower the frequency, the larger the effect of cavitation on the surface to be soldered. Magnetostrictive devices are usually very temperature-sensitive. As the nickel alloys used reach 180 to 200°F (82 to 93°C), they lose their magnetostrictive efficiency.

2. Piezoelectric devices operate using such materials as crystals of quartz and tourmaline or polycrystalline materials such as barium titanate

(Curie point 212°F, or 100°C) and lead zirconate (Curie point 500°F, or 260°C). These materials change dimensions when a voltage is applied across their surface. In polycrystalline materials, this effect is also sometimes referred to as *electrostriction* or *polarized electrostriction*. Here too, the magnitude of the dimensional change is a function of the applied electric field (independent of the sign), of temperature, and of the material itself. Piezoelectric devices for soldering usually range between 20 and 60 kHz. They are more temperature-stable than magnetostrictive devices. The efficiency of piezoelectric transducers is much higher than that of magnetostrictive devices. This makes the price of the generators and the rest of the equipment lower for the same cavitation.

Dictated by the physical size and shape of the transducer, there are certain frequencies at which the unit has mechanical resonance. This results in a large vibration amplitude provided the transducer is driven with a sinusoidal electrical impulse. These frequencies, having a larger degree of efficiency, are therefore used for soldering. Inasmuch as these frequencies usually depend on the load put on the transducer during the specific operation, the generator is provided with a tuning facility. Many devices are tuned by ear, which is simpler than it may appear. However, automatic frequency control, also incorporated in industrial units, is usually achieved by using an additional piezoelectric crystal which translates dimensional changes into currents with a high degree of accuracy. This feedback is used for automatic tuning.

Because of the thermal sensitivity of most transducer materials which lie below soldering temperatures, it is necessary to keep the transducer away from the immediate soldering area and to cool it by air circulation or water jackets. This is achieved using a transmission coupling attached to the end of the transducer which is subjected to the largest vibratory energy. This is transferred through the coupling to the molten solder. The length of the coupling rod is a direct function of mechanical resonance. Stainless steel is an excellent material for transmission couplings because it has relatively poor heat-transfer properties and low damping characteristics. The transducer tip should be made of an extremely hard material to minimize erosion during immersion. Figure 5-39 shows a water-cooled ultrasonic transducer immersed in a solder pot. The transducer consists of a nickel stack brazed to a stainless-steel transmission rod, which in turn is immersed at an angle in the solder pot. Smaller units seldom have water cooling. The solder pot pictured in Fig. 5-40 has a well on top of a vertical-transmission coupling rod with a transducer mounted on the bottom. The well is filled with solder and serves as a solder pot. Figure 5-41 shows air-cooled soldering irons.

The last important component of an ultrasonic soldering unit is the generator, the power unit which drives the transducers and also supplies the heat for soldering. The heat is usually supplied through resistance

Fig. 5-39   Water-cooled ultrasonic transducer immersed in a solder pot.

coils or other conventional methods. The heat is often generated in a component external to the soldering tool. This is one of the problems inherent in ultrasonic soldering. The heat supplies are usually small in order to prevent overheating of the transducers, and large recovery time as well as long soldering times are usually necessary. However, the use of additional heating units eliminates this problem, and ultrasonic soldering is becoming more and more widespread.

Ultrasonic soldering was originally used mainly for soldering aluminum without flux, but additional applications such as wetting silicon, germanium, and magnesium have been found. It is possible to solder nearly any metal without flux using ultrasonics. However, the economics of the problem have to be studied. The price of the equipment, the special effort in keeping the transducer cool, and the slower heat-transfer possibilities are some of the factors which outweigh the cost of the flux material and its application and removal.

Ultrasonics is also a powerful tool for the study of solder properties. The author has successfully used ultrasonics to degas and grain-refine solder specimens under high vacuum for measurement of resistivity, coefficients of thermal expansion, and the like. In addition, the use of ultrasonic devices for detection of cracks has often been suggested as a means of inspecting solder joints, but no satisfactory method is known to the author.

Fig. 5-40   Small ultrasonic solder pot with generator (air-cooled).

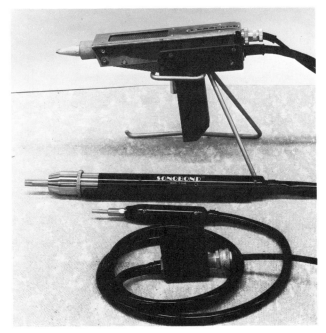

Fig. 5-41   Ultrasonic soldering irons. The two bottom units are forced-air-cooled.

CONSTRUCTION MATERIALS FOR SOLDER EQUIPMENT

5-33 Required Properties Construction materials for solder equipment must fulfill several requirements simultaneously. They must have the physical properties of strength, heat conductivity, and thermal stability as well as corrosion resistance to the soldering alloys at the temperature expected to be used. The last requirement has a double purpose, inasmuch as we do not want to erode the equipment surfaces with a molten solder and we want to avoid pickup of harmful contamination in the solder alloy itself. In addition, the construction material must withstand such additional factors as fluxes and atmospheres.

5-34 The Effect of Liquid Soldering Elements on Materials of Construction *Antimony.* Antimony appears to attack most metals, and for pure antimony, the use of graphite is highly recommended. However, antimonial solders with relatively small percentages of antimony can easily be handled in iron, steel, and stainless-steel containers.

Bismuth and Lead. These two metals can be grouped together because their corrosive effect on construction materials is similar although bismuth is slightly more aggressive than lead. In using bismuth and its alloys, the expansion property of the material after freezing should be included in the considerations because it may deform or break containers unless they are properly designed, especially ceramic materials (see Sec. 3-23 for further details).

Materials of construction which are most resistant are iron and ironbase alloys, high-chrome stainless steels, and the refractory metals tantalum, columbium, and beryllium. The resistance of all these materials up to 1000°F (538°C) is very good. Chromium has only fair resistance up to this temperature. Nickel has very poor resistance to lead and bismuth at temperatures as low as 660°F (349°C).

Cadmium. Molten cadmium will attack copper, nickel, and most other metals, but iron and chromium and some of the refractory metals have a good resistance to corrosion of liquid cadmium.

Indium. Indium is low-melting, diffuses easily, and alloys with most metals. Iron, 18/8 stainless steel, tungsten, tantalum, molybdenum, columbium, and cobalt will withstand indium at normal soldering temperatures.

Silver. As high-melting-point material, silver is present only in small percentages in most solders. In that form it offers few problems to materials of construction. If high temperatures are contemplated, iron and iron alloys should be used.

Tin. This is the most aggressive material in the conventional soldering alloys and will attack many alloys intergranularly. Tin attacks ferrous alloys only mildly up to 800 to 900°F (426 to 482°C). Above these temperatures, the attack is more rapid. Beryllium, tantalum, and sometimes chrome-plated materials show fairly good resistance.

Zinc. Zinc is a relatively active material which will attack most ferrous alloys at elevated temperatures. Columbium, chromium, tantalum, tungsten, titanium, vanadium, molybdenum, and iron alloys are possible container materials for this element.

It is an accepted industrial practice to *weather,* or *blue,* the inside surfaces of equipment which will come in contact with molten solder. This means deliberately tarnishing the materials of construction by the elements of heat. The oxides and tarnishes formed do not wet or react with the molten solder.

In general, it can be said that for most soldering alloys iron and its alloys are the most common and least expensive construction materials, with stainless steel and some of the refractory elements being more expensive but slightly more stable. The use of ceramic-lined containers or equipment is highly recommended for high-temperature applications.

5-35 Spring Material for Use at Elevated Temperatures   In the design of clamping devices, it is often necessary to have some type of spring material which must perform its functions at soldering temperatures. It is imperative that such a material retain its spring properties under soldering conditions. Table 5-7 lists a number of suitable alloys in order of increasing operation temperatures and the heat-treat cycle required.

5-36 Heat Barriers and Heat Sinks   The thermal characteristics of the various heat-transfer materials for soldering equipment will be discussed in Sec. 7-4 and are listed in Table 7-1. Metals and nonmetals are often used for other than heat-transfer reasons, however. In soldering, it is often desirable to prevent the heat from reaching certain parts of the assembly; in that case, heat sinks and/or heat barriers are used.

A heat sink is usually a metallic device which is brought into contact with the assembly being soldered at such a location that its added mass helps absorb the caloric output of the area being soldered and thus prevents the

TABLE 5-7   Spring Materials for Soldering Temperatures

Material	Recommended operating temperature limit, °F	Aging cycle	E	E (torsion)
302 stainless steel	550	None	26.5	10
Inconel	650	None	31	10.4
17-7 PH	700	1 h at 900°F	30.0	
Inconel X	800	20 h at 1215–1325°F	31	11
Rene 41	1000	16 h at 1400°F	25.9	

temperature rise of sensitive parts of the assembly. Heat sinks are usually good conductors with a large heat capacity, such as copper. Figure 5-42 shows several heat sinks designed specifically for the electronics industry. They are clipped between the lead wire and heat-sensitive component. The use of heat sinks on stranded wire is especially important because it stops the wicking action (the climbing of solder in the capillary spaces of the stranded wire) by freezing the solder at the area of contact.

The above-mentioned heat sinks are artificial additions to the assembly and usually fulfill a desirable purpose. However, natural heat sinks in some assemblies which have not been properly designed will rob the soldered area of the heat essential for good wetting. In those cases, good wetting occurs only after prolonged heating or when the whole part reaches soldering temperatures. Common sense in the design can prevent most of these troubles, or a rapid localized heating, e.g., with resistant soldering or induction soldering, should be used.

Heat barriers are materials which are nonconductors, or thermoinsulators. These materials are used when the flow of heat should be restricted to a limited area. They are effective as shields for radiation or convection, in addition to conduction. They include such things as asbestos, various

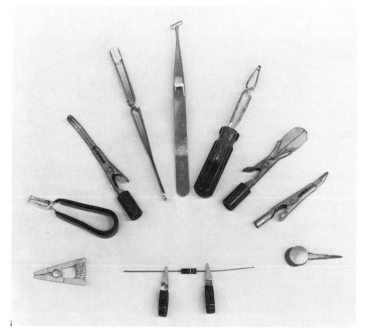

Fig. 5-42   Heat sinks. Note the simple method of application to the resistor (bottom).

ceramic materials, Transite, and to a lesser degree, hardwood and plastic materials (wood is still used for most soldering-iron handles).

POSTSOLDERING CLEANING

5-37 Introduction   The importance of postsoldering cleaning is undisputed throughout the industry, and unless materials and processes are specifically selected to eliminate this operation, serious trouble may ensue.

While it is difficult to give a general guide to the level of cleanliness which any type of electronic assembly must conform to, cleanliness requirements for military and government contracts have been established. Details on methods of monitoring cleanliness and the physical equipment available are given in Chap. 8.

5-38 The Importance of Flux-Residue Removal   The importance of proper flux-residue removal after soldering cannot be stressed enough. To be chemically effective every flux must be corrosive to some degree; otherwise it will be incapable of removing tarnish from the surfaces to be bonded. Manufacturers' claims of noncorrosivity for their fluxing materials have been disproved earlier. Only the residues of water-white rosin or mildly activated rosins can be left on the assembly under special conditions. No blanket approval for activated-rosin residues can be given, however, and each case should be studied in detail. It is safe to say that the best practice for good soldering includes a flux-removal operation.

Proper flux-removal procedures depend on the flux system used. The solvents used for the removal should include materials to dissolve the flux vehicle as well as the fluxing material in its raw form and its decomposed form. We can use the activated-rosin fluxes to illustrate this point. After soldering, we have two major groups of materials, the rosin and the rosin residues, which are solvent-soluble, and the activators and activator residues, which are water-soluble. If only solvents or water are used to clean the assembly after soldering, one or the other part of the flux residue will not be dissolved and will be left on the surface. It is therefore necessary first to remove the rosin and its residues with some solvent and then to follow this with a water-base cleaning operation to remove the ionizable activator materials. We therefore divide the cleaning procedures into the water-soluble and the solvent-soluble materials and discuss them separately. In addition, the efficiency of the cleaning operation can be improved by various mechanical means, e.g., mechanical agitation, ultrasonic cavitation, and vapor degreasing.

5-39 Water-soluble Residues   When both the flux material and its vehicles are water-soluble, the problem is greatly simplified. The work is thoroughly washed in water. One still-water wash, however, is not adequate because it is rapidly contaminated with the materials removed from the work and upon drying leaves thin films of the corrosive fluxing materials that we have tried to remove distributed evenly over all the washed surfaces. This can be remedied in one of several ways as follows:

1. Using several water rinses. The work is introduced into the most heavily contaminated water and is transferred from there to cleaner solutions successively, going into a freshwater rinse as the last operation. This method is called the *countercurrent-cleaning method*. The stillwater rinses used in this method are changed after a certain amount of work has been passed through them in such a way that the most contaminated solution is removed from the system. Successive tanks move up the line so that the second most contaminated tank becomes the first rinse and a freshwater rinse is the last operation. If a continuous stream of water is used, the same basic principle can be applied with the parts moving against the direction from which the freshwater supply comes, thus making the most contaminated water meet the freshly introduced work.

2. Introducing neutralizing agents. Here again the first rinse is a straightforward water rinse, usually with the addition of some wetting agents to facilitate residue removal. The second tank contains a chemical which will react with the corrosive materials in the flux and produce material readily soluble in water. The choice of a neutralizer depends on the type of flux used. Neutralizing agents consist mainly of bases that counteract the acidity of most acid fluxes, e.g., sodium bicarbonate and ammonium hydroxide. The use of ammonium hydroxide is highly recommended because it has a brightening effect on copper and its alloys. In addition, it forms complexes with most metal chlorides which are very water-soluble. If any traces of ammonium are left on the metal, they can vaporize easily without leaving any ionizable residues behind. The neutralizing operation is followed by an additional water rinse. The last rinse is usually followed by a hot-water wash to bring the temperature of the assembly up to a level where the work can easily dry on its own when exposed to the atmosphere and/or be forced-air-dried. Fast-drying solvents like alcohols that are miscible with water are sometimes used to help in the drying operation.

Great care should be exercised when water for the final rinsings is selected. Many water supplies are contaminated with various chemicals introduced to maintain the biological purity of the water. These materials, mostly chlorides, are detrimental to the assemblies cleaned because they introduce a fresh amount of ionizable material. Deionized or demineralized water is essential for the final rinses, especially when electronic assemblies are processed. This processed water is usually expensive and is therefore used in still rinses. The degree of contamination of these rinses is monitored by conductivity cells or the silver nitrate test. However, it is important to remember that all these cleaning procedures do not assure trouble-free assemblies unless adequate protective measures are taken to prevent recontamination from human handling, dust settlement, etc.

5-40 Water-Base Detergents   Many constituents of industrial fluxes are not directly soluble in water. Here the addition of a surface-active agent will ensure the total removal of the flux material. There is no room in a book of this sort to go into the intricacies of detergents and wetting agents. However, most reliable flux manufacturers will specify the type of material which should be used to make their flux residues water-soluble. Above all, it is vital to select a nonionic surfactant because of the inherent residue film left behind.

Many water-base detergents can be effectively mixed with a neutralizing agent, especially ammonia. This eliminates one additional rinsing operation. However, the manufacturer's instructions for rinse temperatures should be carefully followed because many water-base detergents are liable to leave thin residues over the assembly if not removed correctly. The general sequence of operation is similar to that already discussed for water-soluble residues. The first cleaning tank contains the detergent and possibly the neutralizing agent; the second tank contains a water rinse under proper conditions to remove all traces of the detergent; and this is followed by a third rinse, which can be either simple water or distilled and/or demineralized water.

5-41 Water-Base Chemical Reactions   Flux and solder removal can be effected from electronic assemblies with specific chemicals designed to remove known types of dirt. These are normally divided into saponifiers and special additives. *Saponifiers* are intended to remove rosin, oil, and similar nonpolar contaminants with the help of water. These are normally alkaline and react with rosin and similar materials to form a soaplike product which is easily water-washable if not always water-soluble. *Additives* are designed to remove specific reaction products of the flux and its residues with solder and other parts of the assembly. A typical example is the formation of zinc oxychloride when zinc chloride or zinc ammonium chloride is used in flux (Sec. 2-16). The zinc oxychloride develops as a hazy layer of tarnish over the solder joint and in adjacent areas. Here the addition of 1 to 2 percent hydrochloric acid greatly facilitates the removal of the zinc oxychloride and restoration of shiny surfaces. Most flux manufacturers include such information in the data sheets provided with the flux.

5-42 Solvent-soluble Residues   In this group, we find most fluxes formulated around rosin. However, pure solvents will not remove most

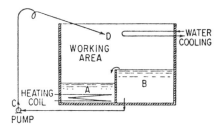

Fig. 5-43   Schematic of a vapor degreaser.

activators and their residues of ionizable materials which cause trouble in electrical assemblies. Therefore, in a typical solvent system, we first remove the rosin and its residues (or for that matter the petroleum jelly or any other solvent-soluble vehicle) and then follow with a water-base neutralizing operation. Or it is possible to blend two solvents which are miscible in each other, where one is polar and one is nonpolar. These blends are normally designed for efficient removal of both types of contamination.

If the asssembly can withstand the temperature of vapor degreasing, this is an extremely successful and rapid method that can be highly recommended. Figure 5-43 shows a schematic of a vapor degreaser. The liquid is boiled in tank *A*, where the heating coils are located, and the vapor given off rises and comes into contact with the cold work introduced into the degreaser. The vapor condenses, and the distilled solvent which is formed dissolves the solvent-soluble materials off the assembly. The solute-rich solvent falls back into the heating chamber, and a fresh amount of vapor redeposits on the surface for additional cleaning. When the part is removed, it is warm enough for self-drying. The unused vapor is condensed on the sides of the equipment and flows into reservoir *B*, which collects the clean fluid. The overflow from this reservoir maintains the level of solvent in the boiling chamber. Various slightly different versions of this equipment are available on the market. Some have an initial ultrasonically activated solvent bath followed by the vapor degreasing operation. Others have a pumping arrangement (*C* in Fig 5-43) which feeds clean solvent from tank *B* into spray nozzle *D* for faster flux removal. Automatic equipment is also available.

5-43 Benefits of Cleaning with Ultrasonics   Ultrasonic cleaning rates special consideration in the removal of fluxes and flux residue as well as in the presoldering cleaning operation. The equipment is similar to that described in ultrasonic soldering and consists mainly of a radio-frequency generator that converts the energy into 20- to 40-kHz frequency. This high-frequency energy is transformed into mechanical vibration of the same frequency through the use of transducers. The various kinds of transducers and their frequency ranges were discussed earlier (Sec. 5-32).

The cavitation formed in the liquid is used to erode the surface of the parts to be cleaned and has a scrubbing action on the surfaces when the proper solvent is used. The flux and flux residues are rapidly removed from the surface. This is the greatest advantage of the use of ultrasonics because it speeds up cleaning times by several orders of magnitude. Furthermore it ensures penetration of the scrubbing action into hard-to-reach places. This makes the ultrasonic-cleaning operation rather economical from many points of view.

Ultrasonic-cleaning tanks can be obtained in a variety of sizes. The transducers are placed inside the solution or mounted externally on the tanks. The selection of the proper cleaning medium is of primary importance. Because of their high surface tension, aqueous solutions have the best cavitation effect. Chlorinated solvents are also preferred to other organic solvents which do not have the same acoustic properties for good cavitation. This applies also to various types of wetting agents which increase or decrease the cavitational effect of aqueous solutions.

The method of loading the ultrasonic-cleaning tank is also important. Single parts correctly spaced will give better results than a randomly arranged lot in some kind of container. When the equipment is misused (wrong solvents, overloading, poor placement of the work in the solution, improper frequency selection, or poor tuning), the process becomes very inefficient. It is therefore recommended that the beginner in ultrasonic cleaning consult a reliable manufacturer to help select the proper equipment. In addition to standard-sized cleaning units, specially engineered systems can be obtained from the manufacturers. These systems are specifically designed for the purpose and may include additional equipment such as continuous cycling and filtering devices, heating units, vapor degreasers, and drying equipment.

Ultrasonic cleaning itself does not automatically clean any assembly but requires careful engineering and selection. The benefits are increased cleaning rates, cleaning of inaccessible places, and in many cases cost reduction.

The use of ultrasonics is often rejected because damage to the component is feared. In electronics, of course, a small part of the assembly such as a transistor may under some unique circumstances resonate to the frequency of the cleaning solution. If the design permits, the amplitude and the force available can exceed those of the material, and failure will occur. However, experience shows this to be unlikely. Many of the major manufacturers of semiconductor devices, where fine wire whiskers are used, incorporate ultrasonic cleaning in their production lines. No specific case where ultrasonic cleaning was detrimental to the assembly has been brought to the attention of the author. On the other hand, ultrasonic-cleaning methods are employed in some automatic sealing operations of electronic components in order to separate the defective units. The average energies involved in ultrasonic cleaning are actually smaller than those in dropping an assembly from tabletop height. It is therefore a suitable cleaning method for most soldered assemblies.

Ultrasonics is mostly used in those parts of the cleaning cycle where scrubbing and mechanical agitation close to the surfaces are most important. This usually refers to the first cleaning operation.

5-44 Protective Coatings   Many materials are applied as protective coatings in soldering over assemblies. Their functions are varied but fall into two major categories.

Solder Resist. These are coatings applied to areas where no solder wetting is desired. An example is a printed circuit with narrow land spacings and a crisscross arrangement where normally bridging would be a major problem. When such a circuitry is selectively coated with a solder resist, so that only areas to which components are to be soldered are left exposed, the possibility of bridging is completely eliminated.

Solder resists are permanent or temporary and are applied by screening, spraying, and brushing. The coatings must be cured before soldering and can then withstand soldering temperatures for reasonable lengths of time without deteriorating. Needless to say, the surfaces before coatings should be extremely clean. It is also important to establish compatibility between the solder resist and the flux to be used. In some cases fluxes (especially amines), which were not compatible with the solder resist caused flaking and/or breakdown of the coating with disastrous results (true for most nonepoxy materials). Solder resists have the additional advantage of presenting a barrier between the atmosphere and the printed circuit, thus eliminating dangers of corrosion and current leakage. The additional expense is balanced by the fact that less solder is used for the operation and that in printed circuits copper circuitry without tin-lead or other protective coatings can be used.

One of the greatest disadvantages of solder resists is the fact that repair work through them is impractical. The solder mask must be removed mechanically by scraping or some other method before soldering can be performed. Most commercial solvents used to remove epoxies will cause extensive damage to other parts of electronic and electrical assemblies.

Conformal Coatings. These materials cover the soldered connection after thorough cleaning and flux removal. The materials are applied in an effort to preserve the cleanliness of the surfaces after soldering, especially for electronic and electrical assemblies. The chemically clean assembly obtained by proper flux removal is an advantage well worth preserving. Perspiration as well as dust and household or factory atmospheres will cause environmental attack on most circuitries. The danger of such an attack is discussed in Secs. 1-14 and 8-10. Conformal coatings differ from solder resists in usually being formulated so that soldering through them is feasible. This is important for later repair work on such assemblies. Conformal coatings also provide a moisture barrier to prevent extreme condensation from affecting the assemblies. In that respect, they are similar to the lacquers applied over electroplated surfaces in decorative platings. They are applied by conventional means of spraying, brushing, or dipping.

The use of protective coatings is not limited to the electrical and electronics industry. The joining of dissimilar metals by soldering or even the galvanic potential formed between solder and base metal offer serious hazards in structural designs. Here protective coatings act mainly as a vapor barrier to prevent the formation of corrosion cells. In this sense, they resemble paint as a protective coating.

SIX
Special Applications

6-1 Introduction In previous chapters, we have treated the subject of soldering in a general way without going into the details of soldering any specific combination. In this chapter, we deal with the most common surfaces soldered which are unique for one reason or another and describe in general terms the problems they cause and how we get around these problems with the proper selection of fluxes, solders, and techniques. We shall try to give an example in each case. The chapter is intended only as a guide and not as a substitute for the proper selection and design of the soldered assembly, outlined in detail in other chapters. Most of the theory, which has been covered before, will be touched on only lightly.

Printed-circuit soldering is a typical example of the use of solder technology in production. We shall review the special needs of this industry and the options available.

The application of special soldering techniques includes such things as high- and low-temperature soldering, soldering to thin films, using expanding solders, and low thermal emfs. Material covered in this chapter involves some repetition of material covered in other sections of the book because of the necessity of presenting material from a different viewpoint.

SOLDERING TO SPECIFIC SURFACES

6-2 Aluminum and Its Alloys   The problem with aluminum soldering is twofold. First aluminum, as a metal, has a tenacious oxide layer which is extremely difficult to remove and which forms easily upon exposure to air. Aluminum oxide is a refractory material which is chemically very inert. Thus, a strong flux is required to clean the surfaces and maintain cleanliness for the soldering operation. Only special fluxes designed for aluminum soldering are effective; however, it is also possible to break through the oxide layer by mechanical abrasion or ultrasonically.

The second problem stems from the position of aluminum in the electromotive series. Under regular conditions, soldering to aluminum with tin-lead solders causes a galvanic potential beyond the tolerable range, and fast deterioration of the joint under humidity conditions results. Table 4-1 indicates that there would be 1.53 V potential between aluminum and tin-lead solder. However, a zinc-magnesium-aluminum alloy could have the same potential as aluminum and thus be suitable. For further details of aluminum solders, see Table 6-2.

Another important consideration in the use of aluminum is its extremely fast heat transfer compared with its low heat capacity. Heating of aluminum is therefore often a problem when localized heat for soldering is required. Another important parameter is the large coefficient of thermal expansion in aluminum, which is actually larger than in most common materials. Unless localized heating can be effected, there is a good chance for heat distortion. Finally, the melting point of aluminum is much closer to that of the soldering alloys than most other base metals, and a heat source must be used that will not bring the aluminum to the melting point in the immediate joint area.

The correct flux for the aluminum alloy also depends largely on the alloying addition found in the aluminum. Many alloying additions contribute to the tenacious oxide film that forms on pure aluminum. The aluminum alloy film is a refractory material which is difficult to attack chemically, and none of the regular fluxes is suitable. In addition, many alloying elements such as silicon and magnesium form similar oxides which are difficult to break. The chemical fluxes which are effective on aluminum contain organic fluorides together with a heat-stable salt such as cadmium fluoborate in a vehicle. They sometimes also contain metal fluorides, inorganic chlorides, and ammonium compounds.

By nature, these fluxes are temperature-sensitive and should not be used over 600°F (315°C). They are used in conjunction with low-melting-point solders and require the special precautions associated with fluorides. Meticulous cleaning after soldering is mandatory because of the corrosive nature of these fluxes. They are usually applied to aluminum alloys with low alloying additions and specifically with little magnesium (under 1 percent) and small additions of silicon (less than 5 percent).

In addition to chemical fluxes, it is possible to use reaction fluxes for soldering aluminum. These materials contain zinc chloride and sometimes tin chloride in combination with other halides. The formulation is designed to have a suitable low melting point for proper soldering and good thermal characteristics to prevent reoxidation. These fluxes penetrate the oxide film so that the salts come in contact with the underlying aluminum. At the soldering temperature (540 to 720°F or 282 to 382°C) these metal chlorides are reduced by the aluminum to form aluminum chloride, which is a gas at these temperatures, and the metallic zinc or metallic tin, whichever may be the case, deposits on the metallic aluminum surface. The evolution of the aluminum chloride breaks up the oxide films on the surface, and the freshly formed film of zinc and/or tin is then available for wetting by the solder. Because of the nature of this type of reaction, it is important that the right reaction temperature be reached. It is lower for the tin chlorides and higher for the zinc chlorides. The proper selection of the reaction flux depends largely on the corrosion resistance of the joint required because the tin in the aluminum fillet has the undesirable side effect of a high galvanic potential described earlier in this chapter.

For further details on the correct matching of chemical- or reaction-type fluxes to the various soldering aluminum alloys, see Table 6-1. Note that there is a definite danger of intergranular penetration by the solder into various types of aluminum alloys, and it is best to consult either a solder manufacturer or an aluminum-alloy manufacturer before deciding on critical solder applications.

A whole range of soldering alloys is available for aluminum materials. Whereas the conventional solders will set aluminum with chemical fluxes, the galvanic potential mentioned earlier is a major factor in the corrosion stability of the joint. A better corrosion stability is possible with different types of solder formulation. Table 6-2 gives a variety of soldering alloys used for aluminum and their relative resistance to corrosion. However, true resistance to corrosion is possible only if extremely high-purity zinc with small alloying additions is used which will have exactly the same galvanic potential as the aluminum itself. For further details refer to tests run by Bell Laboratories.[1]

Fluxless soldering for aluminum is also possible. This is done mainly by forming a pool of molten solder on top of the oxidized aluminum and then breaking through the layer of oxides mechanically or with ultrasonic

[1]G. M. Bouton and P. R. White, A Method for Soldering Aluminum, *Bell Lab. Rec.*, May 1958.

TABLE 6-1   Solderability of Aluminum Alloys*

Alloy group	Typical alloy	Solderability	Recommended flux
1XXX (commercial purity or	1060	Good	Chemical or reaction
higher)	1100	Good	Chemical or reaction
2XXX	2014	Fair†	Reaction
(Al-Cu)			
3XXX	3003	Good	Chemical or reaction
(Al-Mn)			
4XXX	4043	Poor‡	None
(Al-Si)			
5XXX§	5005	Good	Chemical or reaction
(Al-Mg or	5050, 5154	Fair†	Reaction
Al-Mg-Mn)	5456, 5083	Poor†	Reaction
6XXX	6061	Good†	Reaction
(Al-Mg-Si)			
7XXX§	7072	Good	Reaction
(Al-Zn)	7075	Poor	Reaction
8XXX§	8112	Good	Reaction
Al, other)			

*From "The Soldering Manual," The American Welding Society, 1959.
†Susceptible to intergranular penetration by solder.
‡Solderable only with abrasion or ultrasonic techniques.
§Solderability greatly affected by composition.

cavitation. The solder alloy is then in a position to wet the clean aluminum underneath, and the tarnishes float to the surface of the solder. This is especially true for the high-zinc alloys, which have a tendency to penetrate and lift the oxide layer as soon as a small crack in the surface is formed (Fig. 6-1). In some cases, the aluminum solder itself will contain small

TABLE 6-2   Typical Solders for Aluminum*

Composition, %	Temp., °F Solidus	Liquidus	Density, lb/in³	Wetting ability on aluminum	Flux type commonly used	Relative corrosion resistance
100 Zn	787	787	0.26	Good	Reaction	Very good
95 Zn, 5 Al	720	720	0.24	Good	Reaction	Very good
91 Sn, 9 Zn	390	390	0.26	Fair	Chemical, also reaction	Fair
70 Sn, 30 Zn	390	592	0.26	Fair	Reaction	Fair
60 Sn, 40 Zn	390	645	0.26	Good	Reaction	Good
30 Sn, 70 Zn	390	708	0.26	Good	Reaction	Good
10 Cd, 90 Zn	509	750	0.26	Good	Reaction	Fair
40 Cd, 60 Zn	509	635	0.28	Very good	Reaction	Fair
34 Sn, 63 Pb, 3 Zn	338	492	0.34	Poor	Chemical, also reaction	Poor

*From "The Soldering Manual," The American Welding Society, 1959.

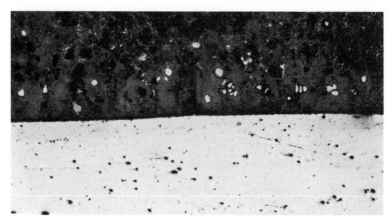

Fig. 6-1   2S aluminum interface with an aluminum-zinc-magnesium solder alloy, no flux used.

particles of abrasives, and rubbing the alloy over the heated surface during the melting will, in effect, achieve penetration of the oxide layer. The cavitation rupture of the oxide layer due to ultrasonic-wave application to the molten solder is described in Sec. 5-43; also see Fig. 6-2.

One unusual method of application worth mentioning here is the abrasion of the surfaces with a rotating grinding wheel loaded with solder. The frictional heat is enough to melt the solder, and the friction itself will remove the oxide layers, bearing clean aluminum to the molten solder. This method has found little use in mass production, however (see also Sec. 3-25).

Fig. 6-2   2S aluminum interface (same as in Fig. 6-1) with a regular eutectic tin-lead solder applied ultrasonically (no flux used).

Although the above material may seem to indicate that soldering to aluminum is difficult, this is misleading because many mass-produced items utilize aluminum soldering. One of these is the soldering of aluminum-base light bulbs or aluminum cooling systems.

6-3 Soldering Beryllium Copper   The problem with beryllium copper is similar to that with aluminum. The beryllium in the alloys forms a tenacious refractory oxide on the surface, especially when heat treating in an uncontrolled atmosphere has been performed before soldering. Oxide scale on beryllium copper can easily be removed by pickling before soldering. For pickling a 20 to 30 percent by volume solution of sulfuric acid (concentrated, 1.83 specific gravity) and water should be used at 160 to 180°F (71 to 82°C). Sufficient immersion time must be allowed to loosen the dark scale completely. This is followed by a nitric acid dip to remove all traces of the black loosened scale, which is mainly cupric oxide, and any red scale, which is cuprous oxide. The solution used for this second dip should consist of 30 percent by volume nitric acid (concentrated, 1.40 specific gravity) and water, at room temperature. After descaling, the parts must be thoroughly rinsed in cold water, followed by a hot rinse to remove all traces of the acid, and then dried by air blast, sawdust, or other available means. A good cold-water rinse between the two pickling solutions is recommended to avoid any carry-over.

A recently descaled surface can easily be soldered with activated rosin fluxes and stronger materials, but an aged surface or one that has light oxide layers requires stronger fluxes. A flux containing lactic acid has been reported to give good results with such lightly oxidized surfaces.

6-4 Ceramic-to-Metal Bonding (Hybrids)   Ceramic materials are non-metallic surfaces to which direct wetting of solder is not possible without an intermediate layer. Usually a thin metallic film is deposited on the ceramic and bonded to it by some means. A good example is the use of a silver-containing glass frit which is fired to the surfaces. The glass wets both the silver and the ceramic, providing the surface with a metallic layer of silver to which bonding is possible (see Fig. 6-3).

Regardless of the method by which the deposits are adhered to the surface and the type of surface to which soldering must be effected, it is important to mention that the thermal shock associated with soldering can seriously affect the structure of the ceramic so that microcracks appear and the assembly fails later upon simple stressing. In order to avoid this, it is possible to preheat the parts in a high-temperature flux and solder them at elevated temperatures.

It is also possible to minimize the temperature shock by reducing the temperature gradient between the hot metal and the ceramic by using a

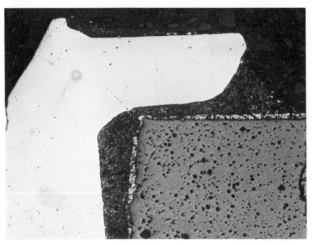

Fig. 6-3 Cross section of silver-fired ceramic surface bonded to a copper header with 97.5/2.5 lead-silver solder (note good feathering on outside of joint).

preheating operation and low-temperature flux and solder and bonding with the minimum temperature required.

The soldering of the particular surfaces deposited on the ceramic is the same as the metal itself and should be looked up under that heading (silver, nickel, gold, palladium, etc.). A word of caution: the layers of metal on the ceramic usually are extremely thin, and there is a definite chance of dissolving most of the metal off the surfaces if the wrong solder alloy is selected.[1] This process, called *scavenging* or *leaching*, can be reduced by alloy and flux selection or process-parameter control.

6-5 Copper and Its Alloys   Copper is one of the easiest metals to wet with solder because of the nature of the tarnish formed on its surfaces. Most fluxes, including water-white rosin, are effective on surfaces of copper provided that the oxide layers are not too heavy. However, the addition of alloying elements to the copper will change the characteristics of the oxide layers so that such additions as beryllium, silicon, and aluminum make the alloy extremely hard to solder. For further details, see Sec. 6-3.

The addition of zinc or tin to copper to form brass and bronze usually results in a slight loss of solderability. In addition, soldering to zinc-containing alloys is a source of serious contamination to a solder pot in dipping processes because the zinc from the base metal causes deteriora-

[1] H. H. Manko, Selecting Solder Alloys for Hybrid Bonding, *Insul. Circuits,* April 1977.

tion of the solder bath. For further details, see Sec. 5-14. In addition, the presence of antimony (no more than 1 percent) and arsenic (no more than 0.02 percent) in the solder as well as the presence of ammonia and ammonia-producing materials in the flux can be detrimental to the brasses because of intergranular attack and embrittlement of the base metal. The use of solder with a low percentage of antimony and free of arsenic is therefore recommended, as well as the use of fluxes which do not contain any ammonia or materials which might break down under high-temperature conditions to give ammonia.

No restrictions on the use of solders other than those mentioned above are found for copper and copper alloys, and ammonia-producing fluxes are the only ones in the family of chemicals which should not be used with brass. However, sometimes the solderability of copper, especially when it has been stored on the shelf for prolonged periods, deteriorates and a chemical etch is required to restore the solderability of the surfaces when mild fluxes are used. Various proprietary materials are available from flux manufacturers; 10% hydrochloric acid, ammonium persulfate (APS), and similar etches are also used (see Table 5-1).

Copper is a relatively soft material, and the possibility of embedding nonmetallic particles is great. These may originate from grinding operations, abrasion with pumice, etc. They reduce the anchorage area, which may result in dewetting. The average flux, however, is not designed to remove them, and only strong chemical etches are capable of restoring the solderability. It is believed that they undercut the copper around the embedded inclusion, thus physically removing the particles. It is suggested that in these cases instead of using the chemical etch the process causing the embedding be changed. The shape of the abrasive (round or sharp), machining speeds, and coolants can be changed around to avoid embedding.

Soldering to copper is one of the commonest operations, and copper leads as well as copper parts are extensively used in electric and electronic circuitry, printed circuits, and similar assemblies. Copper parts soldered together also are used for structural purposes as in plumbing (water-line installation), heat exchangers, automotive radiators, and even for architectural purposes such as roofs and gutters.

6-6 Glass-to-Metal Bonding   This is basically the same as ceramic-to-metal bonding discussed in Sec. 6-4. However, an unusual phenomenon can be obtained on highly cleaned glass (preferably fired to high temperature before soldering), namely, the adhesion of high indium-tin alloys to glass at a specific temperature which is close to the melting point of the alloy itself. This type of bonding is utilized in special instrument soldering and similar application. Because of the high price of the solder and the weak bond obtained unless mechanical strengthening is provided, this is not widespread. See Sec. 3-25 for further details.

The brittle nature of glass and its specific thermal coefficient of expansion usually make it desirable to bond glass to high-nickel alloys and other materials which have a matched coefficient of expansion. In some cases with special techniques it is possible to make a glass-to-metal seal without soldering. However, some of these glass-to-metal seals require the oxidation of the base metal and bonding at 1600°F (871°C) to the molten glass. These leads, if they are to be soldered, will not take the solder alloy even when the appropriate flux is used. In these cases, it is necessary to descale the leads before soldering, using the solutions described in Table 5-1.

6-7 Gold and Precious Metals and Thin Films   The problem in soldering gold and precious metals lies basically in the cost of the material itself and the tendency to use not solid base metals but thin coatings over cheaper base metals. The solder should therefore be one that does not dissolve the precious metal from the surfaces during the short heating cycle of the soldering operation. The normal way to avoid scavenging the metal from the surface is by loading the solder with this material, thus shifting the tendency from a strong rate of solution to a slow rate of solution. However, again because of the high price of these precious materials, this is not always practical. In addition, recent miniaturization has called for the use of extremely thin layers such as are obtained by vapor deposition (on the order of 3000 to 4000 Å) which are deposited on glass, ceramic, or other nonconducting substrates. The use of the right solder alloy is therefore important.[1]

Another problem which stems from the high price of gold and other precious metals is the tendency of manufacturers to supply the precious coatings on the lower part of the specification and thus sometimes supply inadequate porous coatings which barely meet the requirements for good solderability. When base metals are coated with gold or other similar materials, the intention usually is to prevent oxidation of the base metal and improve the solderability of the surfaces. However, if a porous and poorly adhering thin layer of gold is applied over a surface, the oxygen in the air can easily reach the surfaces underneath, and oxidation will take place. Once this surface is soldered with a material which is capable of dissolving the gold, the gold is entirely removed from the surfaces and the base metal underneath is exposed to the solder. This base metal, which is oxidized and unfortunately was never fluxed during the solder sequence, will have a small anchorage area for good wetting, and dewetting usually results. It is important to remember that the wetting process will take

[1] Ibid.

place only in places where the gold has actually covered the base metal. In the other areas, which were oxidized, there will be no wetting at all. The flux which was applied to the coating surface actually acted only on the gold and had no way of penetrating into the pores to clean them for adequate wetting because of the short time of exposure and the sequence of operations in wetting.

In some cases, it is mandatory to use gold or similar materials because of the various processing solutions which electronic devices like transistors are subjected to in the course of the manufacture (etching in transistors). If poor solderability with such components is a problem, it is strongly suggested that these devices be pretinned before use by applying a strong flux and a separate solder bath so that the surfaces are adequately prepared for final assembly; double dipping may be needed to obtain solderable surfaces.

6-8 Iron and Steel   Pure iron is seldom used for structural purposes but presents no problems in soldering. Since iron usually has carbon additions in order to obtain metallurgical properties, we shall discuss the solderability of iron and steel as a function of the carbon content. Pure iron and the very low carbon steels (up to 0.3 percent carbon) have the best solderability. Medium-carbon steels in the range of 0.3 to 0.45 percent carbon have slightly less solderability, mainly because of the addition of other alloying elements. Poor solderability in steel occurs in those having more than 0.45 percent carbon, which in general also have other alloying additions for structural purposes. The cast irons, having between 1.7 and approximately 4.5 percent carbon, have their particular problems, discussed later in this section. They are much more difficult to solder than the steels.

In general, iron and its alloys corrode easily in what is regularly referred to as *rusting*. The rust and heat-treat scales must be totally removed from the surface before soldering to them can be effected. For heavy rust layers, a hot alkaline cleaner followed by an acid dip will give good surfaces for soldering. However, each particular alloy and material should be considered by itself. Most soldering alloys can easily be used in conjunction with iron and its alloys, and the selection depends on the properties of the joint rather than being dictated by the base metal itself. It is normal practice to pretin these metals for good soldering, and the selection of a flux depends on the surfaces to be used. Pretinned surfaces can be bonded by the use of milder fluxes, but untreated surfaces require more corrosive materials.

Most steel used for mass production is precoated with either zinc (galvanized iron) or tin-lead (terneplate) and sometimes with pure tin (tinplate). Other surfaces are plated with cadmium and even nickel for various applications. All these coatings serve a double purpose; they improve the solderability and also give good corrosion protection to the base metal itself.

A discussion of iron and its alloys would not be complete without mentioning stainless steel. Stainless steels are alloys with the addition of nickel and chrome, which are intended to improve the corrosion resistance of the iron. Stainless steels fall into various categories according to the amount of iron and nickel used. However, the mechanism which protects these alloys from environmental and chemical attack also prevents the effective cleaning of the surfaces by fluxes so that the chemical preparation of the surfaces for wetting is difficult. Special fluxes are available for stainless steels, and they are effective for the specific alloy they are designed for. With contamination-free surfaces (foreign materials) the use of these fluxes makes soldering to stainless steel a relatively simple matter. However, the extremely strong chemical nature of these fluxes makes good postcleaning procedures mandatory.

The cast irons, both gray and white, are difficult to solder because of the amount of graphite present in the structure. The graphite itself reduces the amount of anchorage area so that poor wetting results. In addition to carbon, both cast irons contain silicon and other oxides. Electrochemical cleaning, abrasion, and chemical cleaning are some of the methods used for the preparation of cast iron for soldering. Once the surfaces have been prepared, any inorganic flux will suffice to make good joints. A certain amount of discretion should be used in the method of heating to avoid cracking of the casting due to improper heat shocking. However, in general, the soldering of castings is a relatively simple matter.

6-9 Lead and Its Alloys   Soldering to lead and its alloys is a common practice, although it is an extremely delicate operation because of the relatively low melting point of the lead. This melting point is also very close to the soldering temperature of the various alloys used. Pure lead melts at 621°F (327.4°C) and can be used with regular tin-lead solder alloys without too much difficulty. However, since pure lead is extremely soft and lacks the corrosion resistance of some of the antimonial or corroding-grade leads, the solder is often used with lead alloys that have melting points around 450°F. Here, it is extremely important to make sure that the soldering parameters are carefully controlled in order not to melt the base metal itself.

Lead is used primarily because of its corrosion resistance in such applications as chemical piping, roofing, and the sheets surrounding electrical cables for good weathering. Lead and its alloys are also extensively used in the chemical industry for such applications as the production of sulfuric acid and chrome plating. The corrosion resistance of the solder used with the lead should therefore be equivalent to that of the base metal itself or better. It is therefore customary to consider welding rather than soldering in many applications. One should remember here that under welding we consider an application in which the base-metal alloy is used for the formation of the joint. This in effect is an alloy which is normally considered as a solder.

Another term used for lead welding or soldering is commonly known as *lead burning*. In Fig. 6-4, lead anodes are burned to the hooks, for chrome-plating anodes.

Another method associated with lead and its alloys is the *wiping* of solder. In this method, an alloy, usually with 30 to 40 percent tin, balance lead (sometimes 1 to 2 percent antimony), is used because it has a wide pasty range. The solidus of such an alloy is around 360°F (182°C) and the liquidus is around 460°F (238°C). The working range is 100°F (37°C). In that working range, the material behaves like a paste and can be wiped and worked into the joints of the lead in order to make a smooth fillet. This is mainly used in joining lead pipes or sealing lead sheath around electric wires. The soldering temperature here can easily be controlled and will not cause any extensive solution of the relatively low-melting base metal. In order to prepare the surfaces for this type of joining, mechanical abrasion of the surfaces is common. The area is then covered with a relatively noncorrosive flux such as rosin or tallow oil. The solder is heated

Fig. 6-4   "Burning" lead anodes at Alpha Metals, Inc.

in a separate container and ladled out in the correct temperature range to the joint itself, where it is caught in a special tool. The operator then wipes the solder around the joint with an asbestos glove or a similar insulating material until the solder freezes into a tight and well-adhering fillet.

6-10 Magnesium   Magnesium has received little attention in soldering because no effective fluxes for soldering magnesium are available. It is possible to wet magnesium by ultrasonics or friction methods described earlier for aluminum, but this method is not really suitable for mass production. To be soldered in large quantities magnesium must be prepared electrochemically by plating, and the platings are then easily soldered.

6-11 Nickel and Its Alloys   Nickel and nickel alloys are usually used for their corrosion resistance and similar properties. The nickel alloys can normally be soldered with the regular fluxes recommended, unless a heavy oxide or heat-treat scale is present on the surfaces. In some cases the nickel is alloyed with other elements, as described in Sec. 6-8, and then the alloy should be considered not as basically a nickel alloy but as a stainless steel and should be treated as such. For further information on stainless-steel soldering, see Sec. 6-6.

Table 6-3 shows the more common high-nickel alloys and gives a rating of their solderability. Experience has also shown that freshly abraded high-nickel alloys can be soldered with activated-rosin-type fluxes provided the surfaces are covered with the rosin flux immediately after abrasion (Sec. 5-7).

When a color match between the high-nickel alloys or sometimes even the stainless steels and the solder is required, a high-tin alloy such as 95 tin, 5 antimony will give close color resemblance. However, unless these

TABLE 6-3   High-Nickel Alloys*

| Alloy | Composition, % | | | | | | Solderability |
	Ni	Cu	Cr	Fe	Ti	Al	
Monel	67	30	.....	.....	....	....	Good
Nickel	99	....	.....	.....	....	....	Good
Permanickel	98	....	.....	.....	0.40	....	Good
Duranickel	94	....	.....	.....	0.50	4.5	Good
K Monel	66	29	.....	.....	....	2.75	Good
Inconel	77	....	15	7	....	....	Fair
Incoloy	34	....	21	45	....	....	Fair
Nimonic 75	75	....	20	1.75	0.25	0.35	Fair
Inconel X	73	....	15	7	2.50	0.75	Fair
Ni-Span-C	42	....	5.25	49	2.00	0.50	Fair

*From "The Soldering Manual," The American Welding Society, 1959.

surfaces are protected from discoloration by heat treatment or other oxidizing processes the color difference between the base metal and the solder will become apparent. Upon normal exposure to the atmosphere, both the base and solder alloy seem to stay well matched in color.

Soldering to nickel plating is also relatively easy. However, a distinction should be made between electrochemical deposits and electroless nickel. Electrochemical deposits are normally relatively easy to solder provided that the surfaces have not been passivated or oxidized for too long a period, in which case a stronger flux solution is required. Under favorable conditions, nickel-plated surfaces can be soldered with activated-rosin-base fluxes. Electroless nickel, however, is a separate problem. Since a high percentage of phosphor is present in the electroless-nickel deposit, the ratio between the nickel and the phosphor can vary from one plating solution to another and even from the beginning of the production run to the end. Experience has shown that high phosphor content in electroless nickel makes it extremely difficult and sometimes impossible to solder to the electroless-nickel coating. A maximum of 5 to 7 percent phosphor in the nickel coating is therefore preferred. This requires more careful quality control but is well worth the trouble because of the improved soldering conditions.

6-12 Silver and Its Alloys   Soldering to silver really requires no special techniques. Since silver is a noble metal with very little oxidation under environmental conditions it can easily be soldered with most soldering alloys using mild fluxes. Only the silver sulfites and other sulfur-containing products are a real problem in soldering because they will tarnish the silver seriously, requiring stronger fluxes.

Because silver is very soluble in tin and lead the solders used for the tinning of silver surfaces are usually silver-loaded solders. For further details, see Sec. 3-26.

Silver is sometimes applied to nonmetallic surfaces (such as ceramics) in the form of flaked-silver leaflets suspended in a glass frit, which is then fired onto the surfaces, and the glass makes the bond between the nonmetallic material and the silver particles. These silver-fired surfaces are in common use in electronic devices. Here the presence of silver in the solder is extremely important since the layer of silver present in the so-called "paint" is small and scavenging of the silver from the surfaces will result in nonwetting conditions (see Fig. 6-3). The temperature of soldering to silver-fired surfaces should be kept to a minimum in order to reduce the amount of silver scavenging, and the soldering time should be reduced as much as possible. The use of a proper low-temperature flux should be considered carefully when this kind of soldering operation is contemplated. If the use of high-temperature solder is required in conjunction with silver-fired surfaces, low-tin high-lead alloys are recommended.[1] These alloys, having a higher melting point, still do not scavenge much of the silver off the surfaces and can wet the surfaces adequately.

6-13 Tin Tin, like lead, is difficult to solder because of its low melting point. Tin, melting at 450°F, is even more susceptible to distortion than lead. However tin and high-tin alloys such as pewter and babbitt metals are often soldered with eutectic tin-lead solder and similar alloys containing bismuth which have markedly lower melting points than 450°F (232°C), so that soldering temperature can approach this point. Here the color match between the ornamental pewter alloys and the solder is important.

Tin has a light, thin oxide film which forms under normal environmental attack, and this is easily overcome in the soldering operation using the mildest fluxes. Therefore, no extreme difficulty in soldering is encountered. The preparation of tin for soldering usually includes a simple surface-contamination removal (degreasing) and the use of noncorrosive rosin-type fluxes.

Among the interesting applications of tin and its alloys which use solder for joining are the installation of high-purity water lines for distilled and demineralized water in semiconductor-manufacturing facilities, hospitals and pharmaceutical-manufacturing areas, and the chemical and beverage industries. The tin is generally not used by itself because of its extreme softness and ease of deformation, and the application generally involves an outer shell of iron or copper around the pure tin for this type of installation. Other applications involve making pewter objects and manufacturing organ pipes.

PRINTED-CIRCUIT SOLDERING

6-14 The Printed Circuit Printed circuits have taken their place in small and large industries today. They are used for sophisticated electronic equipment like computers and military equipment as well as for everyday appliances like radio and television. One of the reasons for their widespread use is the ease of assembly, the reliability of the soldered connections, and the low cost of production.

The concepts of printed circuits were developed around soldering as the joining method when the tremendous potential of soldering for making multiple reliable connections simultaneously by dipping or wave soldering was recognized. To date, no other joining method can make as

[1]Ibid.

many good reliable solder joints simultaneously for a small investment in equipment.

The numerous methods of manufacturing printed circuits today make a general description impractical. For more information the reader is referred to the more prominent publications in that field,[1,2] but whatever the configuration may be, the basics of soldering do not vary, and the same equipment can be used for most applications.

Although soldering is one of the most versatile production tools in today's industry, it is vital that good design for soldering precede any attempt at manufacturing. In that respect, design is the cheapest investment toward joint reliability which can be realized without additional cost.

After the printed-circuit board has been processed, it can be used immediately for assembly or it can be put in storage. If stored, it is extremely important to check on the solderability of the printed-circuit cards before use. It is possible that the surfaces have deteriorated during storage and are no longer solderable. In that case, it is well worth a short presolder treatment to improve this solderability. In order to preserve these surfaces, a protective coating like that described in Sec. 5-44 should be used. If this material is applied immediately after the last manufacturing step, it will preserve the solderability of the surfaces and ensure reliability.

6-15 Printed-Circuit Design Considerations for Soldering   The single most important advantage of printed circuits from a soldering viewpoint is the fact that all the solder joints on the board can be made in one rapid operation. This, however, imposes certain restrictions on the design which are necessary to assure the above advantages. The following seven rules are meant to help the designer.

1. In a printed circuit which offers high component density, the tendency is to reduce the spacing between the conductors as much as electrical characteristics will allow. This increases the danger of solder bridging (the short circuiting of neighboring conductors with soldered nodules) between adjacent conductors. To alleviate this problem the designer can make sure that all closely spaced lines run parallel to the solder movement across the board, thus getting an additional wiping action which reduces the danger of bridging. Sometimes inherent features of the card such as the widths and/or lengths of board dictate the soldered-dip direction. Precious-metal-plated areas for electric contacts, for example, should always face front and be covered to prevent wetting

---

[1] Publications and manuals of the Institute for Interconnections and Packaging Electronic Circuits (IPC), 1717 Howard St., Evanston, Ill.

[2] C. F. Coombs (ed.), "Printed Circuit Handbook," McGraw-Hill, 1967.

of contact surfaces. In general, however, it is enough to provide one direction for parallel dense circuitry, keeping in mind that conventional laminate materials have an inherent width limitation and that it is therefore recommended that one use the long side of the card as a dip direction whenever possible. Solder-mask coatings are beneficial in correcting problems due to close spacing (see Sec. 5-44).

2. Whereas the shape and size of the line conductors are a function of the current-carrying requirements and the space available, the shape of the land (the metallic conductor to which the component is soldered) depends largely on the designer. Table 6-4 shows some of the configurations of single-hole lands, which are sometimes used in any number of combinations, including several holes through the same land. When the basic form of the land is chosen, such factors as component lead direction and final appearance of the board are important. However, for the electrical requirements only the length of the lead crimped and soldered to the board is important. When 60/40 tin-lead solder and copper circuitry are used, the rule of thumb which gives good results for a single-sided board is

$$l = 9t \qquad (6\text{-}1a)$$

Double-sided boards with plated through holes do not require a bottom crimp when the hole is partially or completely filled; for nonfilled holes

$$l = 5t \qquad (6\text{-}1b)$$

where t is the thickness of the printed conductor and l the length of the solder joint. Mechanically speaking, calculations and tests have shown that properly soldered joints fail in the component or board when subject to tensile or shear loading.

3. The dimensional stability of the board during the soldering operation is an important parameter. Most manufacturers give coefficients of thermal expansion for the x and y directions only, but the expansion in the z direction sometimes is 3 to 4 times greater because of the nature of the laminations. Some materials have been reported to expand as much as 0.001 in during a solder-dip operation (temperature rise of 400°F, or 204°C). On single-sided boards this is not critical, but on double-sided boards with through holes it must be compensated for by the use of either ductile copper plating or straight-through (not crimped) leads.

4. Another factor which affects the soldered joint is the amount of moisture in the board before soldering. This can be in the material itself or come from the etching and plating processes. In either case it will cause sputtering of the solder, gas pockets, and voids. To eliminate the moisture, cards should be baked before fluxing and soldering (see Table 6-5).

TABLE 6-4   Configurations of Printed-Circuit Lands

TYPE	PREFERRED DIRECTION FOR COMPONENT LEAD	SOLDER FILLET CONTOUR WILL BE	REMARKS
TEAR DROP	TOWARD LONG END	EVEN AND ALMOST ROUND	GOOD DESIGN. ENLARGED CONTACT AREA
ROUND	ANY	EVEN AND ROUND	THE UNIVERSAL PATTERN
"D"	TOWARD TIP	UNEVEN	NOT WIDELY USED
RECT.	TOWARD A CORNER OR LONG END	UNEVEN	NOT WIDELY USED
DELTA	TOWARD BASE	UNEVEN	USED IF SPACE VERY LIMITED

Most of the flux vehicle should be driven off by passing the board over heat before soldering. Venting should be provided for each hole because trapped air and resultant gases will prevent good solder filleting.

5. When through holes are used and the solder must rise by capillarity to the top side of the board for metallic continuity, proper hole-to-lead spacing is of utmost importance. A. 0.002- to 0.006-in clearance is recommended for optimum strength. With automated insertion equipment this clearance is 0.012 to 0.015 in, which is good for repeated hole filling.

TABLE 6-5   Printed-Circuit Presolder Baking Temperatures*

	Temperature		
Equipment	°F	°C	Time, h
Recirculating oven	225–250	107–120	1–2
	160–180	70–80	3–4
Vacuum oven, 1 torr	120–130	50–55	1.5–2.5

*These temperatures have been established empirically as sufficient for wave soldering and represent approximately one-third of the time needed for total moisture removal. They should be a convenient starting point for production testing and may be reduced according to need. Use the lower times or temperatures with thin materials up to 0.062 in and the higher times or temperatures for thicker laminates. Multilayer board with over four layers will require the maximum indicated.

6. After the joint is fully designed, it is advisable to consider the heat capacity of the assembly and how it affects each particular joint. Because it is essential that the joint area be allowed to heat up to the wetting temperature during the soldering cycle, large heat sinks near the joint should be avoided. It is therefore advisable to distribute the assembly so that the heat capacity of each joint area is equal or similar to that of all the others which are soldered simultaneously.

7. Many types of coating are available on the market for printed-circuit use. Some are used as a solder mask to prevent solder from wetting the coated areas. These are usually good heat- and solvent-resistant organic materials which are left on the board. Other coatings are applied after final assembly operation as coatings. They are used to prevent environmental precipitation and outside contamination from affecting the electrical characteristics of the board. Whichever coating is applied, the designer should make sure that it is applied to a clean surface. Otherwise the advantage of the coating is lost.

Assuming a good design, let us discuss the general flow of the printed circuit through the production line. In each case, we shall discuss in some detail the materials recommended for proper use.

6-16 Outline of Printed-Circuit Assembly and Soldering Line (Fig. 6-5) In this section we consider the various options used in industry. The latest trend to emerge is the move to *straight-through* unbent leads, intended to simplify field repair, provide better stress distribution, and lower manufacturing cost.

Assembly is rarely feasible on automatic insertion equipment alone, because there are a variety of components that are difficult to automate. On the other hand, automatic insertion is more economical for large runs of uniform design. Computer control further adds to the versatility of this equipment. The automatic equipment also trims the lead to preset length and can bend (crimp) it at the same time at no added cost.

Manual insertion is more versatile in relation to component shape and mix but may be more costly for uniform long runs. It is used alone or to augment insertion equipment. Bending and cutting the leads after hand stuffing is expensive and may damage the boards and/or the components. To reduce these costs, it is possible to "preprep" the components by shaping and cutting the leads before assembly. In this sequence, bending the leads is not recommended for economic reasons and repairability.

As a result of economic pressures, in-line trimming of component leads, inserted straight through as received, has developed (see Fig. 6-6). Cutting leads in the range of 0.030 to 0.090 in underneath the board surface is common. Tighter tolerances are feasible but require control of warpage, conveyor, and process. In any case the components must be supported

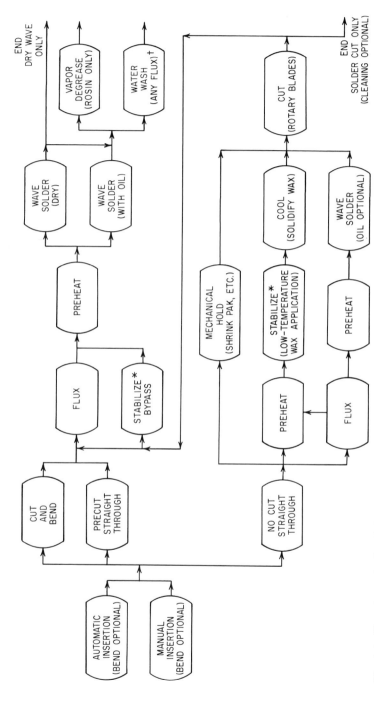

Fig. 6-5 Flow chart of printed-circuit-assembly options. Note trend toward straight-through leads for ease of repair.

*Trade name Hollis Engineering Co.
†With the proper additive to wash water.

253

Fig. 6-6   Rotary cutters for lead trimming. *(Hollis Engineering, Inc.)*

during trimming for accuracy and blade life and to prevent component dropout.

Since the accuracy of trimming is directly related to board flatness, cold holding the component is desirable because it reduces the amount of precutting temperature and its associated warpage problems. Mechanical means, such as shrink-pack plastic and web spraying, have been used. A more economical in-line process utilizes low-melting organic compounds like wax. In this Stabilizer[1] process a quasi joint of wax is made which supports the lead during rotary cutting. The boards then proceed to the wave-solder application without wax removal. The wax is compatible with the solder process and can be used with standard fluxes, or it can be activated (no additional fluxes needed).

For economy some manufacturers prefer to solder the leads before cutting and trim the excess length later *(solder-cut system)*. There is exposure to joint damage here if solderability is not adequate and/or hole-to-wire-diameter ratios are out of control. In addition, the lead ends are left exposed, and while this is no problem with copper-base alloys, iron-based

[1]Trade name of Hollis Engineering Co., Inc., Nashua N.H.

materials (Kovar, 52 Alloy, etc.) will rust. For these reasons higher reliability levels can be achieved by resoldering such assemblies *(solder-cut–solder system)* although this causes increased heat damage, etc.

Flux application has been discussed in Sec. 5-8. For printed-circuit soldering foam, spray, and wave fluxing are most common. The industry has been dominated by rosin fluxes, but more recent trends are toward organic-acid formulations, encouraged by the availability of good water-washing equipment (Secs. 5-39 to 5-42).

The use of adequate preheat for printed-circuit boards cannot be overstressed. Preheating was discussed in Sec. 5-9, and its benefits to the flux were explained. In addition preheating raises the assembly temperature before soldering. This makes it possible to have less contact time with the wave to reach the same thermal balance.[1] Thus we can speed up the production line and reap the technical advantage of less metallurgical interreaction in the solder fillet and less heat deterioration of the plastic materials in the laminate and components.

For top-board temperature as measured on the laminate upon leaving the preheater see Table 6-6. To achieve the higher top temperatures or when high production speeds are required a top preheater may be necessary (Fig. 6-7).

Wave soldering of printed-circuit boards requires an analysis of the impact of the wave on the board (Fig. 6-8). The process can be divided into three distinct parts:

1. *Point of entrance (see enlargement in left circle of Fig. 6-8)*. Because the board moves in the direction opposite the solder, this point has the largest differential speed. At the point of impingement, therefore, we get the largest turbulence and washing action. This serves to remove the pre-heated-flux-tarnish combination from the base metal and brings the solder into contact with the board conductors. When wetting temperature is reached, wetting occurs instantaneously. If the conductors were plated

[1]H. H. Manko, Understanding the Solder Wave and Its Effect on Solder Joints, *Insul. Circuits*, vol. 24, no. 1, January 1978.

TABLE 6-6   Top-Board Temperature for Average Speed and Component Density

Board type	Temperature	
	°F	°C
Single-sided*	175–200	80–90
Double-sided	210–250	100–120
Multilayer, up to 4	220–250	105–120
Over 4	230–270	110–130

*Also true for most flexible circuits.

before soldering, the wave turbulence will help wash away these layers. In the case of fusible alloys like tin-lead or pure tin, the mechanism is a *melt-wash* combination. For soluble platings like silver or gold the mechanism is a *solution-wash* combination.

2. *Passage through wave (between entrance and exit points in Fig. 6-8).* This zone is a thermal *heat-transfer region* needed to supply both heat and solder to the leads and the plated-through holes if any. Remember that the lead is a bigger heat sink and will require longer dwell time to reach wetting temperatures. In addition the immersion depth in the wave does not push up the solder to the top, for obvious reasons. Enough time in the wave must be allowed to let surface energies at wetting temperatures draw the molten alloy to the top for good filleting.

3. *Point of exit (left circle enlargement in Fig. 6-8).* To understand the effect of this critical wave area we must quickly review the forces which act on the molten solder fillet. Surface energies in wetting (Chap. 1) will retain the molten solder in the joint, while gravity (or the solder weight) will tend to pull it down. The balance between these forces explains the need for proper hole-to-wire ratio and also depends on solderability. In order to isolate the point of exit from these forces at equilibrium the exit must be at a static part of the wave. This is ensured by matching the withdrawal speed of the board as closely as possible to the speed of the receding wave.

The use of oil in the wave has its greatest impact on the point of exit. By improving the solder surface tension and excluding the air, we can control the shape of the solder fillet and reduce solder consumption, resulting in contour fillets that are easy to inspect. Dry waves (no oil) tend to leave more solder,[1] due to the dross skin which is formed as soon as the solder

[1]C. D. Bernard, Horizonal vs. Inclined Conveyor Wave Soldering, *Circuits Manuf.*, September 1977.

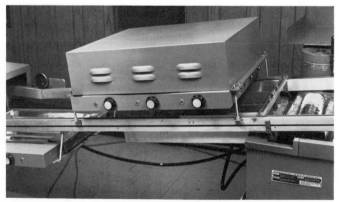

Fig. 6-7   Top preheater radiation type. *(Hollis Engineering, Inc.)*

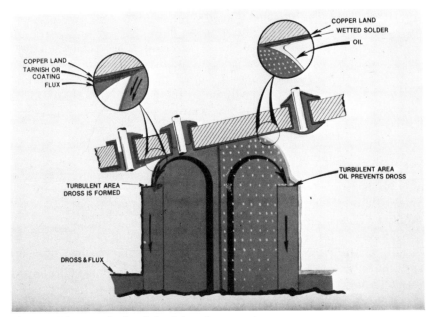

COPPER LAND
WETTED SOLDER
OIL

COPPER LAND
TARNISH OR
COATING
FLUX

TURBULENT AREA
OIL PREVENTS DROSS

TURBULENT AREA
DROSS IS FORMED

DROSS & FLUX

Fig. 6-8   Diagram of printed-circuit board transversing wave.

surface contacts the air. Oil also prevents dross formation on the solder reservoir in the pot, which is an important economic consideration. Oil cannot be used, however, when washing is not feasible since oil residues should not normally be left on the boards. In many cases, savings on dross losses in dry systems can offset the cost of the use of oil and cleaning.[1]

Rosin flux removal can be effected by vapor cleaning in solvents (Sec. 5-42) or water washing with additives (Sec. 5-41). When properly planned, both cleaning processes will remove oil and wax residues as well as the rosin. Organic-acid fluxes on the other hand must be cleaned in water-wash systems, with or without additives. Remember that the clean board may get recontaminated, and the use of protective films (such as conformal coatings) may be advisable (Sec. 5-44).

SPECIAL SOLDERING TECHNIQUES

6-17 Thermal-free Solders   Two dissimilar metals which have a junction at a point that has a different temperature from the rest of the wire will cause an electric potential. This phenomenon is utilized by thermocouples, where materials are chosen which can generate a large emf

[1]H. H. Manko, Understanding the Solder Wave and Its Effect on Solder Joints, *Insul. Circuits*, vol. 24, No. 1, January 1978.

(electromotive force) for small temperature changes. Thermocouples are then used to measure temperature gradients.

In a solder joint we have a similar situation. The metal leads (base metal) and the solder alloy are mostly of dissimilar composition, and therefore under certain temperature-gradient conditions an appreciable amount of current can be generated in the circuit. The use of special solders to prevent this effect has been suggested for circuits where extremely low currents are used and the cumulative effect of many joints might cause misleading effects in the circuitry. This is specifically true for such delicate equipment as computers and low-current devices.

When a low-emf solder or thermal-free solder is used, it is matched to one specific base metal. Specific alloys are available for all base metals used in the electronics industry. The importance of using thermal-free solders should be determined according to the application at hand.

Soldering with thermal-free solders requires several precautions.

1. Since the thermal emf depends largely on the purity of the alloy applied, only clean soldering irons with tips which have been pretinned with the same thermal-free solder should be used.

2. Only wires pretinned with the same thermal-free alloy should be used.

3. Only fluxes containing no metallic ions should be used.

4. The assembly should be carefully cleaned, and all thermal emf joints should be marked with some kind of ink, paint, or marker so that any repairs or replacement of components can be made with the same type of solder.

In addition, some of the alloys which must be used will freeze in a dull gray appearance, resembling that of disturbed joints. Here inspection criteria should be centered around proper flow and low contact angles on both components of the solder joint and not on the appearance of the fillet, which in this case is an inherent property of the solder alloy itself and not a disturbed joint.

6-18 Low-Temperature Soldering   In Chap. 3 various low-melting-point alloys were discussed. The problems involved in low-temperature soldering are pointed out in various parts of the book and summarized here. First, the correct flux should be used in conjunction with these solders, keeping in mind that the temperature of activation for the normal fluxes is relatively high and that the average flux is therefore not suitable for low-temperature applications. The organic-acid fluxes are usually adequate for this type of an operation, although in some cases inorganic-acid fluxes must be used. Should neither of these fluxes be adequate at the low soldering temperatures, it is always possible to preclean the surfaces with a chemical etch (see Sec. 5-5) followed by a good rinse and then solder with rosin-type fluxes which just prevent reoxidation during heating and promote the spreading of the solder over the previously cleaned surfaces.

If we want to retain the advantages of low-melting-point solders and low soldering temperatures, we must use proper tools. Using a low-melting-point alloy with a high-temperature iron will defeat this purpose. In Chap. 7 we show how it is possible to reduce the tool-tip temperature by lowering the input voltage. The same can be done on soldering pots and other resistance-heating devices. Other methods of soldering can also be adjusted so as to give the proper temperature for the wetting operation. The low-melting-point solders should be soldered at temperatures of 60 to 100°F (15 to 37°C) above their liquidus. In some cases this temperature may be even lower if a sufficient dwell time is allowed. Low-temperature soldering operations in piggyback style (one solder operation on top of another where the first solder is not molten in the second soldering operation) can easily be produced using these simple safeguards.

6-19 High-Temperature Soldering   The problems in high-temperature soldering are reversed from those in the previous section. The high soldering temperature required usually exceeds the temperature stability of most organic fluxes, and an inorganic salt flux is therefore preferred. Again, if this type of material cannot be used, a pretinning operation coupled with temperature-stable nonactive flux can offer the proper solution. If organic materials are exposed to too high a temperature, they become charred; and the black layer, which is hard to dissolve in ordinary solvents, may interfere with wetting or be objectionable in the function or appearance of the assembly.

Soldering equipment for these higher temperatures is easy to obtain. However, the higher melting point requires a larger temperature difference between the soldering temperature and the liquidus of the alloy. Here, a temperature gradient of 100 to 150°F is normal. Also, the heat source has to be relatively larger in order to ensure short soldering times. The choice of the proper equipment was covered in Chap. 5.

The higher the soldering temperature the faster the base metal dissolves in the solder alloy if there is a mutual solution. In addition, the tarnishing of the adjacent areas to the bond and the tarnishing of the solder alloy itself, if not properly protected by the flux, are also a problem. The dull appearance of high-temperature solders which contain high percentages of lead is sometimes misconstrued as disturbed solder joints. This, unfortunately, cannot be easily corrected, and a dull appearance on high-temperature solders should not be considered a sign of poor wetting or incorrect cooling. The high-lead alloys will not have a tendency to dross quite so fast as the higher-tin-content alloys at the elevated temperatures.

High-temperature soldering operations are sometimes used to strip wires of thermal-sensitive insulation such as the various polyurethane coatings (see Sec. 5-6). If this operation is critical, the nature of the wire underneath should also be considered. Copper wires, for instance, should be soldered with low-tin-content solders since they have a tendency to dissolve rapidly in tin at elevated temperatures.[1]

6-20 Expanding-Type Solders   As mentioned in Chap. 3, alloys containing more than 50 percent bismuth have a tendency to expand after solidification, which is not due to the thermal coefficient of expansion. This expansion of the solder can be utilized to advantage in those applications where there is a natural shrinkage of the materials surrounding the solder joints. Thus the expansion of the solder will compensate for the shrinkage of the assembly material. This is particularly true in printed-circuit applications where the soldering operation would cause a drying out of the material surrounding the solder joint holes and the tendency of the material to shrink upon drying (see Fig. 6-9).Using expanding solders here will, in effect, eliminate any cracking in the hole. Studies run on bismuth-containing solders for printed-circuit boards have indicated that the joints are as good as those made with tin-lead. However, adequate

[1]H. H. Manko, "Soldering Fine Copper Wire, *Electron. Packag. Prod.*, vol. 6, no. 2, February 1966.

Fig. 6-9   XXXP printed-circuit board with tube socket lead. Around the pins note the cracks attributed to material shrinkage (solder, eutectic tin-lead).

Fig. 6-10 Same XXXP board as in Fig. 6-8 soldered with expanding solder. Under the same conditions no cracks were found.

compensation for the lower conductivity of the bismuth alloys must be made in all design calculations for the printed-circuit boards (see Fig. 6-10).

In addition to the use of expanding solders in conventional printed-circuit boards, these materials can also be used in structural joints where the added pressure on the parts, especially a wire through a hole, will help strengthen the assembly mechanically.

Bismuth alloys have been applied by hand-soldering techniques as well as by dipping and wave applications. However, this property of solder expansion in the solder pot should be kept in mind in setting up such equipment. It is definitely a hazard to let the solder freeze in a solder pot. It freezes on the surface first, locking the liquid underneath in place, and when it expands, it has been known to push the bottom out of the solder pot. To avoid this, either the solder should be removed from the solder pot before cooling or the solder pot should be shaped to relieve the stresses (sloping walls, etc.).

SEVEN

Hand Soldering and the Soldering Iron

7-1 Introduction Before soldering irons were heated by electricity they were heated by flames, coals, stoves, and other direct methods. Only in 1893 was the first patent granted on an electrically heated soldering iron. At that time most soldering irons were of the heavy variety used for large sheetmetal work.

With the advent of the electrical industry, the smaller electrically heated soldering iron came into its own. Today irons in a large variety of sizes and shapes are available to modern industry, the latest development being a generation of temperature-controlled tools.

Approximately 760,000 soldering irons are sold yearly to the electronics industry (excluding sales for sheetmetal work, hobbies, home use, etc.), and it is estimated that there are over 2 million irons in use. In spite of the inroads that wave soldering and other automatic methods of manufacturing have made, there will always be a large number of soldering-iron applications that cannot be replaced. All touch-up, repair, and replacement on printed-circuit boards is done with soldering irons, and so is all point-to-point wiring, installation in main frames or chassis, etc. The role of the soldering iron is thus relegated from the more robust initial assembly of low-cost components to the realm of delicate operations on high-cost equipment. This further accentuates the need for a better understanding of soldering irons.

In spite of its early history, little engineering analysis of the soldering iron can be found in the literature, yet requirements for this common everyday tool are becoming more and more stringent. The nature of delicate electronic assemblies requires that the thermal characteristics of an iron be carefully balanced and that the iron be properly matched to the job. This can spell the difference between success and failure.

Excess heat in any soldering operation is detrimental from many points of view. Plastic materials in the immediate joint area deteriorate, the metallurgical interreaction between the solder and the base metal is accelerated, exposed metallic surfaces tarnish and may anneal, and the soldering flux may volatilize, char, and become inefficient, exposing the solder joint to excess drossing. The common misconception that extra heat will improve solderability has been disproved throughout this entire book. The need for solderability restoration by different means has been analyzed. Restricting the amount of heat, however, may result in poor connections; hence, the proper tools, equipment, and procedures are important.

The following study is an analysis of the thermal characteristics of soldering irons. It establishes a mathematical model and a method of adjusting and matching the thermal characteristics to the job at hand, reviews the materials and construction details of industrial irons, gives sample calibration curves for industrial models, and outlines the proper use and maintenance of irons and their tips.

A discussion of the soldering iron would not be complete without a review of hand-soldering practices, where a wide gap exists between the artist and the artisan on the one hand and the trained operator on the other. A misconception exists that hand soldering is an uncontrolled process and inferior to automatic mass-production techniques. This assumption is based on unnecessary but sad experiences and observations. A well-trained and well-motivated work force using correct equipment and materials and having properly designed assemblies is more reliable than any automatic piece of equipment. A schooled operator observes each connection as it is being made and is the best judge of good and marginal wetting. In that respect, the operator sees more than any inspector can after the fact. If we learn to use the human brain like a controlling computer over an operation, we can achieve levels of quality that cannot be matched by any unintelligent piece of equipment no matter how sophisticated it is.

To achieve this we need one basic change in attitude. Presently operators are left to their own devices to do the best they can with the materials given them. In many cases they feel guilty if a solder connection does not come up to the quality standards imposed and try to rework the connection or otherwise hide the failure. This often results in too much heat deterioration and hidden faults, which lower the overall reliability of the equipment. What is actually needed is a different emphasis altogether. The operators must be told that unless a solder connection is formed within given and specific conditions, they must alert their supervisors. The burden is thus taken from their shoulders and placed on the materials-control group. They must monitor incoming solderability and have methods of restoring surfaces which have lost this condition.

Methods of soldering have evolved over the years, and the details have changed as more and more about the process of soldering has been learned. In our discussion we concentrate on the latest techniques, which obviously have to be adapted to all kinds of special situations.

Training the operator in the techniques of hand soldering is relatively simple, but it constitutes only a small part of the overall training required. The bulk of the training is associated with the unique workmanship standards of each organization or industry. (This is discussed in greater detail later in this chapter and in Chap. 8.)

7-2 Construction Details   Although the soldering iron is only a means of carrying heat to the solder joint, it is a tool which incorporates many engineering concepts. The electric iron consists of a heating coil, usually made of nichrome wire, which produces heat through its high resistance (I^2R). The size and length of the wire determine the resistance of the heating element and thus its output, which is expressed in watts; however, soldering irons should generally not be classified by their watt output because this is a misleading quantity in most applications. Additional information such as iron efficiency, iron heat content, and maximum tip temperature should be included in the rating or classification of irons. A more thorough discussion is presented later in the chapter.

The heat developed in the coil must now be transmitted to the working tip. This is achieved by the use of metallic conductors, mostly copper or copper alloys. In addition to transmitting heat directly to the working tip, the metallic conductors also serve as a caloric reservoir to be called upon when the heat has dissipated from the tip during the soldering operation. (For further details see Sec. 7-15). Selection of conductors takes into account the following properties:

1. *Heat conductivity.* A short time lapse between heat formation in the coil and the availability of the same heat at the working tip is of vital importance.

2. *Specific heat of the conductor.* This determines the heat content of the iron. It is always coupled with the conductor volume, which in turn is dictated by the application.

3. *Oxidation resistance of the surface.* The amount of tarnish formed on the surfaces of the conductors reduces the overall conductivity of the iron and thereby decreases its efficiency. The conductor material should therefore be able to withstand soldering temperatures without excessive oxidation.

The working tip is the subject of Sec. 7-14, but the considerations which applied to the conductors from the heating coil to the working area apply also to the tip. An additional restraint is that the tip material itself should be wettable with the solder alloy and not dissolve in it or form metallic compounds in the working-temperature range.

A nonmetallic handle is incorporated in the soldering iron which enables the operator to handle the iron without being exposed to the high soldering temperatures. Many handles are made of wood, which is an excellent material for this application, but in recent years various types of plastic material have been used, mainly to lower the price. Sometimes a cork sleeving is put over the handle to further lower the temperature of the outside of the handle making contact with the operator's hands.

Current is introduced to the soldering iron through a flexible electric cord. The flexibility of the cord contributes largely to the ease of handling and should not be overlooked when new irons are evaluated. The heat stability of the cord insulation is also important to safety; patched cords should not be permitted.

All soldering irons must be grounded for the safety of both the operator and the assembly. Voltage-sensitive devices like FETs and CMOS, which require special handling for electrostatic shock prevention, require extreme precaution in hand soldering. For that reason we must separate the soldering irons into constantly operating types that can be grounded directly without any danger to the devices and switching irons, which may have voltage spikes generated during the switching cycle. Careful study of this parameter is highly recommended, and vendor specification sheets should be consulted.

Figures 7-1 and 7-2 show cross sections of two well-designed soldering irons. Let us survey Fig. 7-1 from the handle toward the tip. The cross section shows the wooden handle with a metallic anchor embedded in it and a cork insulation sleeve around it. The wire is introduced through a hollow at the base of the handle, and a grounding screw is available on the terminal strip. The two current-carrying wires terminate at similar screws on the same terminal strip. A stainless-steel case is threaded into the metal anchor to hold the heating assembly and the tip firmly in place. The stainless-steel jacket serves as a low-heat-conducting protective sheet around the element and its components. Stainless steel is less prone to attack by the soldering fluxes than most metals and has adequate strengths to absorb mild stresses and blows. A brass heat-reserve jacket located between the sleeve and the heat element itself serves to increase the metallic volume within the iron. A heat-transfer sleeve made of bronze is located inside the heating element, and the soldering tip is fastened into this heat-transfer sleeve with a setscrew shown on the side.

Figure 7-2 shows the cutaway view of a fully temperature-controlled soldering iron, which has an accurate thermistor sensing element located in the unit and protruding all the way into the tip, continuously monitoring the temperature of the iron. A light indicator on the stand indicates when the current is on and when it is off. Under normal idling conditions, this unit will switch on and off approximately 50 times a minute, keeping its temperature within 5°C of that dialed. The element here is 100-W, which has enough power to regenerate any heat drain within a relatively short time. The biggest advantage of this type of iron lies in the fact that the selection of the tip enables the user to adapt it to delicate applications on the one hand without any danger of too much heat while heavier load applications can still be satisfied with larger tips. Unfortunately, these irons are the highest-cost units and can only be justified for delicate work or where one iron has to do more than a single type of job. They are by far the most accurate irons available.

The construction of soldering irons is different for almost all manufacturers and is usually protected by some kind of patent. In some cases, the element is included in the soldering-iron tip and the number of oxidizable heat-transfer surfaces is thereby reduced. This has a great disadvantage, however, in that the element must be discarded whenever the tip life has been surpassed or the element burns out. In general, models having a greater portion of the soldering tip inserted into the heating element are preferred to those having a smaller section of the tip screwed into the casing. Press-fit tips over the heating element should be avoided because they are not reliable and the heat-transfer properties are not reproduci-

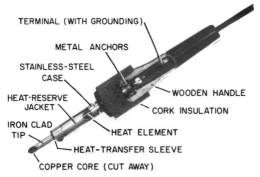

Fig. 7-1 Cutaway view of a medium-sized soldering iron. *(Hexacon Electric Co.)*

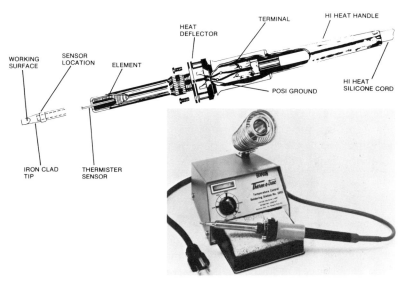

Fig. 7-2   Cutaway view of fully temperature-controlled iron (100 W). The tip maintains temperatures within 5°C of any setting dialed between 260 and 382°C (500 to 720°F). Suitable for precision soldering of anything from delicate to robust applications. *(Hexacon Electric Co.)*

ble. In practice, the type of iron and its final applications dictate, to a large extent, the construction and the details of coupling the tip to the iron itself.

Since the better models of irons are all available with grounding facilities, the use of a transformer to lower the iron voltage is no longer necessary. The 6-V irons were made primarily to minimize electric-shock danger to the operator and damage to electronic components due to current leakage. The use of 110-V irons with grounding is recommended for good heat characteristics and easy handling.

7-3 Available Industrial Irons   To complete the picture we look at the soldering iron just as a source of heat and see how it can be engineered to the items to be soldered. There are several approaches: truly temperature-controlled irons, which maintain a narrow temperature band, even under load conditions; the temperature-limiting irons of varying design; and the constant-load irons, which undergo a much larger temperature variation under load conditions.

The truly temperature-controlled solder irons, which have a large element, normally 100 W, are matched to a wider range of applications than their partially or uncontrolled counterparts (Fig. 7-2). The temperature is maintained through an efficient heat-sensing and regulating system, e.g., a thermistor. The heat-up time is very rapid, and they recover the specified temperature within the normal time it takes to move from one connection to another even under high-speed mass production. The use of this device does not necessitate any special calculations for picking the size and the shape of the tip.

The temperature-limiting irons, which unfortunately are often referred to as temperature-controlled irons, fall in a different category (Fig. 7-9). While they still use large heat-generating elements, the temperature-sensing mechanism is much more sluggish. The tip-temperature variation is therefore much larger. For instance, magnetically switching soldering irons fall into this category. It is difficult to generalize on the calculations for these irons without knowing the exact thermal response of the sensing mechanism, the element size, efficiencies, and similar characteristics. The calculations we shall make for the constant-load irons described below can be adapted to this category (see the mathematical model in Sec. 7-5).

Finally, we must consider the constant-load soldering iron, which is the most widespread and most economical iron in the market (Fig. 7-8). This is a constantly "on" iron which must be carefully selected for an application if the tip temperature is to be in a specified range (see Table 7-6). The mathematical model (Sec. 7-5) deals in great detail with this type of tool. Matching this iron to a specific task is covered in Secs. 7-11 to 7-14.

Since the constant-load-iron temperature is voltage-dependent, it is often coupled with a regulating device for temperature adjustments. Figure 7-10 shows a schematic set up for an entire workbench. Individual voltage regulators are also used for single solder systems.

7-4 Tips for Soldering Irons   The purpose of a soldering tip is to convey heat from the iron to the work. Its shape and size are mainly determined by the type of work required. In addition to its thermal characteristics, a soldering tip should have a long life in the environment in which it is used; therefore, the solid metal at the tip should not erode rapidly and dissolve in the liquid solder.

In the past, most soldering tips were made of copper, which is suitable because of its good thermal conductivity, high heat content per volume, and low cost. Unfortunately tin-lead soldering alloys (specifically the tin) attack the copper in the tip and dissolve it during the operating cycle of the iron. As a result, copper tips can never maintain their shape for long and must be periodically reshaped with a file. This gives the copper tips a relatively short life (experience shows that copper tips have worn out after 2 to 8 h of use), and although the cost is only a fraction of that of the more modern tips, considering labor costs and reshaping time, the modern tips are far more economical in the long run.

TABLE 7-1   Comparison of Heat Characteristics of Metals

No.	Metal or alloy	Heat conductivity cal/(cm²)(cm)(°C)(s)	% of copper	Measured at °C	Specific heat per volume, cal/(cm³)(°C)
1	Aluminum	0.461	50.0	0	0.56
2	Brass	0.204	23.0	0	
3	Copper	0.920	100.0	0	0.81
4	German silver	0.070	7.6	0	
5	Gold	0.744	81.0	0	0.60
6	Iron (1%C)	0.1085	11.8	18	0.82
7	Nickel	0.140	15.2	18	0.92
8	Silver	1.096	120.0	0	0.59
9	Tantalum	0.130	14.0	17	0.60
10	Tungsten	0.383	41.5	0	0.62

Let us look briefly at other possible metals for soldering tips and discuss their relative merits (Tables 7-1 and 7-2).[1] Silver and gold have excellent conductivity but a lower heat content per volume than copper; however, they are too expensive to be considered useful soldering tips, especially since they also readily dissolve in molten tin alloys.

Brass and other copper alloys have a lower conductivity and erode in the same fashion but at a slightly slower rate. Nickel-silver does not erode so fast but has less than 8 percent of the conductivity of copper.

Tungsten and tantalum are not wetted by most solder alloys at soldering temperatures and are therefore used as nonwetted heat-transfer tips for special applications in miniature soldering. Of the two, tungsten is far more useful because of its better conductivity and higher heat capacity per volume.

Iron and nickel, in spite of their low conductivity, are wettable and offer the greatest resistance to erosion. This, along with a specific heat per volume which matches that of copper, makes them very useful as a tip material. In order to increase the conductivity of the tip and at the same time retain the erosion-resistant properties of the surface, ironclad or nickel-clad copper bits are used. These clad copper bits make up a large percentage of the modern electric soldering-iron tips.

Aluminum is in a class by itself. Its conductivity is half that of copper, and its specific heat is one of the lowest. In addition, once the aluminum is wetted, it will contaminate the solder and cause a large amount of grittiness.

[1]This discussion refers mainly to tin-lead alloys and most high-tin solders. Erosion problems of other alloys should be considered separately.

TABLE 7-2   Some Industrial Models of Soldering Tips

Manufacturer's name	Catalog No.	Core	Platings from core to outside, in	General observation	Trade names
Hexacon Electric Co.	HT 866D	Brass	Iron (~0.0028) pretinned	Bare threads	Durotherm
	HT 6030	Tellurium copper	Iron (~0.008) pretinned	Stainless-steel shaft† (0.003)	Durotherm
	HT 603X	Tellurium copper	Iron (~0.008) pretinned	Iron shaft† (0.001)	Xtradur
	HT 603	Tellurium copper	Iron (~0.008) pretinned	Iron shaft (0.001)	Hexclad
American Heater Co. (American Beauty)	B-3	Copper	Iron (~0.003) copper* silver*	Plating on thread	Superior
	177 PAR.	Copper	Iron (~0.007) pretinned	Iron shaft† (0.006)	Paragon
Ungar Co. (Imperial)	6319	Copper	Iron (~0.007) copper* gold*		
	6407	Tellurium copper	Nickel* copper*		
General Electric Co.	6A-214	Copper	Iron (~0.02) pretinned	Heater included with tip	
Oryx Co.	Type J	Copper	Nickel (~0.0003)	Nickel flash on shafts, nonvisible inside	

*Flash plating.
†Specially treated.

Note that in Table 7-1 the heat capacity (or the specific heat) of the metal is given in calories per cubic centimeter per degree Celsius. In other words, this is the capacity of the volume, not of a certain weight. These figures were specially computed for this purpose and are easy to compare for given tips made of various metals where the volume rather than the weight is constant. Table 7-1 lists in alphabetical order some of the common materials that might be considered for soldering tips. The conductivity is given in both absolute units and as a percentage of copper conductivity. The table is also useful in designing special tools for soldering.

Figure 7-3 shows a cross section of an ironclad copper-core soldering tip. Two areas of the tip were microsectioned at ×25 magnification and are shown superimposed on the line drawing in their respective positions. The tip itself shows an outer layer of solder (which is difficult to see at this magnification because it is so thin), an iron coating, and the basic copper tip. The microsection of the shaft shows the copper core with the iron around it. The well at the back of this particular soldering tip is necessary for the various plating procedures.

Let us analyze the reason for the various coatings on this tip. As shown earlier, the copper core is the most suitable material for a soldering tip except that it erodes rapidly in the molten tin-lead alloy. The tip of the soldering iron is therefore coated with a layer of iron that has approximately the same heat content but a much lower heat conductivity. Because it does not react with the molten solder, this iron plating gives the tip long life. Continuous abrasion of the work surfaces against this iron skin, however, eventually wears through it, and once the copper is exposed, the tip has to be discarded. Iron by itself is difficult to wet with solder unless careful procedures are followed; therefore, the manufacturer will pretin these tips up to a certain depth, as shown in Fig. 7-3. If an ironclad tip without pretinning is used, it should be fluxed when cold and immediately upon reaching soldering temperature should be wetted with tin-lead (preferably by dipping in a solder pot). Otherwise heavy oxidation of the iron plating will make the wetting difficult and sometimes impractical. For maintenance details on the soldering tip, see Sec. 7-17.

The shaft of the tip does not necessarily require the iron plating because it will not be wetted. It could be coated with any metal which has

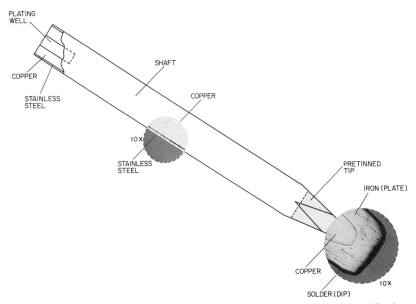

Fig. 7-3 Diagram of a soldering-iron tip. Note the two areas of ×10 magnification showing tip and shank. *(Hexacon Electric Co.)*

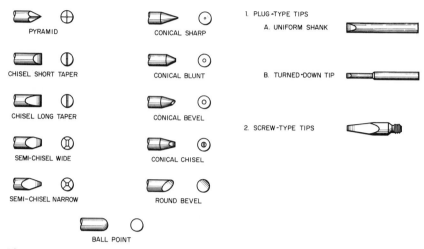

BASIC TIP CONFIGURATIONS

PYRAMID

CHISEL SHORT TAPER

CHISEL LONG TAPER

SEMI-CHISEL WIDE

SEMI-CHISEL NARROW

BALL POINT

CONICAL SHARP

CONICAL BLUNT

CONICAL BEVEL

CONICAL CHISEL

ROUND BEVEL

BASIC TIP TYPES

1. PLUG-TYPE TIPS

A. UNIFORM SHANK

B. TURNED-DOWN TIP

2. SCREW-TYPE TIPS

Fig. 7-4 Basic configurations and basic types of common soldering tips.

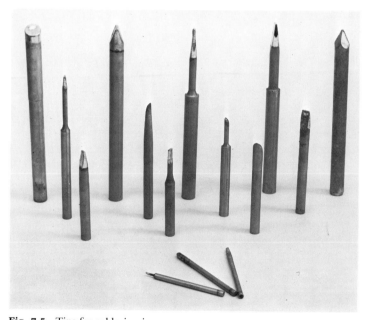

Fig. 7-5 Tips for soldering irons.

good oxidation resistance so that the tips will not develop an oxidation scale rapidly at soldering temperatures and bind inside the iron. Some manufacturers go to the extent of coating the shaft with a thin layer of precious metal like gold or silver. Experience has shown, however, that the extremely thin coatings of precious metals applied do not preserve the shank from heavy oxidation over prolonged periods.

The hole at the end of the tip is a well, made specially for plating purposes, and has no particular significance in the operation of the tip.

Figure 7-4 shows some of the basic configurations and the basic types of common soldering tips. Figure 7-5 is a photograph of the actual soldering tips for standard irons. It does not show screw tips.

7-5 A Mathematical Model for Soldering Irons   It would be simple to set up a mathematical model to calculate the caloric requirements of a solder joint if the system were truly adiabatic (in this context adiabatic refers to any change in caloric content to or from the environment; in other words, it excludes changes in which there is a heat gain or heat loss to the surrounding area). In such an isolated system the differential heat (between tool and work), the contact area, and the contact time would determine the caloric heat transferred from the soldering tool to the joint. In reality, however, our system is far from being adiabatic, and we must include the heat dissipation into the surrounding areas, which is a time-dependent variable.

7-6 The Basic Axioms   In order to set up a realistic mathematical model we first postulate a set of axioms and make some general assumptions justified by practical experience.

Axiom 1: Prolonged heating is deleterious.

Since it is possible to achieve a solder joint with various rates of heating, let us first set up the reasons why extended heating is objectionable. In Chap. 3 we discussed the metallurgical reactions between the solder and the base metal and showed that they are in the large part undesirable. The reactions are all time- and temperature-dependent and can be drastically limited with the limitation of time. The metallic surfaces interfacing with the solder are not the only materials suffering from heat damage. Plastic materials are even more prone to heat deterioration. Glass insulators, coatings, component markings, and printed-circuit-board laminates should be protected from prolonged heat exposure. We must consider heat damage to the components and devices which may change value. Finally we must avoid the spread of tarnish on unprotected metallic surfaces and the possibility of annealing the base metals below their useful range.

All this can be achieved by shortening the heating time and/or increasing the heating rate. These actions by themselves are not free from additional problems, however, as the next axiom shows.

Axiom 2: Excessive temperatures are objectionable.

One method of shortening the soldering time would be to increase the temperature differential between the tool and the work, i.e., using a higher soldering-iron-tip temperature. This has inherent problems, however, as it will destroy the flux in the core of the solder, not allowing it enough time to spread over the surfaces for the fluxing action described in Sec. 5-8. In addition, the solder may melt prematurely and because of its density may replace the flux before it gets a chance to react properly, thus destroying the proper sequence of events. Finally, excessive temperatures may accelerate the time-related problems described in Axiom 1 beyond the protection that a shorter time may afford. As a result, it is necessary to keep the tip temperature of the soldering iron within certain limits in order to obtain proper hand-soldering results.

A much more acceptable method of shortening the heating time at a relatively lower temperature is by increasing the contact area between the tool and the work. This is limited by the configuration of the work and is not always possible, although certain manual techniques may help in this respect (see Sec. 7-18).

Axiom 3: Joint quality depends on a relaxed operator.

We must incorporate the human operator into the process. This is possible only when the industrial conditions provided are compatible with our scientific aims. The relaxed operator is one whose motions have been properly planned by the design engineer and who is using comfortable tools suitable for the task. The solder engineer who plans the operation must make sure that the conditions are right for the time-temperature-material combination to give the right fluxing-wetting-spreading sequence for easy joint formation.

The importance of operator training cannot be overemphasized. With the planning mentioned above and the correct training procedures we can use the intelligence of the human operator to execute high-quality solder joints often exceeding those made by automatic means. The key here is the trained operator who recognizes soldering difficulties and brings them to the attention of the supervisor for corrective action. We cannot allow ourselves the luxury of blaming operators and/or having them blame themselves for faults in the materials, tools, or design of the work. As a result they try to hide their mistakes under excessive amounts of solder.

Axiom 4: The prerequisites for hand soldering must be supplied.

It goes without saying that the materials, design, and methods (process) must be correct for the application. This includes such items as good solderability, correct flux and solder for the application, adequate geometric design and configuration for easy joint formation, proper tools (irons and tips), appropriate workmanship standards, and a training program for the operator. Without those prerequisites our model would suffer from the deficiencies described earlier.

The implications we derive from our axioms are as follows:

Axiom 1: Keep time short.

Axiom 2: Keep temperature low.

Axiom 3: Assure that operator is in a relaxed mode.

Axiom 4: Be sure that design, materials, and methods are correct.

7-7 Understanding the Sequence of Events in Joint Formation   We must define the times and steps involved in making the solder joint and the temperature levels reached in each case. Table 7-3 gives the 12 steps in the sequence in joint formation related to a soldering iron. The time involved between steps 1 and 12 is the *total time* for the task t_{tot}, whereas the time to make the solder joint falls between steps 3 and 8 and is labeled *time per joint* t_j; steps 5 to 9 are the critical time t_{crit} when the solder is molten. T_{mp} is melting temperature and T_i initial temperature.

Let us follow the steps in the sequence in which they occur. In our case we assume that the soldering iron has reached its maximum idling temperature ($T_i + 40$) after the previous cycle during the idling time (defined

TABLE 7-3   The 12 Steps in the Sequence of Joint Formation

	Step	Description	Temperature, °F
1	Iron removed from holder	$T_i + 40$
2	Tip sponge-wiped	T_i
3	Tip and core solder touch work	$T_i \rightarrow T_{room}$
4	Flux melts out of core and reacts	T_f
5	Solder melts, creating a heat bridge	T_{mp}
6a	Work and solder reach wetting temperature	T_w
6b	Tip temperature drops	$T_i - \Delta T$
6c	Solder fillet formed	T_w+
7	Core removed first	$T_{room}+$
8	Iron removed second	$T_j \rightarrow \Delta T + dT$
9	Fillet solidifies	T_{mp}
10	Work cools	T_{room}
11	Steps 3–8 repeated n times per task	$T_i - n(\Delta T - dT_{room})$
12	Iron returned to holder between tasks	$T_i + 40$

(Bracket groupings: t_{tot} spans steps 1–12; t_j spans steps 3–8; t_{crit} spans steps 5–9.)

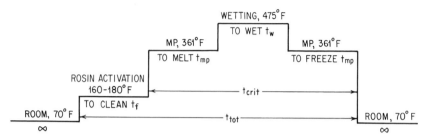

Fig. 7-6   The temperature-time cycles. The number on the top of the line represents the temperatures, and the numbers on the bottom represent the time in seconds. $T_{crit} = 1 \pm \frac{1}{4}$ s.

as the time between two successive tasks). When the iron is removed from the holder and sponge-wiped to remove the charred flux and dross, it will lose approximately 40°F (22°C) and then be ready for soldering (T_i). As soon as the tip and the core solder touch the work (they are both at room temperature), the soldering iron will start raising the temperature of both. Next the flux starts to melt out of the core and react with the surfaces at a flux temperature T_f. When the surfaces have been prepared by the molten flux, they are ready for wetting with the molten solder; thus the next sequence in the operation is the melting of the alloy itself to create a heat bridge T_{mp}. The heat bridge may start to conduct at the solidus temperature, but it is effective only at the liquidus temperature, i.e., the melting point of a noneutectic alloy. Three things happen simultaneously: (1) the work and solder reach wetting temperature T_w; (2) the tip temperature drops by ΔT and (3) the solder fillet is formed. As soon as all the solder required for the joint has been liquefied, the core solder is removed. This is followed by the removal of the soldering iron. The fillet is then left to solidify, and the work cools back to room temperature T_{room}. In the meantime, steps 3 to 8 from Table 7-3 are repeated at least n times per task (where n is the number of joints per task) before the iron is returned to the holder. Figure 7-6 shows a typical temperature cycle for a eutectic tin-lead solder with an activated-rosin core being applied to a copper surface. Note that here we have a specific critical time of $1 \pm \frac{1}{4}$ s.

7-8 Theoretical Basis for a Mathematical Model   Although we recognize that the system is not adiabatic, we must start out with this assumption. Later we shall show that the heat losses to the environment are proportional to the tool and work and thus can be neglected in our model. Under adiabatic conditions the first and second laws of thermodynamics make it possible to determine the temperature changes in the hot tool and the cold work when they are first brought into contact. To help our thinking, however, it is much simpler to consider the entire system to consist of two sets of masses converted into their *copper mass equivalents*. This term was specifically developed for our model and will represent the various parts of the system. Thus we can view the various parts of the tool and the work as units of copper mass at predetermined temperatures, which is the key to our calculation.

7-9 The Work   Let us consider the work and what makes up its copper equivalency. For simplicity we shall take an axial component soldered to a printed-circuit board with a plated through hole. The solder used is eutectic 63/37, and the fillet is ideal top and bottom (see Fig. 7-7). The mass of the joint M_j is equal to the mass of the lead M_{lead} directly affected by the soldering operation plus the mass of the printed-circuit board M_{pcb} plus the mass of solder in the joint M_{solder}

$$M_j = M_{\text{lead}} + M_{\text{pcb}} + M_{\text{solder}} \tag{7-1}$$

The mass does not tell the entire story, however; we must consider the caloric consumption Q by the work during the soldering process. We must also consider the temperature change ΔT, which in this case is uniform from room temperature to wetting temperature. This enables us to calculate the caloric consumption of the work if we consider the following three components:

1. The heat used to raise the temperature of the task
2. The heat used by the solder, which includes heating it to melting, the heat of fusion, and the heat used to raise the temperature of the liquid solder to soldering temperature
3. The heat used by the flux

Actual calculations indicate that the heat used by the flux is negligible compared with the rest and can be neglected.

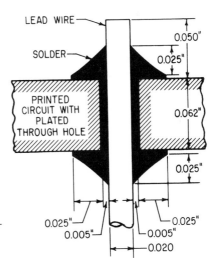

Fig. 7-7   Schematic of solder fillet in a plated through hole.

When calculating the caloric requirements of the task (without the solder) we use the formula

$$Q_{task} = fM_{task} \times \text{specific heat} \times \Delta T \qquad (7\text{-}2)$$

where ΔT is the change from T_{room} to T_w and f is the conversion factor equating everything to copper. For the lead wire and the copper land $f = 1$. For the laminate we can neglect the effect, and $f = 0$.

For the printed-circuit solder joint our mass would consist of the mass of the lead plus the mass of the printed-circuit board

$$M_{task} = M_{lead} + M_{pcb} \qquad (7\text{-}3)$$

An actual calculation of the joint depicted in Fig. 7-7 is summarized in Table 7-4. In order to derive these values an effective lead length of 0.250 in was used. This also accounts for the mass of the tinned coating.

Next we must consider the solder fillet, where we must separate the heat of fusion needed to melt the solder from the heat needed to raise the temperature. Simple spatial calculations are used to determine the volume of solder in a joint. Inspection criteria are used to define the joint geometry. Table 7-4 also reports the values calculated for solder in the joint depicted in Fig. 7-7.

The temperature rise ΔT used in the calculations is the difference between ambient (room) temperature and the melting temperature. For Table 7-4 we used

$$\Delta T = T_w - T_{room} = 260 - 20 = 240°C \qquad (7\text{-}4)$$

When we calculate the copper equivalency of the solder, which includes the heat of fusion, we find that it is nearly the same as the solder volume. Thus for all practical purposes we can include the volume of the solder in the volume of the copper provided the wetting temperature stays the same.

In summary, the amount of heat needed for the task can always be translated into a function of the metal content of the joint (as copper) and

TABLE 7-4   Analysis of Straight-through Lead in Plated Through Hole

Zone	Volume, cm³ × 10⁻³ Actual	Cu equivalent	Heat required, cal	%
Lead wire	5.15	. . . .	1.064	72.5
Printed-circuit board	0.405	. . . .	0.084	5.7
Solder*	1.54	1.55	0.321	21.8

*Includes the heat of fusion.

the temperature differential (between room and wetting). The calculation is much easier when discrete terminals or pins are considered for joining since they can be weighed accurately. The solder and wire are then added to the weight in the same way as for the printed-circuit board. Remember, to add a factor to the work in order to increase it to the effective length. The heat conductivity of the metal will determine the size of this factor, poor conductors like Kovar having a smaller effect.

7-10 The Soldering Iron   From the thermal point of view, the soldering iron consists of three distinct parts:

1. The tip which acts as the conductor of heat and must be defined in those terms
2. The active element, which is the source of heat
3. The inactive case and handle, which serve as insulators

We shall treat each component separately, stating the theoretical basis for the mathematical model and the practical application to actual soldering tools.

When an iron is matched to the task, the temperature fluctuations within the instrument are minimal and for all practical purposes we can consider the system in a thermal steady state. This simplifies all the calculations by making them independent of time. Any inaccuracies which occur as a result of this assumption are compensated for when the actual *efficiency factor* of an iron is physically measured. (A calculation of the actual variation due to temperature drop in matched iron-work systems indicates a variation of less than 1 percent.)

The Tip. The soldering-iron tip is the conductor of heat from the element to the task. The heat is transmitted through conduction, following Fourier's law, which states that the instantaneous rate of heat flow dQ/dt, where Q is the heat amount flowing during the time t, is equal to the product of three factors:

1. The area A of the section (taken at right angles to the direction of heat flow)
2. The temperature gradient dT/dx, or the rate of change in temperature T with respect to the length of the task x
3. The thermal conductivity K, which is a proportionalty factor unique to each material

$$\frac{dQ}{dt} = KA\ \frac{dT}{dx} \tag{7-5}$$

The conditions for this equation are (1) that a thermal gradient exists in order to induce conductivity from task temperature to wetting temperature and (2) that a steady state exists where the temperature is not a function of time. Under these conditions, and specifically in a steady state, the formula can be further simplified to read

$$\frac{dQ}{dt} = \frac{Q}{t} = q \qquad q = KA \frac{dT}{dx} \qquad (7\text{-}6)$$

where Q = quantity of heat, cal
t = time, s
A = area, cm^2
T = temperature, °C
x = length of conductor path, cm
K = thermal conductivity at temperature T
q = steady-state heat-flow, cal/s

For the soldering-iron tip we can adapt this formula as follows. The average tip is made of ironclad copper, but we can neglect the outer coating, taking the entire tip as if it were made of copper. The heat-transfer impedance due to the plating will be compensated by the efficiency factor later. The formula is now

$$q_{\text{tip}} = KA \frac{dt}{dx} = K \frac{\pi}{4} D^2 \frac{T}{x} \qquad (7\text{-}7)$$

where K = 224 for copper
D = diameter of tip, cm
T = tip temperature, °C
x = effective length of tip, cm

The heat-conducting length of a tip consists of two parts, the length of the exposed tip and the length of the tip inside the element. For all practical purposes the external part is the critical component, but the internal part can be taken as

$$x_{\text{int}} = \frac{\text{element length}}{2} \qquad (7\text{-}8)$$

Thus the total effective length x is

$$x = x_{\text{ext}} + x_{\text{int}} = x_{\text{ext}} + \tfrac{1}{2} \text{ element length} \qquad (7\text{-}9)$$

The formula is not complete without including an efficiency factor E, which accounts for (1) heat loss from the external part to the surroundings, (2) conduction restrictions due to outer coating (iron) and wetted or oxidized outside of the tip at the point of contact, and (3) a geometric factor since tips are machined and shaped in various configurations. These are easier to measure on an actual tip than to calculate geometrically.

It is easy to find the true conductivity of a soldering-iron tip experimentally. The actual heat output of an iron is measured first with a defined solid piece of round copper filling the element cavity exactly and having a thermocouple embedded at the end. This is then compared with an actual soldering-iron tip with a thermocouple at the wetted interface. If we include the measured efficiency factor E for the tip, we get the instant rate of heat flow for an actual tip

$$q_{tip} = E_{tip} \ K \ \frac{\pi}{4} D^2 \frac{T}{x} \qquad (7\text{-}10)$$

The Active Element. This component of the soldering-iron system must be viewed only as the heat generator.[1] The conductivity of this part will again be incorporated in the efficiency factor. From Joule's law the caloric output of the element is

$$Q_{elem} = \frac{I^2 R t}{j} \qquad (7\text{-}11)$$

where I = current, A
$\quad R$ = resistance of element, kΩ
$\quad t$ = time, s
$\quad j$ = 4.2 cal/s
Since most soldering irons are rated in watts, we can convert this formula to read

$$Q_{elem} = E_{elem} \frac{W}{j} \qquad (7\text{-}12)$$

where E_{elem} is the measured efficiency of the iron and W is the rating in watts. Note that once again we are independent of time.

The efficiency rating of the iron is calculated from the first value measured earlier for the tip divided by the theoretical heat output of the iron.

The Inactive Case and Handle. The case, handle, and other inactive components of the iron do not contribute to the heat transferred, but they constitute part of the heat loss from the element. Because of the way we have calculated the efficiency factor, the interaction of these elements with the iron is included in the efficiency factor and can be neglected. The same holds true for the heat capacity of the entire assembly, excluding the

---

[1]For temperature-controlled irons with a switching mechanism the calculation is very clumsy. It is necessary to measure the percentage of time the element is actually on and to include this in the calculation. This is relatively easy for the idling condition of an iron but may be more complicated when a large task is being soldered with this type of iron, necessitating an integrator.

tip. While this is an important factor in the recovery rate of the iron, it can be neglected in this mathematical model.

In summary, the heat supply of the iron depends on the heat output of the element. This is directly proportional to the wattage of the element and the efficiency factor of the iron, which depends on the heat losses.

The conductivity of the tip, on the other hand, is geometry-sensitive and depends on the materials of construction and the existing temperature gradient.

Knowing the heat requirements of the work from Sec. 7-9, we are now in a position to equate them to the heat output of the iron. This theoretical model is important for understanding the heat flow and balance in hand soldering. A simple empirical procedure is actually used to match the iron to the task.

7-11 The Iron to Fit the Task   As mentioned earlier, the wattage of a soldering iron does not really determine the temperature of the tip. Most reputable soldering-iron manufacturers will provide approximately the same tip temperature for a variety of iron sizes designed for a particular application, i.e., miniature electronics, electrical structural, etc.; thus a group of soldering irons intended for printed circuits will have approximately an 800°F tip temperature regardless of the size, which can vary from a tiny tip to a massive configuration (for further details see Table 7-6). Irons intended for nonelectrical work have a much higher tip temperature, however, usually coupled with a larger-sized iron because of heat requirements for mechanical joining. It is therefore impractical to compare soldering irons according to their wattage but much more feasible to refer to them by their maximum tip temperature.

It has been made clear that the wattage of a soldering iron does not tell enough about the properties of the iron. In addition to the wattage, we should be given the efficiency of the iron, defined as that quantity of the energy generated as heat which actually reaches the working tip and the solder area itself; the heat content of the iron; and its recovery rate. These three parameters, which are unique for every model of soldering iron, can be used for comparing and evaluating all irons.

Maximum Tip Temperature. The maximum tip temperature is the temperature at equilibrium; at this point the heat generated in the coil and the heat lost to the surroundings are equal, and the iron is maintained in a steady state. A well-designed soldering iron should have the same maximum tip temperature for the given rated voltage that is constant for any particular model.

Heat Content of Iron. The heat content is a measure of the caloric content of the soldering iron that is available for the soldering operations. It depends on the specific heat of the various metallic conductors present between the heating element and the work. The amount of heat stored in any of these conductors is a direct function of the temperature and the volume of the material. If the heat content of the iron is too small, a single solder joint can lower the tip temperature of the iron so that the operator has to wait for the iron to recover its tip temperature, thereby losing production time.

If the solder-joint configuration is matched to the iron, consecutive solder connections can be made without a recovery period, and the amount of heat available to the solder joint is then limited to a reasonable amount.

Recovery Rate. The recovery rate is the amount of time necessary for a tip to regain its maximum tip temperature after the heat is drained out of the tip during the working cycle. It is indirectly a function of the wattage, the heat content, and the work load of the iron and is limited by the steady-state conditions, which are approached in a decreasing asymptotic manner. Without a reasonable recovery rate, the iron is useless for production.

The wattage, heat content, and work load are naturally proportioned so that a high tip temperature or a large heat content will necessitate a slower recovery rate and, on the other hand, a small heat content at a low temperature will make possible a faster recovery rate. When an iron is considered for a specific job, all three characteristics of the iron should be examined and the desired balance found for the given conditions.

Figures 7-8 and 7-9 indicate the great variety of soldering irons available on the market. (These are only representative models of the lines manufactured by various manufacturers.) We have considered it very important to establish a method whereby a group of basic soldering irons can be used under a variety of conditions. Two reasons for this are that (1) it is not feasible for a user to have a large variety of soldering irons in stock in order to accommodate a variety of jobs requiring various heat conditions and (2) it is impractical to try to buy one iron to do all jobs. The need for a basic group of irons also implies the need for irons having various tip temperatures because of the variety of soldering alloys available.

Because the heat source of the soldering iron is a resistance type of element, it is possible by changing the voltage of the input to change the caloric supply in the iron. A study based on this fact was made of the changes in the temperature of soldering irons as a function of the input voltage (refer to TTV curves, Fig. 7-11). The results of such a study made it possible to adapt a basic group of soldering irons for a wide variety of needs.

Little equipment is needed to carry this out in practice. A variable autotransformer (Variac or Powerstat) can be easily used to effect the changes in voltage. The average Powerstat can handle 5 to 10 soldering

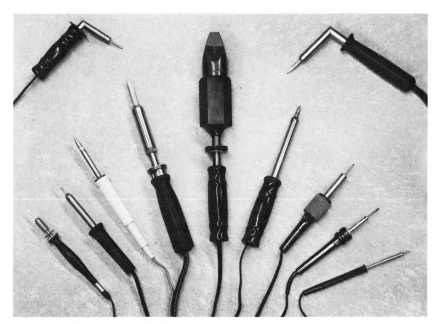

Fig. 7-8 Soldering irons, 110 V.

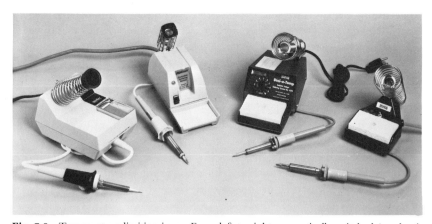

Fig. 7-9 Temperature-limiting irons. From left to right magnetically switched (mechanical); bimetal switched (mechanical); voltage-controlled (nonswitching); and a dual-heat unit.

irons at the same time. Since there is always a possibility that the existing line voltage will vary from 5 to 10 V over the working day, for a critical application it is suggested that a voltage stabilizer be interposed between the input line and the Powerstat. Figure 7-10 shows such an arrangement. Note that the number of receptacles is limited only by the equipment rating. It is easy to figure out the amperage used by each individual iron from its wattage and the input voltage (refer to Fig. 7-12 for conversion of resistance to wattage which can be used to establish uniformity in some models or ratings for irons not marked; note that the phase shift is not accounted for). Thus the number of irons per Variac or Powerstat can be established. The photograph in Fig. 7-10 shows how easily a single soldering iron can be regulated.

Figure 7-11 shows a typical family of TTV curves (tip temperature vs. voltage). A certain amount of deviation can be anticipated from these lines since manufacturers will guarantee only ±5 percent in the wattage rating of their irons; however, experience has shown that the more reliable soldering-iron manufacturers maintain closer tolerances, and the differences in temperature due to the change in wattage are usually negligible. It is therefore recommended that the optimum temperature be chosen from the TTV curves, considering also the possible changes in temperature during the selection. Not only the wattage of the soldering iron is

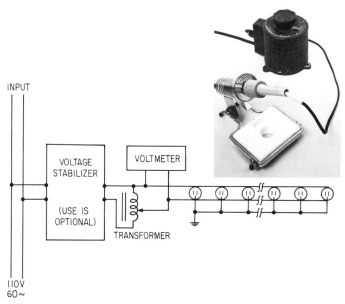

Fig. 7-10 Regulating the voltage for the production line. Note the number of receptacles. which is limited only by the equipment rating.

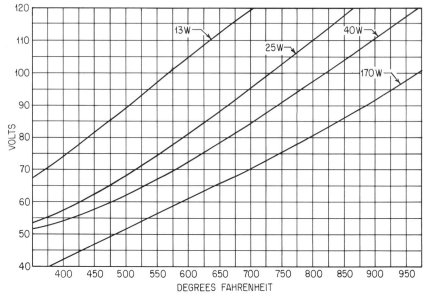

Fig. 7-11 A typical TTV curve for brand A soldering iron (maximum tip temperature vs. input voltage).

important but also the type, model, and catalog number. These data should therefore be included with the TTV curves in addition to the ratings of each of the irons tested. The test procedures are described in detail in Sec. 7-13, test 2.

7-12 Evaluating Industrial Soldering Irons   A short survey of the literature revealed an astounding lack of engineering information on industrial soldering irons, and it was necessary to devise a series of tests to evaluate these tools. The tests themselves were designed to obtain the maximum data with the simplest equipment and most straightforward procedure; thus an additional test could be run by anyone to obtain information on a particular iron. These tests are described in Sec. 7-13.

Based on the same considerations and tests, a practical approach to matching soldering irons to specific production tasks has also been developed. Figure 7-12 shows a soldering-iron analyzer commercially available. This unit, which can be obtained at various levels of sophistication, enables the user to run specific tests on the production floor. Thus we shall distinguish between laboratory tests and evaluation and industrially matching the iron to the task. It is recommended, however, that the reader read both sections since they complement each other rather than duplicating information.

7-13 Laboratory Tests and Evaluation   To help engineering and laboratory personnel select an industrial iron to be adapted for their specific in-house application a series of laboratory tests is vital to weed out the poorly designed or constructed irons, although it is impossible to evaluate element life without actual production runs. Thus, reproducibility of thermal conditions and a spectrum of temperature characteristics suitable for a group of soldering applications can be well predicted. These tests include

1. Checking iron wattage
2. Measuring tip temperature vs. voltage
3. Measuring tip temperature as a function of geometrical configuration
4. Measuring the efficiency factors of the iron and tip
5. Testing handle temperature
6. Checking current leakage

Remember that soldering-iron temperature depends closely on voltage, which is an independent factor that must be monitored. Variations of voltage at test locations and industrial buildings will affect the tip temperature of any soldering iron (unless the tool is temperature-controlled), its

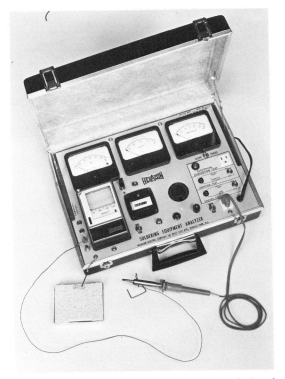

Fig. 7-12 Commercial soldering-iron analyzer designed to evaluate irons on the production line or in the laboratory. (*Hexacon Electric Co.*)

recovery rate, and other characteristics. It is recommended that the voltage be closely checked and monitored, and if variations exist, a voltage stabilizer should be considered.

Checking Iron Wattage. The wattage in a soldering iron is a measure of the number of calories developed in the element. Although the wattage by itself is not indicative of the thermal characteristics of the soldering iron, it is one of the major parameters used to classify them by models and manufacturers. As mentioned earlier, manufacturers state only that the wattage of their irons falls within 5 or 10 percent deviation. Experience has shown that some manufacturers stay well within these limits. It is well to check the wattage of new irons to establish the deviation in a lot and thus to avoid extreme changes in tip temperature in soldering tools. This could easily be made part of the incoming inspection on new irons and should not raise objections from iron manufacturers. Figure 7-13 gives the iron wattage as a function of the cold resistance. This curve is useful in checking uniformity of tools but does not account for the power factor. It is not really essential to find out what the absolute wattage of a particular iron is. It is more important to make sure that a group of irons used for a specific application all fall within close limits in their cold resistance. In this

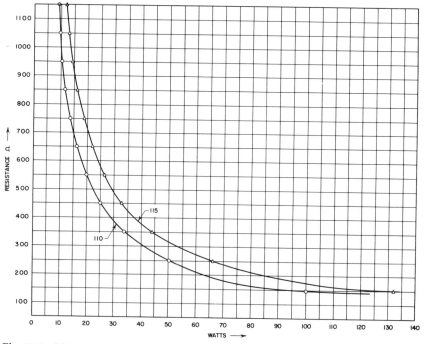

Fig. 7-13   Wattage vs. resistance in soldering irons.

test, a regular ohmmeter is used, and the resistance of the iron is measured from the two prongs of the plug. It is not necessary to take the iron apart for this type of measurement.

Measuring Tip Temperature vs. Voltage. The maximum tip temperature of a soldering iron should be constant for every specific wattage at a known voltage. This relationship is described on the TTV curves. Figure 7-11 shows a typical family of TTV curves for a particular manufacturer. Once a set of soldering irons is selected for use, such a family of TTV curves is all that is required for intelligent control of soldering temperatures in this operation.

For an individual soldering iron the purpose of this test is to determine the tip temperature of the particular model at a given voltage. If this information is known, an iron can become a versatile tool; i.e., an iron with a high wattage can be connected to a voltage regulator, the input current reduced, and connections made where a large amount of heat at low temperature is needed. In order to obtain this information, a V groove was filed in the tips to remove existing iron plating (or any other coating) and to position a thermocouple for brazing. The groove was located ¾ in from the end of the tip so that it would not interfere with the work area if these tips were later used for production-type tests.

Iron-constantan thermocouples were obtained 25 ± 0.5 in in length to accommodate all the tips used in the test. The leads of the thermocouple were spot-welded. The spot-welded bead was then seated in the V groove, and sufficient braze material was applied to fill the V. Caution was taken to control the amount of brazing alloy in order to avoid the possibility of heat sinks and have uniform samples (Fig. 7-14).

The tips were inserted in their respective irons. Six irons at a time were placed on a test stand. The irons were tested in pairs, three pairs to a run.

Voltages of 40 to 110 V were applied to the irons in 10-V increments. The irons were allowed to reach equilibrium before each increase. Equilibrium temperature was recorded for each iron for a given voltage. The input to the irons was regulated by a voltage stabilizer and voltage regulator.

A voltmeter was in the line at all times so that any fluctuation in the input could be detected and to help in making critical adjustments. The temperature readings were taken on a recorder at a speed of 180 in/h. A 12-point switch was incorporated into the circuit so that the readings of the irons could be made without having to disconnect the iron after a reading. An ice bath was used as a cold junction for the thermocouple.

A measurement of the change in temperature due to the aging of heat-transfer interfaces in the iron is well worth making. It was found that irons do change their heat characteristics after several hours, but it appears that once an iron has been thoroughly heated and used for several hours, its

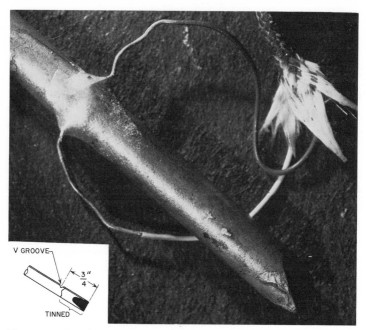

Fig. 7-14   Attaching the thermocouple to the tip.

heat characteristics stay constant provided the maintenance rules and regulations specified by the manufacturer are followed. In the better irons, the change in temperature recorded between "new" and "used" was no more than 20°F (11°C).

Measuring Tip Temperature as a Function of Geometrical Configuration. From the mathematical model it is obvious that tip geometry has a large effect on the tip temperature, and thus it is necessary to obtain additional information on any proposed configurations in production. The test itself is very simple and follows the guidelines in the section on measuring tip temperature vs. voltage.

Thermocouples are simply embedded in a variety of irons, but the thermocouple must be placed at the extreme tip of the iron, making it unsuitable for production testing.

The information obtained from these tests can be used in conjunction with the test below to establish the efficiency of every geometric configuration of the tip.

Measuring Efficiency Factors. The efficiency factors are a function of all the corrections discussed in the mathematical model. We shall concern ourselves with two factors, that for the iron and that for the tip. Both require one basic test for calculation purposes, as follows.

The iron to be tested must be in the "used" condition so that thermal characteristics will not change with the initial heat-up. This aging of the iron must be done with a tip in place even though the test itself does not call for the use of a tip. In the test a well-defined copper plug is made to fill the entire element cavity. The copper must not extend outside the barrel of the iron. A thermocouple is placed at the exposed part of the copper plug flush with the outside of the element. The iron is then allowed to idle, and the temperature of the copper dummy is measured.

By comparing the plug temperature with an actual tip temperature one can easily calculate the efficiency of the tip. The efficiency of the iron itself can also be interpreted from these measurements inasmuch as the theoretical heat output of an element can be related to the measured output.

In order to simplify the test, it is possible to cut down an actual soldering tip and place the thermocouple in the exposed cut. Differences in tip material, especially on the walls, will cause some variations, but they may be disregarded.

Handle-Temperature Tests. The irons were allowed to idle on 115 V for over 200 h before the temperatures of the handles were measured. Tempil sticks (temperature-sensitive crayons) were used for measuring. The temperature of the handle was recorded in a vertical and horizontal position. The purpose of the test was to determine the actual temperature and the effect of the idling position. The results verified that the horizontal position is cooler than the vertical position because of heat convection. The actual temperature difference was 50 to 100°F (10 to 38°C), depending on the iron. Another factor which maintains low handle temperatures is the use of cork sleeves over the handle. Hatchet-shaped irons and the use of heat baffles also help maintain cool handles.

Recognizing Electrically Induced Damage. Irons must be grounded for operator safety and to prevent static electric charges. Grounded units, however, may also be the source for electrically induced damage to the circuitry and components. Let us review the areas of concern.

STEADY-STATE LEAKAGE. The heating element of an iron is insulated with materials (mostly mica) that will permit minute current leakage in the microvolt range. Since this leakage increases with temperature and gets worse with the aging of the materials, irons should be periodically checked using an oscilloscope to assure that this parameter is kept below 5 mV.

TRANSIENT VOLTAGE SPIKES. Some irons are controlled by mechanical switches that produce an arc at operating voltages. Each arc is like a broadcast, transmitting voltage spikes through the metallic tip to the work. Voltage-sensitive devices have been shown to fail because of this factor.[1] A

---

[1]J. D. Keller, Latest Soldering Irons Need Updated MIL Specs, *Assem. Eng.,* vol. 17, no. 6, December 1974.

recording oscilloscope, sensitive in the nanosecond range and capable of measuring up to 200 V is required.[1] Spikes of 150 V peak to peak have been recorded on common industrial temperature-controlled irons.

MAGNETIC AND RADIO-FREQUENCY INTERFERENCE. Coils, transformers, and magnetic switching devices also can cause damage to sensitive work. Delicate instruments, meters, and the like may be temporarily or permanently distorted. Special shielding precautions thus are necessary when soldering near or producing such equipment.

In general, select the type of equipment which fills the thermal requirements of the task to be soldered. Then check it out against the dangers listed above. There is standard equipment available to meet all these results, but the cost increases with complexity.

7-14 Selecting an Iron for an Industrial Task   The number of variables involved in the determination of the iron size required for a job are numerous. After reviewing the mathematical model, however, one can select an iron by the following guidelines:

1. Determine the type of iron according to the cost (temperature-controlled, temperature-limited, constant-load irons, etc.).

2. Pick a tip geometry according to the task to be soldered (consult Sec. 7-15 for specific steps).

3. Select an iron from the preferred vendor's catalog to provide desired tip temperature. Table 7-5 may simplify this selection.

4. Obtain a soldering iron with a slightly higher wattage for production testing purposes. Use an analyzer as described below to zero in on the ideal iron parameters (wattage, tip length, etc.).

In practice, the skilled person can pick the right iron for the job from experience alone; however, to make this task easier, Table 7-5 should be consulted. It lists the six general categories into which irons can be divided. It is suggested that a small iron with low wattage be tried first, and only if it is found inadequate should a larger iron be used. If the smallest iron still has too much heat, use a reduced voltage, as described in the TTV curve.

[1] L. Lapanne, Hand Soldering: A Look Ahead, *Quality*, vol. 17, no. 1, p. 20, January 1978.

TABLE 7-5   Iron Classification

No.	General classification	Wattage
1	Miniature	To 25
2	Small electrical	25–35
3	Medium electrical	35–45
4	Large electrical	45–60
5	Light duty	60–100
6	Heavy duty	100 up

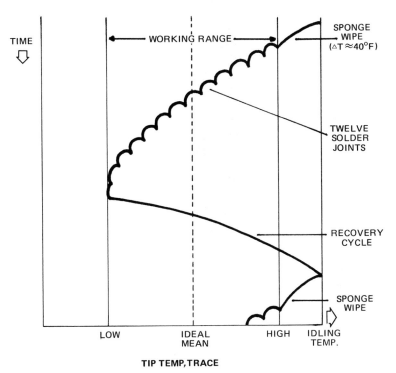

TIP TEMP, TRACE

Fig. 7-15   Typical analyzer trace during production tests.

The use of an analyzer or similar instrumentation of one's own design is intended to help establish production parameters. This can most easily be explained from the type of curve obtained on an analyzer recorder. Remember that the analyzer helps establish tip temperature as a function of time under actual production conditions.

Figure 7-15 shows a typical trace during production testing of a soldering iron. It shows the temperature drop due to the wet sponge wiping (approximately 40°F or 22°C) and the tip-temperature variations as the operator moves from joint to joint and from task to task. This temperature range from top to bottom is considered to be the thermal working zone of the iron and should be maintained within the limits indicated in Table 7-6. This curve also enables the engineer to establish the recovery time between two consecutive tasks and to ensure that the iron can return to its original idling temperature so that successive operations follow the same pattern.

When using the analyzer in Fig. 7-12 one merely plugs the soldering iron into the outlet provided while at the same time connecting the tip thermocouple. The unit is then switched on, and with the help of the voltage regulator one dials in the wattage desired for the test. The iron is then allowed to idle, and the idling temperature is read off on the appropriate meter. Once hot, the iron is handed to an operator under standard working conditions, and the strip-chart recorder is turned on. The operator is instructed to solder for a standard period of time. If after approximately 10 soldering cycles (consisting of at least 10 tasks), it appears that the iron is either too hot or too cold, an adjustment in wattage will bring the iron into the desired range. Note that for special refinement, a time counter is mounted on this particular unit to enable the observer to note any increase or decrease in soldering speed, which is an indication of soldering ease.

When the best working conditions for the particular iron and tip in question have been reached with the help of the analyzer, the wattage is carefully noted. Soldering irons with that specific wattage should then be ordered for that specific application. If only small variations in temperature are needed, however, it may be possible to effect a change in tip temperature without purchasing special elements of unique wattage which are not shelf items. Remember, that shortening the tip and/or decreasing its diameter will increase tip temperature while lengthening the tip or using a larger diameter will lower tip temperature. Neither of these adjustments affects the heat output of the element.

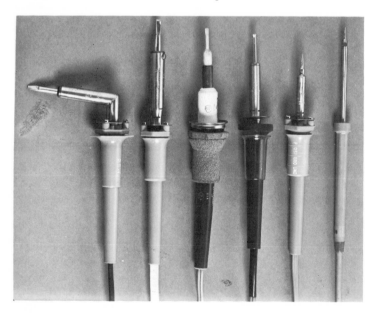

Fig. 7-16   Relative size of irons ranging from miniature to medium-sized constant-load units. Despite variation in tip size and weight, overall length is fairly uniform.

The circuitry shown on the bottom right-hand side of the analyzer in Fig. 7-12 is used to establish the efficiency of grounding and element insulation. Such circuitry is often provided in a separate box for iron maintenance purposes.

The reader is reminded that the above test is basically intended for constant-load irons without transformers or temperature-regulating devices. While the accuracy of these tools can also be verified with a solder-iron analyzer, the adjustment of tip temperature is not easily effected by such simple means as a variation in voltage.

One iron property which is tied in with the third axiom does not yield to laboratory testing; the shape of the iron, its balance with tip and cord, and the fit to the hand are entirely subjective. The feel of the iron, however, becomes apparent when it is handled, and the right technical model must pass this added evaluation. The irons depicted in Fig. 7-16 are typical for the industry. Note that the handle length is quite uniform but weight and balance are not.

7-15 Selecting a Soldering Tip   The selection of an iron divorced from the tip is wrong. A matched tip to the job as well as the iron is essential. A variety of tips is shown in Fig. 7-5. The following rules are presented to serve as a guide:

1. Select the configuration which will give maximum contact of the tinned areas on both tip and work. This will greatly help the heat transfer.
2. Consider the best approach for the work when selecting your configuration. Straight as well as hatchet-type irons are available.
3. Keep the taper of the tip as short as possible for good heat transfer to tip. Long tapers have small cross-sectional areas for heat flow.
4. Select the shortest reach to minimize wobble and shorten heat pass from the element.
5. Decide on the proper heat supply and remember that:
    a. Diameter of the shank determines the heat transfer from the elements; the larger the better.
    b. Diameter of the tip determines the heat transfer to the work. Turned-down tips have higher temperature but smaller caloric content. Larger tips have lower temperature but larger heat reserves.
6. Avoid screw-on or press-fit tips; they have limited heat transfer and poor reproducibility.
7. Combinations of tip and element in the same unit are usually more expensive because they have to be discarded when either fails.

7-16 Soldering Aids in Industry   Many forms, shapes, and sizes of soldering aids used in conjunction with irons are now commercially available. We shall limit our discussion to the most widely used aids in the electronic industry.

Heat Sinks and Barriers (Fig. 7-17). As the name implies, a heat sink is used to keep heat from reaching a temperature-sensitive device. Copper is generally used to make up the mass of the heat-sink material because of its large heat content and high thermal conductivity. Aluminum, having good conductivity but a small heat content, is also used, especially if there is a possibility of damage due to the physical weight of the clamp and because of the reduced production cost. Many types, shapes, and sizes are available, the usual design being a spring clip (similar to an alligator clip) that will easily fasten on the part. For further details see Secs. 4-5 and 6-36 as well as Figs. 4-1 and 5-42.

Antiwicking tools (usually in the form of tweezers) are heat sinks with an additional purpose. They are used to prevent the liquid solder from climbing up the capillary spaces formed in a stranded wire, which would result in charring or melting of the insulation and stiffening of the wire, which defeats the purpose of stranded wire and weakens the joint. They are usually covered with a nonwettable metallic coating such as chrome. The stranded wire is completely encased in the metal sheath of the tool, and the opening in the metal jaws varies according to wire size.

Heat insulators are used to prevent the surrounding area from robbing heat from the joint. These barriers have the opposite role of the heat sinks. Their construction is much the same, with an insulating material

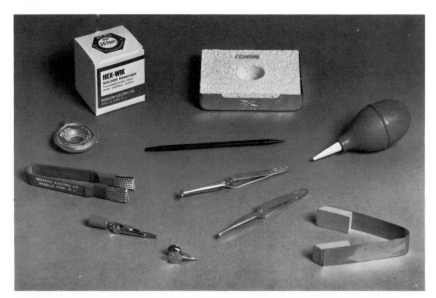

Fig. 7-17   Soldering aids. The sponge has a hole in the center to contain excess solder and dross. The carton and spool contain fluxed wick for solder removal. (*Hexacon Electric Co.*)

replacing the metal mass of the sink. This results in faster heating of the work area and shorter soldering times, which are very desirable.

Nonmetallic and Metallic Probes. Probes and forks are tools used to straighten and wrap wires or hold down parts to be soldered. Nonmetallic probes are used if there is danger of physical damage or in order to avoid a heat sink (low conductivity). They may be essential to avoid nicking or similar deformation of the wires, which considerably reduces the strength of the joint. The size of a probe is about the same as a pencil, but a pencil should not be used because of harmful graphite. One end of the probe is usually shaped to a dull point and the other in a fork or similar design.

Brushes. Perhaps the most widely accepted means of applying flux is by brush. Brushes range in size from 2 to 3 in down to just a few bristles. It has been found that in a miniature assembly the stiffer and usually cheaper "acid brush" works much better than the soft sable.

The second role of brushes is to clean foreign particles from the joint area before fluxing. Manufacturers often make a combination brush and probe. These tools should be restricted to use before the application of solder. If they must be used later, care should be taken to avoid damage to the soldered fillets.

Cellulose Sponges. Sponges to wipe and maintain a clean tip are finding more and more use in the industry. If supplied, they will keep the operator from using another material that might hurt the tip more than help it (charring of rags, wear due to abrasive materials, etc.). The sponge should be kept soaked with water at all times; if pressed on top they should show signs of water; otherwise they are not wet enough.

In miniature soldering irons it may be necessary to wait a few seconds after cleaning the iron with a sponge before the connection is made, this waiting time being dependent on the recovery of the iron from the quench it received from the wet sponge.

The tip should not be overwiped, and under no circumstances should it be wiped dry. After the tip is cleaned, it is a good practice to apply fresh solder immediately if it will not be used.

Holder or Cage. The holder has a great effect on the idling temperature of an iron, the handle temperature, and similar characteristics (Fig. 7-18). While strict safety regulations exist on the quality of holders (one must be able to touch a stand from the outside while a hot iron is inside without getting hurt), there are many technical aspects which are also important:

1. The angle with which the cage holds the iron should be approximately 45° so that convection will not cause the handle to get too hot for comfort. This angle is not always convenient for iron return at the completion of a task and there is a tendency to hold irons at a perpendicular position for that reason. This should be discouraged because the handle gets too hot and deteriorates.

2. The mode in which the cage holds the iron must be adjusted so

Fig. 7-18   Iron holders designed for operator safety. The two on top are for bench mounting (over or under). The two below are table-top models. *(Hexacon Electric Co.)*

that the tip does not touch the cage. Tip contact causes heavy heat sinking, and affects tip temperaure. Most cages are arranged so that the tip is entirely suspended in free air.

3. The cage must be designed so that the tip slides in and out without any mechanical jarring or knocking. This is vital in order to lengthen element life.

4. The cage must be constructed so that solder drippings can easily be cleaned out, since they are unavoidable.

Fig. 7-19   Temperature-control stand. *(American Electric Heater Co.)*

5. The cage must be positioned so that solder dripping will not fall on the operator, the work, or other sensitive areas.

Temperature Stand (Fig. 7-19). Temperature stands provide an automatic means of controlling the soldering-iron temperature. The stand has a temperature-sensitive device, usually a bimetal, that will turn the power to the iron on and off during idling time as it is needed.

7-17 The Maintenance and Care of a Soldering Iron   Neglected soldering irons not only impede the operator but enhance conditions for unreliable connections. The following suggestions will help to prolong their useful life and get the maximum benefit from the tool.

1. Under no circumstances should irons be allowed to heat up without a tip inserted in the barrel. This allows the element to overheat and causes short life together with excessive oxidation inside the barrel.

2. When a new iron is heated for the first time, liquid activated-rosin flux should be added to the tip or the tip should be tinned in a solder pot. If a solder pot is not available, cored solder may be used to tin the iron. The solder should be applied when the tip is just hot enough to melt the alloy. When the iron is cooled, the tip should be cleaned and well wet with plenty of solder. This preparation will permit simple reheating.

3. When a used iron is reheated, the tip should be treated the same as a new tip unless it was cooled as described above. If, because of neglect, a used tip fails to tin, acid flux and/or emery paper will have to be used to remove the iron oxides, and the above tinning procedures can then be followed.

4. During long idling times, solder should occasionally be added to the tip to prevent dewetting, which will occur at the high tip temperatures when the solder oxidizes in the air and no longer protects the iron. Once the iron oxidizes, activated-rosin flux cannot retin it.

5. No matter what manufacturers' claims may be, the iron should never be given a physical shock, even to remove burned oxides. To clean the tip, it should be gently wiped with a cellulose sponge. Care must be taken not to overwipe the tip, because oxidation of the iron will result if all the solder is removed.

6. When cold, the tip should be removed from the barrel and the bore cleaned by gentle tapping. Depending on the type of tip and coating, this cleaning should be repeated either once a day or once a week. Tips with antifreezing features may be left in the iron for relatively longer periods.

Various elements are customarily manufactured for the same handle and casing. Since only these elements are interchangeable and the casing with the marking stays the same, a quick check of the wattage by a resistance measurement is recommended in a new iron. For convenience, the resistance across the male plug is measured and converted into wattage (refer to Fig. 7-13). If an iron becomes damaged, or if an element burns out, the iron need not be discarded and only the defective parts need be replaced. To replace a damaged part of an iron the manufacturer's instructions should be consulted.

7-18 Hand Soldering   For the purpose of this discussion let us define what good hand soldering consists of. *Given the right materials (core solder), solderable parts, and well-designed joints, an operator should be in a position to complete the soldering operation in the critical time within 1 ± ¹/₄ s.* Remember that the critical time is defined as the time in which the solder is molten (see Sec. 7-7). As indicated in Sec. 7-1, it is up to management to set up manufacturing logistics which will enable the operator to notify the supervisor of any difficulties rather than try to hide mistakes.

To achieve this we have set up the *Ten Commandments* of hand soldering:

1. When touching the work with the iron, try to contact the part of the greater mass with the flat (larger) part of the tip while trying to touch the part of the smaller mass with the side or edge (smaller part).

2. Immediately on tool-to-work contact, create a fast heat bridge by placing the core solder in the gap. This will melt the flux and then some solder, which will permit localized wetting for rapid heat flow.

3. Once the core solder starts melting, draw it around the joint in the direction of flow.

4. Remove solder first after sufficient metal has been added to the joint.

5. Next, remove the iron, but only after the solder has reached the desired contour as dictated by surface tension.

6. Allow the solder to freeze without vibration to avoid disturbed joints.

7. Always place the heat (soldering-iron tip) on the side opposite insulation and heat-sensitive components.

8. Do not melt the solder on top of the soldering iron and carry it to the work (puddling).

9. Do not place the soldering iron on top of core solder or interpose core solder between the work and tip; flux will spit and not run onto the connection.

10. Do not pull or push on solder joint for inspection.

Specific assemblies may require additional hand-soldering instructions. Any method is acceptable provided it enables the operator to solder the assemblies within the permissible time of 1 ± ¹/₄ s. Experience has shown that the tip temperatures given on Table 7-6 yield the best all-around results without damaging the work. Note that the table is specific for eutectic and near-eutectic alloys and that the use of a fully temperature-controlled soldering iron permits a much lower soldering tip temperature.

TABLE 7-6   Tip-Temperature Ranges For 63/37 and 60/40 Tin-Lead and 63/36/2 Tin-Lead-Silver Solders

Type of work	Temperature	
	°F	°C
Thick and thin film microcircuits	540 ± 20	280 ± 11
Flexible printed circuits	575 ± 25	300 ± 14
Fine copper wire	735 ± 15	390 ± 8
Multilayer printed circuits*	820 ± 20	435 ± 11
Standard printed circuits*	820 ± 40	435 ± 22
Terminals and lugs*	810 ± 50	430 ± 28

*Subtract 100°F (55°C) for irons with fully controlled tip temperatures.

The sequence for hand-soldering operation is as follows. In the morning the operator should inspect the cold iron for external damage, rattle, and insulation of the cord. This is the time when the tip is removed (daily), the barrel is tipped over to remove loose debris, and any loose material is wiped off the tip. The tip is then reinserted all the way to the bottom and fastened in place. The operator then inspects the tip for wetting and adds more solder while it heats up, either by wrapping it around the cold tip or adding it to the hot tip as soon as the flux and solder will flow.

During the working day, the well-wetted tip is kept idling in the cage. When a task is to be soldered, the tip is removed from the cage and sponge-wiped, which provides steam cleaning of the tip. The tip is then used on the work consecutively (wiping is needed only if excessive flux or dirt accumulates on the tip during the task soldering). When the task is completed, the iron is returned to the stand without wiping. If the tip is to stay in the stand for long periods of time, more solder is added. If the tip is to be used within the next minute or two, it is returned without additional treatment. This process is repeated all day long until the end of the workday, when a generous amount of solder is melted on the tip and it is left out to freeze for a repeat of the operation on the following workday. Under no circumstances is a tip to be wiped dry, since this will start a dewetting process. A dewetted tip must be retinned with the use of strong fluxes or an abrasive material, which should be done only in a tool crib or under supervised conditions.

In addition to keeping the work station clean, the operator should rinse out the sponge once or twice a day under running water. The sponge is kept moist enough so that when it is depressed a pool of water will form. It is better to use a sponge with a hole in the center and to start the wiping motion from this hole. This way, any excess solder drippings will accumulate in the hole and not on the surface of the sponge; then there is little danger that they will be picked up during subsequent soldering operations.

A word of caution on personal hygiene and safety is needed here. There is no danger of lead inhalation during hand-soldering operations, but it is possible to ingest the poison. To explain this, one can run a simple experiment using a white paper towel or napkin and a piece of core solder. Draw the core solder through the white paper exerting mild pressure and note the black strip or line it leaves behind. No amount of rubbing will remove it all because it forms constantly. This black material is a poisonous lead oxide. It rubs off, not only on paper, but also on the hand, workbench, and other surfaces that come in contact with solder. It is therefore necessary to wash carefully before eating, drinking, smoking or chewing. This is why no eating, smoking, or drinking is permitted at the work station, and it is anticipated that future government specifications will reflect these regulations.[1]

7-19 Training and Workmanship Standards   Training for manual soldering has to be custom-tailored to the workmanship standards of the company, the product, and the industry (Sec. 5-2). Here are suggestions for organizing a workable training program:

1. Obtain one each of every soldered joint in your product. Then make samples of these joints to show the desired configuration, one that is marginally acceptable, and one that is unacceptable.

2. Prepare visual aids, such as slides, photographs, or drawings of the above samples. Provide written descriptions of each sample, orienting them to the type of visual aid.

3. Tie the visual aids and written descriptions into a workbench training program.

Training a new employee can be expected to take about 40 h. Semiannual retraining, which is desirable, should take about 10 h. Inspectors should receive identical training, plus an additional 20 h.

A typical training program for operators and inspectors should include these elements:

1. Give a statement endorsed by management that describes the company's cost and quality policy, with emphasis on the need for quality.

2. Describe and explain soldering processes. Teach the proper use of the soldering tools and materials required for your operations.

3. Describe your own workmanship standards and the hardware involved in your operations.

4. Tie the above elements into a workbench training program.

Formal training certificates are issued by many companies at the conclusion of solder training programs. Personnel departments are effective

[1]Occupational Exposure to Lead: Proposed Ruling, Department of Labor, Occupational Safety and Health Administration (29CFR pt. 1910) docket H004, *Fed. Regis.*, vol. 40, no. 193, Oct. 3, 1975.

sponsors for solder training and certification programs. The formality of this arrangement establishes management's interest. Furthermore, because soldering training programs have been found to improve employee morale and job satisfaction, responsibility for such programs fits into the basic objectives of personnel departments.

EIGHT

Inspection and Quality of Solder Joints

8-1 The Role of Quality Control in Soldering

Quality control and inspection are an integral part of making reliable solder joints. Their primary function is the continuous monitoring of all the parameters which contribute to the reliability of the solder connection. The responsibility of the inspection runs through the entire process from incoming inspection of materials and equipment through in-process control of the process parameters through final inspection of the product before it leaves the plant. Quality control, however, does not inspect the quality into the product; this must be built into it through careful planning and processing. Quality control merely monitors it to make sure that each one of the steps leading to the reliability of the end product is carried out properly by the various functions within an organization.

Quality-control standards, or workmanship standards, must be carefully compiled for every type of assembly (Sec. 5-2). There are no universal standards possible within such a diversified industry as this. Setting up quality standards must be based on knowledge, data, and facts in order to be useful. Copying standards from other organizations is hazardous since it may lead to poor reliability and inflated costs. Chapter 4 is useful not only when a new assembly is contemplated but also when existing assemblies are scrutinized and workmanship standards are set up. Specific instructions can be found in Chap. 4 for setting up the quality criteria.

Final inspection is not the entire function of quality control. In-process inspection is just as vital, and Chaps. 5 to 7 give the basic parameters on the processes so they can be monitored continuously. The present chapter concerns itself with methods of monitoring final quality and some of the peripheral tests needed. We shall also spend some time discussing incoming inspection, solderability testing, cleanliness monitoring, etc.

QUALITY-CONTROL SCHEME

8-2 Three Steps to Ensure Reliability   In previous chapters we have laid the foundation for higher reliability in soldering. Remember that quality must be built into the product and cannot be inspected into it later. There are three basic steps for achieving reliability, and each step requires some degree of inspection and quality control by itself. Then only if all three steps are carefully monitored will final quality control become meaningful.

8-3 Step 1: The Material Selection and Incoming-Material Inspection   One of the most important considerations is matching the flux to the base metal. If we are using an inadequate flux, the metal will not wet properly, and no matter how much time we spend trying to make a single connection, we shall never be able to get a reliable joint if we have not selected the right material to begin with. The flux is one part, and the correct solder alloy is another. Not all solder alloys have the same properties, and it is important that we select the right alloy to match the application. Material selection is covered in Chaps. 2 to 4.

As for the control of the material selected, it is easy to determine whether the flux and the base metal are properly matched by a simple solderability test. The procedure for this test will be discussed later in this chapter.

Another problem is the inspection of incoming materials. Whereas wet analysis is used to establish the major constituents of a solder alloy and spectrographic analysis suffices to reveal information about the levels of purity, the check of soldering fluxes is more difficult. Since fluxes are generally of a proprietary nature, the full formula cannot be obtained from the manufacturer. However, in essence, it is not necessary to have the full formula from the manufacturer as long as he supplies such major items as flux density, so that the solid content of the flux, if it is a rosin, can be established, and minimum chemical requirements, so that it is easy to determine the presence or absence of certain agents that are either desirable or undesirable in the formulation.

The Principles of Incoming-Material Inspection[1]

It is always the right of the purchaser to specify the quality of the product obtained from the supplier. In soldered assemblies no purchasing specification is complete if it does not have a solderability clause. Although many people would like to include solderability in the "good workmanship" of the specification, it really requires close surveillance. Unfortunately, the users of components and printed-circuit boards will specify in great detail the dimensional, electrical, and other parameters of the circuit without considering the necessity of having an assembly which will adequately solder with the fluxes intended for the final assembly operation.

Many industry and government specifications are available for use with the printed-circuit industry. They are discussed in detail below. Suffice it to say that mutual agreement on the quality and solderability of all surfaces can be reached, and then the role of incoming inspection becomes simple in that respect.

All too often the printed-circuit-board manufacturer will be allowed to use various materials on a printed-circuit board which are basically solderable when newly prepared. Upon aging, however, they lose their solderability because of improper material selection. Here again the specification must be clear in order to avoid this unhappy circumstance. It is therefore suggested that for all printed-circuit boards as well as components, proper solderability specifications be incorporated into the overall purchasing specification. If none exists, it might be wise to write a solderability specification for use in conjunction with all such specifications.

Also requiring surveillance under incoming materials are the fluxes, the solders, and soldering chemicals. Although these are normally bought under proprietary formulations which make a complete material inspection impractical, it is possible to obtain from any reputable manufacturer enough technical and chemical parameters to ensure quality and uniformity of the materials. A general outline of this is given in Table 8-1 so that the readers can easily decide on which checks are adequate for their particular applications and include them in the quality-control manual. These tests are described in some detail below. Special tests on specific products should always be set up in conjunction with the supplier to make sure that the specification is not unreasonable and would cause a slowdown in deliveries and/or increase in price.

Density Check. The density is a unique property of the material which gives a good indication of the amount of solids in the formula in conjunction with the type and quality of the solvents used. The density of the material is the ratio of the weight of a predetermined volume of this material compared with that of water. Thus the density is an indication of the weight of the material per volume.

[1]Adapted from H. H. Manko, Chap. 14 in C. F. Coombs, Jr. (ed.), "Printed Circuits Handbook," copyright 1967 by McGraw-Hill, Inc. Used by permission of McGraw-Hill Book Company.

TABLE 8-1 Recommended Tests for Incoming Materials*

Material	Density	Water-extract resistivity or ion content	Acid titration	Surface tension	Viscosity	pH	Wet Sn analysis	Spectrographic analysis General	Spectrographic analysis Special	Special
Fluxes:										
w/w rosin	x	x	...	x†	x	...	...	...	...	Acid number of rosin
RMA rosin	x	x	...	x†	x	...	...	...	...	Acid number of rosin
RA rosin	x	x	x	x†	x	...	...	...	...	Acid number of rosin
Water-base organic acid	x	x	x	...	...	x	...			
Water-soluble organic acid	x	x	x	x†	x	x	...			
Inorganic acid	x	x	x	...	...	x	...			
Chemicals:										
Etches	x	x	x	...	...	x	...	...	...	Effect on polished surfaces
Protective coatings	x	...	...	x	...	...	...			Sulfide aging
Solder resists	x	...	...	x	x	...	...			
Strip coatings	x	...	...	x	x	...	...			
Conformal coatings	x	...	...	x	x	...	...	...		Dielectric strength
Cleaners	x	x	...	x	...	x	...	...	...	Capacity as polar solvent
Oil	x	...	...	x	x	...	...			
Sn-Pb solders:										
ASTM-A	x	...	...	...	...	...	x	x	...	
QQS-571	x	...	...	...	...	...	x	x	...	Sb

*Adapted from H. H. Manko, chap. 14 in C. F. Coombs (ed.), "Printed Circuits Handbook," copyright 1967 by McGraw-Hill Inc. Used by permission of McGraw-Hill Book Company.
†For foam fluxes mostly.

The density of metal can be used for tin-lead ratio if no third alloying constituent is present. Figure 8-1 shows the change of density of tin-lead alloys as a function of lead content. This is normally checked by weighing the solder first in air and then in water. The density is computed by dividing the weight in air by the loss of weight in water. Balances for specially cast slugs are available which are directly calibrated in tin-lead ratio. In most cases, density for solutions is an excellent and simple method of determining the uniformity of incoming products. With soldering alloys density is also an important tool, although it requires some additional work to determine whether the material is of ASTM grade A quality (without antimony) or of QQS-571 quality, which contains antimony to prevent tin pest.

The measurement of liquid density is usually accomplished through the use of a hydrometer, a calibrated float which is submerged under the liquid level according to the density of the material. The stem, which protrudes out of the liquid, can be used in conjunction with a scale to read the density. Density readings require a specific temperature, as density in most solutions is a direct function of the temperature; otherwise erroneous results might be obtained. The density has to be measured in a container (usually a cylinder) wide enough not to cause friction with the hydrometer or sticking to the side walls.

Figure 8-2 gives the density of water-white rosin in pure isopropanol. This is the type of flux required for the inspection of solderability under the various tests described below. Note that the density is expressed in grams per cubic centimeter at a temperature of 77°F, a convention throughout the soldering industry.

Water-Extract Resistivity. This test is a measure of the ionizable materials present in a rosin-base flux. The test utilizes the conductivity of distilled or demineralized water before and after a predetermined amount of flux has been added and the ionizable materials have been extracted from this flux through a brief boiling period. The boiling serves the double purpose of helping extract all the ionizable materials and removing any absorbed carbon dioxide from the solution, which would lower the insulation properties of the water, thus lowering the water-extract resistivity.

Roughly speaking, rosin-base fluxes with no activators have an extremely high extract resistivity value of 200 kΩ·cm and above whereas the mildly

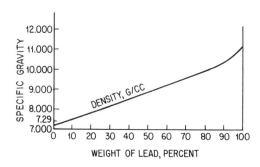

Fig. 8-1 Density of tin-lead alloys.

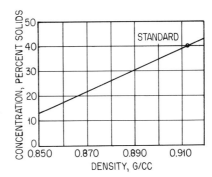

Fig. 8-2 Density of water-white gum rosin flux in isopropanol expressed in grams per cubic centimeter at 77°F.

activated rosin fluxes have a water-extract resistivity of 100 k$\Omega\cdot$cm and above and the activated-rosin fluxes fall in a range anywhere from 20 to 50 k$\Omega\cdot$cm. For full details of this measuring method, the reader is referred to MIL-F-14256C specification, where this test is described in detail. It has been the author's experience that this is an extremely valuable tool in the plant quality control over the production of fluxes since it is sensitive to the ratio of activator to rosin and alcohol in the formulations. The test requires great skill and clean work area and glassware and should not be attempted without adequate preparation.

This test is useful for rosin formulas only, since in many industrial applications, the flux residues are left on the surface without being removed and thus the electrical properties become important. However, this test has not been found suitable for any of the other types of fluxes since their water-extract-resistivity range would be extremely low and on the opposite side of the scale from the rosin fluxes, which makes it relatively meaningless.

The ion content of these chemicals can also be used as measured by the methods described under cleanliness monitoring (Secs. 8-22 and 8-23).

Acid Determinations. These determinations must be suited to the type of material checked and follow normal analytical procedures. It is customary to check activated-rosin fluxes, for instance, calculating the results as if all the ions were chlorides. This gives an easy quality-control handle over the amount of activator added in each particular lot of fluxes and makes it possible to inspect incoming materials for their activity.

For organic-acid and inorganic-acid fluxes as well as etches, the technical information supplied by the manufacturer of these materials should be adequate to determine the type of chemical titration required to establish the quality and uniformity of the formula.

Surface Tension. This is usually one of the important characteristics for foam fluxes and/or organic coatings which are expected to give certain physical characteristics to the total assembly. The surface tension itself is usually measured with a capillary tube and is also a temperature-sensitive test which requires extreme care to obtain reliable information. This type of test is not usually performed unless trouble occurs in the performance of the material.

Viscosity Measurements. The viscosity of a flux can become important if the flux is used in some mechanized method of application which greatly depends on the fluidity of the material. Viscosity measurements are temperature-dependent and therefore should be carried out at the same temperatures at all times. Viscosity information, similar to surface-tension information, is normally checked only if trouble develops and not as part of the incoming inspection.

pH Readings. The pH is a measure of the acidity, alkalinity, or neutrality of the solution and is generally used for a quick check of organic and inorganic fluxes and etches. This is a good incoming-inspection test.

In some cases where chlorinated solvents are used for cleaners, the pH check is also important. The chlorinated solvent, by itself, will exhibit no pH reading because it is a nonpolar solution. If the chlorinated solvent is inadequately stabilized, it may break down to form some acidic materials which might be harmful to the assembly. It is possible to measure the pH of the material by mixing the cleaner in a 1:1 ratio with distilled or demineralized water. The two liquids will not mix, and after shaking well, they will separate. It is now possible to check the pH of the aqueous part of the mixture, which will contain all the ionizable material. This will quickly reveal whether the material is properly stabilized and whether breakdown of the chlorinated compounds has occurred to give hydrochloric acid.

Analyzing Solder Alloys. Soldering alloys are usually checked by wet analytical methods for their major constituents. In the case of tin-leads, which are the commonest materials used for printed circuits, a wet-tin determination is used to establish the amount of tin in the formula. The density of the solder also establishes the tin-lead ratio but not quite so accurately as wet chemistry.

Spectrographic-analysis methods are used to establish impurities as well as the presence of antimony in the formula. The presence of antimony is required in all solders meeting QQS-571 specification.

Special Procedures. Various items have unique tests specific for them, e.g., the *rosin acid number check,* which determines the quality of the rosin used for rosin flux; the effect on a polished surface of the etches, to determine the severity of the attack on the base metal and the cleaning power, humidity, aging characteristics for protective coatings; dielectric strength; heat resistance; electrical properties, etc. These usually become apparent when a specific property of the material is sought after for the assembly at hand.

8-4 Step 2: The Geometrical Design and Workmanship Standards The second major step in the assurance of the quality in a solder joint is its proper design. The joint must have adequate current-carrying capacity and strength. Without these properties, there is always a danger of overheating of the joint or mechanical failure under stress. This is discussed in detail in Chap. 4. The other geometrical consideration, even more important for visual inspection, is the fact that a joint must be inspectable. Configurations which do not allow for visual inspection should be rejected at the design stage. It is also at this stage that we can set up our dimensional requirements so that the inspector has fixed guides to go by (Secs. 5-2 and 7-9).

8-5 Step 3: The Process   In Chaps. 5 and 6 we carefully reviewed the equipment and soldering techniques available on the market. Adequate soldering control must be established over the procedures to eliminate the possibilities of variations in such parameters as soldering temperatures, times, density of flux, and cleaning efficiencies. Here again, the common practices used throughout the process industry are required to maintain close surveillance over the equipment.

Once we have made sure that the proper steps have been taken for the control of the three basic steps for good inspection, we can take a closer look at the final inspection itself. Here we have the finished product. It is now too late to make any changes in materials, design, or process. All we can do is make sure that the assembly fulfills the requirements of the quality we have designed into it. We do this three ways: (1) we check the quality of the wetting, (2) we check the physical dimensions, and (3) we check the cleanliness of the assembly.

Many people complain that solder connections can be only visually inspected, but this is an advantage not a shortcoming. Soldering and brazing are the only joining methods responsive to visual examination, and the advantage far outweighs the necessity for inspector training. To begin with, visual examination permits 100 percent inspection, while other methods require sampling. When it comes to joint reliability, it is better to deal with human frailty than with projection from a small sample to an entire run. Speed is another important factor. When inspection methods involve cumbersome techniques and complex equipment, there is a strong tendency to bypass the sampling schedules set up by a statistical expert and to cut corners because there just is not enough time. With visual inspection, examination is as quick as a glance. And finally, although it is true that visual examination depends to a great extent on inspector training, the same is certainly true, and more so, when it comes to the use of destructive methods. Not only must inspectors be trained in the use of the equipment and running the test but they must also be trained in evaluating the results.

Visual inspection should be compared not only with destructive testing but also with some nondestructive testing methods such as electrical methods. The big drawback with electrical methods is that they are time-dependent. They rely on aging before a defect develops. Fresh clean surfaces wrapped around each other will give a good electrical connection as long as there are no oxide layers between the clean surfaces. With aging, however, with humidity, even at room temperature, the picture changes entirely. Production speeds do not allow us sufficient time to age our components before checking them. Therefore, electrical tests for 100 percent testing of soldered connections are not feasible.

Various mechanical methods have been suggested. The big problem here is that mechanical methods cause undue stresses in the assembly. It is poor practice to have inspectors poking at electrical connections to see whether they are strong. If they were strong to begin with, there is a good chance that they will be damaged after such stressing. The undesirable stresses introduced by mechanical joint inspection will tend to cause a lack of reliability.

And finally, there are such methods as x-rays, radiation, and ultrasonics. These methods are excellent if they have a very uniform part or joint to evaluate, but as soon as there are variations in the configuration, even the experts recommend other testing methods.

Let us repeat, therefore, that the soldering method permits 100 percent visual inspection, and 100 percent visual inspection is the best guarantee to 100 percent reliability; but we should not forget that quality cannot be inspected into a connection. Inspection methods separate the good from the bad but do not change the bad to good. Quality and reliability must be built into the joint. The formula for reliability, therefore, is the proper material selection and continuous inspection of incoming materials coupled with proper design and the rejection of any configurations that do not live up to the concepts of the design. When added to intelligent process control this formula makes it possible to secure 100 percent reliability through 100 percent inspection. Reliability in soldering does not just happen; it must be planned for, supervised, and ensured through continuous and thorough final inspection. With these safeguards, soldering is the most reliable, time-proved, and versatile electrical joining method offering all the benefits of speed, dependability, and economy.

FINAL INSPECTION

8-6 The Benefits of Visual Examination   In final inspection, we deal with the finished product. It is really our assignment to make sure that all the concepts of good soldering have been adhered to by the various people responsible for the finished product. This includes workmanship to ensure proper wetting, processing parameters to avoid damage to the assembly, and cleanliness checks to ensure freedom of corrosion. Thus we make sure that the assembly fulfills the requirements of the quality we have designed into it.

Under ideal conditions, a well-designed soldered assembly where materials were selected properly, ground rules of design were followed, and proper processing techniques were selected should have a prototype assembly. The sample would then be tested in a metallurgical laboratory destructively, using some of the techniques described later in this chapter. These would include such checks as strength (mechanically), current-carrying capacity (electrically), and cross sectioning to check the soundness of wetting and fillet formation. Once the prototype assemblies are found adequate by such an exhaustive destructive mechanical and electrical test, the visual inspection of the production lots will have more meaning. They then indicate whether the quality of the solder joint is equal to the quality of those joints which were tested for actual performance in the laboratory. This type of information is vital only on the prototypes; but any change in materials such as base metals, solders, or fluxes and/or any variations in process or configuration of the joint, its performance or function should be reevaluated in an exhaustive, destructive laboratory test. Thus theoretical considerations are combined with practical determinations to give the assembly the best possibilities for 100 percent reliability.

8-7 The Contact Angle   As we have seen in Chap. 1, the dihedral angle can be used as a measure of the degree and state of wetting of the solder system. The two extreme conditions would be (1) total nonwetting, where θ equals 180°, and (2) total wetting, where θ equals 0°. Partial wetting will occur between these two conditions. This concept of partial wetting needs further consideration, especially if we remember that the system seldom reaches true equilibrium during soldering. Normally the soldering time is too short and the system is frozen before equilibrium is reached. In this case, the wetting angle reveals additional information. It gives an idea of both the direction in which the wetting is going and the stage of wetting reached. Let us break this range of 0 to 180° into three separate conditions.

1. $\theta < M$ (theta is smaller than M). This indicates the condition of good wetting. M, the marginal limit of the dihedral angle, is usually arbitrarily set at 75° or less when extremely high quality is required.

2. $90° > \theta > M$ (theta is less than 90° and greater than M). This indicates a condition of marginal wetting, and unless special conditions exist, this type of wetting is not acceptable.

3. $\theta \geq 90°$ (theta is equal to or larger than 90°). This indicates *nonwetting* or the condition termed *dewet*. The solder has frozen before wetting or in the process of dewetting the surface to be soldered. In the case of dewetting, the driving force moves in the direction of nonwetting and the rate of dewetting is a direct function of the dihedral angle.

For quality-control purposes, M should be clearly stated as the criterion for good wetting and sound solder joints. For further information on the use of the dihedral angle in the inspection of solder joints, see Table 8-2 and Sec. 1-4.

TABLE 8-2 Solder-Joint Inspection Chart

M = marginal limit of the dihedral angle, usually 75°

No.	Type	Diagram	Dihedral angle	Description
1	Good wetting		$0-M°$	Solder is feathered out, indicating a small dihedral angle; solder surface is bright and smooth, with few or no pinholes (see Fig. 8-3)
2	Poor wetting		$M-90°$	Solder makes a large contact angle; solder surface is not continuous; irregularly round, nonwet areas are exposed (see Fig. 8-7)
3	Dewetting and nonwetting		$90-180°$	Solder does not completely cover the surface, having a large contact angle; solder appears as droplets or balls, either having withdrawn from previously wet adjacent areas or never wetting them at all (see Fig. 8-4)
4	Insufficient heat		Usually large	Solder solidified before adequate wetting occurred; usually flux was not properly activated and work is still covered with tarnish, in which case solder can be pried loose; solder appears smooth and continuous
5	Rosin		Usually large	Bond achieved through a layer of solidified flux, usually rosin type; in its worst form, this joint has no metallic or electrical continuity and has little physical strength; solder fillet is continuous
6	Disturbed joint		Usually small	Joint appears frosty and granulated because of movement during solder solidification; in its worst form, it is the fractured joint (see Fig. 8-6), also referred to as a cold joint

These considerations of the angle involved in wetting must be qualified by saying that the angle formed between the solder and the base metal, at the periphery, is actually a function of the amount of solder and base metal available. To illustrate let us take the two extreme cases. An infinitely large surface of base metal with a limited amount of solder will give a true picture of wetting. On the other hand, if there is a limited small surface with a large supply of solder, the solder by necessity will form a large fillet which actually "bulges out at the seams"; here the contact angle bears no relation whatsoever to the wetting conditions (see also Fig. 3-18).

Wetting conditions tie in very closely with inspectability. An excess of solder on a limited surface masks the actual inspectable areas, and we therefore obtain erroneous results, which speak further against excess-solder joints.

In addition, remember that the soldering process is performed in a limited time, which does not give the solder–flux–base-metal system a chance to reach equilibrium. The contact angle is therefore only an indication of the type of wetting obtained and does not give absolute values of the systems at hand. Under ideal conditions and prolonged time the surfaces, solders, and fluxes would as a rule give a much smaller contact angle than those obtained during actual soldering.

To demonstrate how the contact angle can be used to evaluate the quality of the joint, a few actual examples are shown. Figure 8-3 shows a well-wetted joint on a printed-circuit card. Figure 8-4 shows a dewetted surface. Under dewetting, we observe the phenomenon whereby the solder first wets the surfaces and then draws back because of improper wetting, leaving behind a thin coat of solder over the base metal. The solder itself then "balls up" on the surface. This condition is usually the result of improper surface preparation and should not be accepted for a quality joint.

Fig. 8-3 Good wetting.

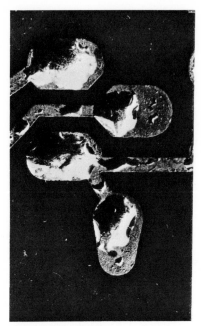

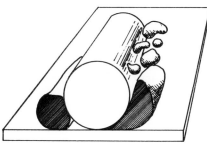

DEWETTED

Fig. 8-4 Dewetting.

Fig. 8-5 Nonwet area in portion immersed in solder. Note circular exposed area in wetted film. Nonwetting is usually accompanied by dewetting, which is also visible.

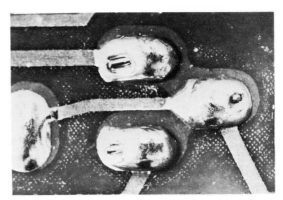

Fig. 8-6 Poor wetting on leads; only the printed-circuit board is well wetted. Note that the solder resist limited the flow to land areas.

A similar condition in surface energies and poor wetting conditions is the nonwetting of surfaces. Here the solder never comes in full contact with the base metal. Areas, normally circular in shape, of the base metal are plainly visible through the solder (Fig. 8-5). This too is unacceptable quality and will be included with dewetting in the balance of this text.

Since there is always more than one surface involved in joining, both surfaces in a bond should show good wetting. This is not the case in Fig. 8-6, however, where only the printed circuit was wetted and the transistor leads are not.

Figure 8-7 shows a disturbed joint (also called a cold joint), i.e., a joint which has been disturbed during the freezing period. Cracked, uneven surfaces result and in the ultimate condition can lead to a fractured joint. Again, this is a doubtful joint and should not be accepted as quality

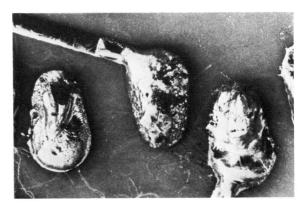

Fig. 8-7 A disturbed, or cold, joint.

soldering. The contact angle in this case may be deceiving since the joint itself could have good wetting, but most cold solder joints have so much surface variation that even an estimate of the contact angle is impossible.

8-8 Acceptance Limits   In order to accept or reject a solder joint, it is customary to specify a certain amount of magnification for the inspection. The average human eye cannot always distinguish clearly the fine differences between various contact angles and states of wetting. Low-power magnification is recommended for the initial inspection. Viewed under 2- to 10-power magnification, the part can be categorized as definitely in one or the other group of solder joints. However, if any doubt exists, a higher magnification on the order of 150 to 200 times is necessary to determine borderline cases.

Because of human error and a certain amount of lack of reproducibility of the same solder joints, some deviation from the clearly defined groups can be expected in every joint. It is therefore necessary to specify that part of the area joint which does not conform to the specified quality. A typical specification might therefore read that "if more than 5 percent of the surface is classified as unacceptable wetting, the part will be considered a reject."

This percentage of nonconforming area depends largely on the type of assembly and the quality expected of the joint. The figure 5 percent quoted above is for high-reliability assemblies. Up to 10 and 15 percent of the area can be specified for the commercial electronic circuitry. An even higher percentage can be specified when we know beforehand that a certain amount of damage to the areas to be soldered will be imposed by the presoldering equipment. Things like piercing platings, heavy scratches and gouges, or areas covered by fixtures during precleaning operations will contribute to large areas of poor wetting or similar unacceptable conditions. When such a physical cause is present, affected areas are satisfactory as long as the total amount of good wetting in the joint is enough to take care of the physical and electrical properties including the safety factor. Usually, in these cases, a certain amount of compensation in design is required. However, when no physical reason exists for poorly wetted areas, they serve as an indication of the efficiency of the wetting process in the solder joint and should not be dealt with lightly. It is well to study the cause of such recurring deficiencies and try to eliminate them by changes in either design or manufacturing procedures.

Another area that requires attention is the amount of mechanical security required in soldering as seen through the solder fillet. In Chap. 4 we discussed such items as overlap and wrap connections and classified solder connections by their mechanical security before soldering. The amount of overlap and wrap-around required can easily be set up according to those base rules. However, for convenience, more than the minimum necessary is often desired to hold the parts to the assembly. When a solder fillet is inspected in its final shape, the requirements should not be too stringent, and there is an economical advantage in maintaining the ground rules specified earlier. Thus there is no advantage in inspecting a solder joint that has 300 percent of the minimum current-carrying capacity and strength required for 100 percent perfection. Small imperfections to as much as one-third of the solder joint can still be safely passed and will not require expensive reworking. The rule of thumb therefore is as follows: *The minimum requirements of the solder joint including the safety factor should be clearly spelled out. A 100 percent perfection in each joint is not mandatory as long as the fault does not exceed one-half the safety factor.*

Fig. 8-8 Contours visible, good feathering.

Fig. 8-9 Same joint as in Fig. 8-8 but with excess solder. This is a doubtful joint. (Note that one of the wires was deliberately pulled out of joint to demonstrate lack of inspectability.)

8-9 Inspectability and the Size of Fillets The term *inspectability* requires discussion. It originates from the fact that we are using the external appearance of the solder joint to evaluate its quality. It is therefore essential that the criteria which determine the quality be plainly visible. In that respect, a fillet which is well feathered, continuous, and smooth can be defined as a good soldered connection. However, it is also important that the contours of the two surfaces be visible through this fillet; otherwise we are masking the properties we are looking for. Figure 8-8 shows two wires inserted and looped around a terminal, and the wires are plainly visible. This is, by definition, a good soldered connection. Figure 8-9 shows the same basic configuration with an excessively large solder fillet. Here the contours of the wires are not visible, but in spite of this, the fillet is smooth, continuous, and feathered out on the terminal. This connection has no inspectability and is therefore a poorly soldered connection.

Other violations of the inspectability requirement can be corrected by the simple application of common sense. No soldered connection should be located in an area which is not visible and/or accessible for a mirror type of arrangement. It is the explicit responsibility of the quality inspector to reject any excess solder joint on the premise that the quality of the joint is doubtful and could therefore, under certain circumstances, lower the reliability of the assembly.

Careful consideration of the parameters of the solder joint indicates that the mere addition of solder mass does not improve the quality, strength, or current-carrying capacity. The size of the fillet should by all means be large enough to fill up the solder area; but for inspection purposes it is imperative that the contours of both parts joined by the solder be plainly visible for inspection. This was a basic assumption in our design considerations (Chap. 4).

8-10 Cleanliness Although cleanliness is a general inspection criterion and applies not only to solder joints, it is important to remember that traces of fluxes or flux residues indicate poor postcleaning procedures. In addition, many solder operations are by nature contributors of small solder droplets. These small metallic particles can be dislodged eventually in the assembly and can cause short circuits and other undesirable side effects. When an assembly is improperly cleaned after soldering or fluxes are not removed at all, the job of visual inspection is physically hampered. Cleanliness is therefore one of the items that should be specified in any proper solder-inspection procedure, even if the assemblies do not need to be cleaned after soldering.

It is true that a segment of the soldering industry does not clean after soldering, but these joints are subject to the same hazards of corrosion and current leakage. It is therefore vital to check uncleaned assemblies, on a low AQL basis, for the presence of dangerous contaminants, which may result from excessive exposure, improper materials, or processing faults. Their presence signals the need for changes and some cleaning of affected lots.

Recent developments in high-reliability low-current circuitry (especially printed circuits) have emphasized the importance of cleanliness of electronic assemblies because of the danger of corrosion. Let us see what corrosion really means and why it is undesirable. We are concerned because corrosion can damage conductors. It can increase circuit resistance, and high resistance is undesirable. It can also cause a physical failure of the conductors by weakening and embrittlement. In addition, corrosion products themselves can cause current leakage. Current leakages are particularly bad because they are not consistent, since humidity changes in the atmosphere cause variations and sometimes intermittent occurrence of the current leakage. Corrosion products can also cause contamination throughout the whole system in the form of nonconductive deposits on mechanical contacts, relay surfaces, etc. Corrosion is definitely a problem, but it is important to remember that corrosion is not always caused by fluxes. For further details on the source of corrosive materials and the mechanism of corrosion, see Secs. 1-12 to 1-14, and 5-37 to 5-43.

Cleanliness is easily checked after the cleaning procedures outlined are followed. As shown earlier, the danger to the assembly stems from the presence of ionizable material, mostly chlorides. The easiest check and the most thorough one is with the use of a conductivity cell and alcohol with demineralized water, either leaching the ionizable materials off the surfaces in premeasured amounts or submerging the whole assembly in a container and measuring the resistivity of the solution. This indicates the presence of ionizable materials. However, since chlorides are the major contributors in the corrosion mechanism and are the most abundant form of ionic contamination on the assembly, many places have standardized on a simple silver nitrate test to establish the presence of chlorides.

Cleanliness after soldering is important because soldering is usually the last step in a long process of assembly. Since recontamination of the surfaces is easy, however, the necessary precautions must be taken; otherwise we defeat the purpose of cleanliness and the check for corrosion products in the final inspection.

THE QUALITY LOOP

8-11 The Elements of the Quality Loop The flowchart in Fig. 8-10 depicts the entire quality loop. It shows the interaction between the quality functions, the process itself, and the management functions. In the diagram, the diamond-shaped boxes denote a quality control (decision)

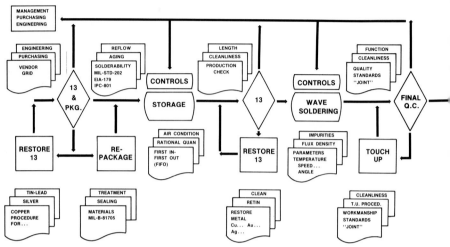

Fig. 8-10   The quality loop. Note that 13 is the number of letters in the word *solderability* and is used as a shorthand notation.

check. The rectangular boxes denote an operation associated with quality control, and the storage and wave-soldering operations, which are under manufacturing controls, are listed separately. Each operation and function shows the necessary associated documentation.

The material flow goes from incoming inspection through storage to wave soldering and final quality control, while the information generated by the various inspection stations is fed back in the opposite direction to management, purchasing, and engineering. Only a total effort encompassing all these functions can result in a low-cost, high-reliability product.

The tests themselves, their relative merits, and similar information will be found in Secs. 8-16 to 8-28. Here we discuss only the interface between the various activities since we have already gone through the test requirements for incoming inspection of soldering materials. Our discussion here is centered entirely around those parts of the assembly which are to be joined (the base metals). This chapter applies to structural as well as electronic assemblies.

8-12 Phase 1: Purchasing and Receiving   Engineering specifications to cover all properties needed for ease of soldering are a first step in the right direction. Purchasing specification in turn must be very clear in regard to those properties, and often an organization finds it easy to set up a vendor grid, on which the merits of the various suppliers are rated. This then gives an order of preference to their materials with the idea of a minimum of touch-up and rework needed, commensurate with the appropriate price. Vendor grids are relatively easy to set up since the manufacturing organization normally has a good memory for parts which have given them trouble in the past. Purchasing must be sensitive to these requirements because rework, repair, and field maintenance outweigh any possible savings in component price.

When the parts are received, they should be checked for solderability (Secs. 8-16 to 8-20). Plain solderability check, however, will yield limited information, and it is recommended that the parts be aged to simulate storage conditions since they are seldom used right away. Whenever solder-plated surfaces have been reflowed, they require additional checks for the quality of the reflow (judged by freezing characteristics) and cleanliness for reflow-media removal. The frequency with which parts must be checked for solderability depends on the experience of the company with the components and their vendor grid.

Parts that fail the solderability test should be returned to the vendor for corrective action. If time does not permit this or return to vendor is not feasible, the solderability must be restored in-house. Restoration of solderability is a simple matter (see Secs. 5-3 to 5-7), and the steps normally follow this sequence:

1. Degrease surface to remove foreign contamination
2. If inadequate, follow by hot-water rinse
3. If inadequate, follow by chemical treatment
4. If inadequate, use very aggressive flux and retin or replate parts

After each of the above steps has been completed, the solderability is checked again. Most parts can be used without going through the entire procedure.

The preservation of solderability depends on the storage conditions. In most properly designed solder systems the flux is perfectly capable of removing normal tarnish, which results from the exposure to air (oxygen), humidity, and temperature. Undesirable pollutants such as sulfur, organic materials, plating and etching chemicals, etc., may cause attack on the surfaces not removable by the normal flux. Gross dirt, from handling, sedimentation out of the air, etc., must be avoided since it too may cause an interference layer between the flux and the base metal. If the storage area cannot be kept free of the second category of materials, parts should be stored only in protective packages. If the packages that might originate with the supplier are inadequate, they should be replaced with approved containers during receiving or incoming inspection (see Sec. 5-3).

8-13 Phase 2: Storage and Handling Once parts have been found to be solderable and properly packaged, they move on to the storage area, where they are kept until required in production. The controls over storage are relatively simple and include the logistics of "in and out" and cleanliness precautions. We are not concerned with inventory keeping and similar stocking operations.

The advantages of first-in–first-out (FIFO) are obvious, and the benefits of rotating the parts in stock for solderability shelf life follow the same concept. Good planning, however, will dictate that the quantities issued to production will follow some rational control. It is suggested that the quantities issued match the production needs for a convenient cycle such as a production run, work day, or work week. Component storage in the production area is usually under more adverse conditions, and the solderability of the surfaces may deteriorate.

Storage areas should be kept free of dust and airborne contamination as much as possible. If the conditions are especially bad, filtered air under positive pressure may keep the storage area clean. Humidity and temperature control are luxuries seldom needed for the storage of components, especially if they are packaged in approved containers, plastic bags, etc.

Handling is another source of undesirable surface coatings, e.g., perspiration, lubricants, and miscellaneous carry-over from other operations. While it is not feasible to have operators wear gloves or similar protection, proper motivation and training may teach employees to handle parts by the edges or surfaces not to be soldered, so that a minimum amount of solderability deterioration will occur. Handling equipment, conveyors, and similar devices must be kept clean and free of harmful materials, like silicone oils, known to destroy solderability.

8-14 Phase 3: Presoldering Checks   Even though surfaces to be soldered have been checked for solderability before storage, they must be double checked just before final assembly. While all precautions were taken to preserve solderability during storage and possibly handling, this is mandatory because touchup on unsolderable parts is difficult. Preproduction checks are often done by assembling a small number of actual units and running them through the production equipment with all appropriate production parameters rather than using a laboratory solderability check. Any parts that fail this preproduction evaluation are then sent for solderability restoration before rejects are produced.

8-15 Phase 4: Soldering and Final Inspection   The role of quality control during soldering has already been highlighted. Such processing parameters as temperature, soldering time, density of flux, etc., must be monitored continuously. This is often left to the responsibility of the machine operator, but even then they should be periodically monitored by the quality control department.

We have finally reached the last stage of quality control, where we look for the workmanship standards of the joint, cleanliness, and other details. It is important to weed out at this point joints which cannot be corrected by standard touchup techniques (poor wetting, dewetting, and nonwetting) and to follow touchup procedures.[1] Remember that unnecessary heating of the solder joints in a touchup procedure may well accelerate metallurgical reactions and lower the stress-coupling ability of the joint, thus shortening its life or lowering its reliability.

SPECIFIC QUALITY TESTS

8-16 Solderability and Aging Tests   Solderability is defined as *the properties of a base metal to be wet by molten solder under specified conditions of time, temperature, and environment.* The environment in our case would be the flux. It is important to specify the wetting conditions because the test is used at various levels (Sec. 8-20).

Phase 1: Incoming Inspection (See Sec. 8-12). Here we are interested in weeding out not only the bad parts but also the marginal ones. Since parts are seldom used right away, we must also simulate storage through aging. Thus the ideal test would age surfaces to be checked (see Sec. 8-19) and impose restrictive conditions of weak flux, low temperature, and short time.

Phase 3: Presoldering Check (See Sec. 8-14). At this stage we are concerned only with the results of the production run. The parts are therefore checked under simulated production conditions of flux, time, and temperature. Even better is an actual sample lot of assembled units run through the production line.

Solderability tests must be divided according to the surfaces checked. There are separate industry and government specifications for component leads (mostly round wire) and flat surfaces (mostly printed-circuit laminate) but this division is not really sufficient. We should concern ourselves with the nature of the base metal too and check:

1. Bare base metal (requires no special discussion).

2. Coated base metal (pretinned) with a fusible alloy such as pure tin, tin lead, etc., applied by electroplating, hot coating, etc. Testing of the outer surface of this group yields only limited information. The solderability test must also evaluate the interface between the coating and the metal underneath.[2]

3. Coated base metal with a soluble plating such as gold, silver, cadmium, zinc, etc. Here again the surface solderability information is of limited value because the dissolved plating will leave a new interface to be wetted.[3]

[1] H. H. Manko, Eliminate Poor Solderability; Don't Bury It under Solder, *Insul. Circuits,* vol. 22, no. 2, February 1966; H. H. Manko, Printed Circuit Touch-up, *Circuits Manuf.,* vol. 18, no. 7, July 1978.

[2] H. H. Manko, Solderability: A Prerequisite to Tin Lead Plating, *Plating,* July 1967.

[3] Ibid.

Fig. 8-11 Visual solderability test standards for flat surfaces. Coupon on left shows first class wetting. The second shows a small amount of dewetting at the lower edges. The third shows complete dewetting while the one on the right shows nonwetting and dewetting simultaneously.

    4. Coated base metal with a nonfusible, nonsoluble plating, such as nickel, iron, etc.; here the requirements are the same as in item 1 for bare metal and require no special discussion.

Only the dip test (see Sec. 8-17) is universal in this respect and can evaluate bare and coated specimens. Numerical methods (see Sec. 8-18) require additional checks to verify the solderability of fusible or soluble coatings.

    One other way to classify solderability tests is through their method of evaluation. The quality of wetting is recognizable visually. Thus a large part of the test uses a subjective comparison with visual wetting standards (see Fig. 8-11). This evaluation is hard to quantify, and numerical methods have also been developed. As we shall see, they also have their drawbacks.

8-17 The Dip Test   This is by far the most popular and widespread method. It has specific documents issued by government agencies (MIL-STD-202 method 208 and MIL-STD-883 method 2003, etc.), industrial organizations (EIA STD-178, IPC STD 801, etc.), and individual companies. These tests specify hand or machine dipping, types of surface preparation (steam aging, etc.), kind of flux to be used (w/w Rosin, RMA, etc.), and the composition of the solder alloy.

    In principle the test consists of the following sequence. The part is used as received or aged and fluxed by immersion. Flux excess is drained and the volatiles allowed to evaporate. The part is then dipped into molten solder that has been skimmed for dross removal. The rate of immersion is usually 1 in/s (see Table 5-4). The specimen is allowed to dwell in the solder for a predetermined time established for adequate heat transfer and withdrawn at the same rate as dipping. The solder is then allowed to freeze on the surfaces without vibration. Any flux residues are washed off, and the condition of wetting is observed and/or compared with visual standards.

8-18 Numerical Tests   Several pieces of commercial equipment derive a numerical value equated to solderability.

1. The Meniscograph measures the surface energies of the specimen during a dip cycle (see Fig. 8-12). The principle of the test utilizes the change in force exerted on a surface immersed in solder.[1] While the specimen is still not wet, we get a buoyancy effect due to the greater density of solder (true for most base metals) and the fact that the surface area of the solder is mechanically increased. After the flux and heat from the solder both induce wetting, we get a force in the opposite direction. The wetted specimen is drawn into the solder. These changes in force are recorded on a chart. Figure 8-13 shows six actual test traces indicating

[1]D. Macay, The Meniscograph Method of Solderability Measurement, Central Research Laboratories, HIRST Research Center, Wambly, England, 1970.

Fig. 8-12   The Meniscograph. *(General Electric Co., Wembly, England.)* This unit is connected to a fast chart recorder not shown. *(Hollis Engineering, Inc.)*

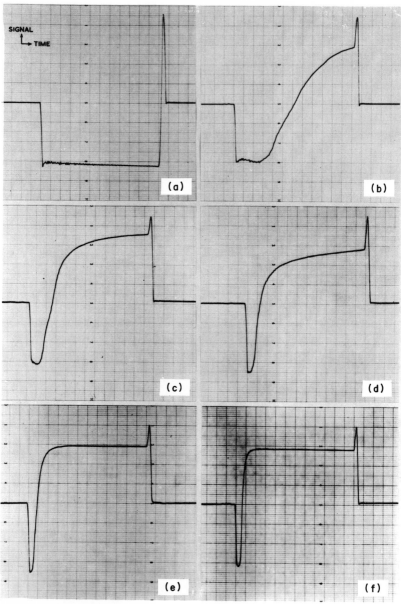

Fig. 8-13 Typical Meniscograph trace according to Duncan Macay, General Electric Co., Ltd. Note that (*a*) has no solderability while (*f*) has excellent fast wetting. Other graphs show stages in between.

total nonwetting (Fig. 8-13*a*), slow wetting (Fig. 8-13*b*), through to excellent wetting (Fig. 8-13*f*). The time recorded is an indication of the systems interactions and can be translated into a solderability rating. The test is being adapted as an IEC standard.[1]

2. The globule test assesses solderability by a time measurement. It utilizes a molten sphere of solder (the globule) of predetermined volume. The fluxed test wire is introduced into the globule, and the time needed to wet and surround the specimen is recorded. The test is popular in Europe and is part of the IEC test Ta, Solderability.[2] Good statistical correlation between test results and production results have been developed by G. Becker of L. M. Ericson[3] in Sweden who has also adapted this method to integrated-circuit solderability testing.[4]

3. The wetting-time test.[5] In this test the specimens are mounted on a rotary arm, which dips them into a solder pot. A timing circuit measures the time from first contact with molten solder to wetting. This time is used to assess solderability.

8-19 Aging of Surfaces   When components are not used immediately, it is desirable to simulate deterioration in storage. These procedures are called *Accelerated aging techniques* and must become a part of the solderability specification. DeVore[6] has studied this subject in detail and evaluated the various available techniques. While various tests have special applications and merit, steam aging, as described in MIL-STD 202 method 208, appears to be the most universal and is highly recommended by the author. The test procedure calls for suspending the test surface 1½ in above boiling distilled water, with a cover to contain most of the steam. A 1-h exposure to this environment is enough for most applications, but longer exposures may be specified (see MIL-STD 202 method 208 for more details).

8-20 Solderability Interpretation   The need for solderability interpretation is very important since this is the only one of the five soldering parameters which is not controlled. Figure 8-14 shows the reliability balance and lists the five variables that normally affect a soldering opera-

[1]International Electro-technical Commission 50C WG3 (Secretariat) 217, restricted communication to committee members, Nov. 22, 1977.

[2]Basic Environmental Testing Procedures for Electronic Components, *Int. Electrotech. Comm. Publi.* 68-2-20, pt. 2, 1968.

[3]G. Becker, Numerical Definition of Component Solderability Reduces Soldering Defects, *Assem. Eng.*, January and February 1970.

[4]G. Becker, How to Pinpoint IC Solderability Problems, *Assem. Eng.*, vol. 20, no. 2, February 1977.

[5]C. J. Thwaites, *Electr. Manuf.*, vol. 8, no. 5, p. 18, 1964.

[6]J. A. DeVore, Solderability of Terminals and Aging, *General Electric Co. Final Rep. Tech. Inform. Ser.* R64 ELS-44, May 1964.

tion, the more constant of these being the time, temperature, and solder alloy (provided it is not contaminated). They can be controlled together with the flux activity by the using organization. The quality of the surfaces in terms of solderability, however, depends on the supplier, storage, handling, and other variables discussed earlier. From the balance we can see that a minimum level of solderability is required for every type of flux to be used. If the solderability level falls below the minimum acceptable flux activity, bad results can be guaranteed.

The solderability tests described in Secs. 8-17 and 8-18 are strictly functional and provide no clues to the origin of the fault. Additional diagnostic testing is still necessary to correct the solderability of existing surfaces while preventing recurrence in future production.

From a practical standpoint, users of components are advised to incorporate accelerated aging into their test schemes. Experience has shown that solderable parts, as measured by any method, stay solderable if stored correctly. Unsolderable parts also remain unusable with time and can be weeded out by all tests without difficulty. The troublesome parts are those which are of marginal solderability when received, and here the Meniscograph offers the best chance of early detection.

Producers of solderable surfaces, however, face a different problem. They must not only predict future solderability but must be able to analyze the cause of poor quality to rectify and prevent it. This can be done if we remember the definition of solderability as a surface property. Current analytical tools, based on the scanning electron microscope, allow the interpretation of these surfaces (see Fig. 8-15). Further analysis by Auger Electron Spectrometry allows idenficiation of the contaminant in the surface layer only.[1]

---

[1]G. Tissier and C. Legressus, Surface Phenomena in Relation to Solderability and Solder Joint Reliability, *Int. Packag. Conf. Proc., 1977.*

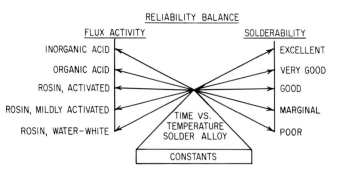

Fig. 8-14   The reliability balance shows the interaction between the five soldering parameters of time, temperature, solder alloy, flux activity, and solderability.

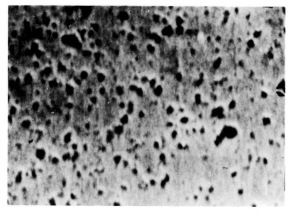

Fig. 8-15 Scanning electron micrograph of flat-pack lead reveals a heterogeneous surface with micropitting. This in turn caused poor solderability. (*G. Tisser—C. S. F.—Thomson, France.*)

8-21 Cleanliness of Surfaces and Ionic Contamination   The need for flux removal has spurred the practice of cleanliness monitoring (see Sec. 5-38). Flux removal is not the only part of the soldering process that requires contamination-free surfaces. As we have seen, soldering depends on wettability of the base metal first with the flux and then with the molten solder. While this phase of the process can be monitored through solderability, the condition of the final surface requires additional tests. Other peripheral processing steps that also require close control over surface contamination include electroplating, etch-resist application, solder-mask screening, conformal coating, etc.

The three most common forms of dirt are removed in different ways (see Secs. 5-39 to 5-42) and can be detected as follows:

1. Particulate matter present on the surface is easily checked by microscopy.

2. Nonconductive dry or wet films must be dissolved in appropriate solvent and identified.

3. Ionic contamination is measured as described in Secs. 8-22 and 8-23.

Groups 1 and 2 may interfere with surface phenomena, but group 3 also poses corrosion and current-leakage danger.

8-22 Ionic-Contamination Measurement   The solution of ionizable materials in a polar solvent like water changes its electrical characteristics. The conductivity of the solvent is in proportion to the amount of ionic content. This gives rise to a simple method of measurement. Known surface areas are exposed to a polar solvent of predetermined electrical characteristics. The resultant change in conductivity (or its inverse, resistivity) is used to assess the cleanliness of the specimen (see Sec. 8-23 for instrumentation).[1]

The basic difficulty lies in the interpretation of the results. While we can develop a number of ions per unit measured, this by no means tells us the location of the contaminant on the surface. One fingerprint, for example, may be the only contamination on the surface. Our test is unable to indicate the location or distribution of ions on the part. It yields a statistical figure dividing total contamination into surface area. The second problem associated with this test comes from the lack of correlation between the number of ions per unit area and the functional behavior of the circuit. Even if we assume the contamination to be evenly distributed on the surface, we are unable to come up with universal limits. The variations between assemblies in materials, design, and electrical functions necessitate setting up specific limits for each application. While this sounds very negative, it is really meant to show the test limitations as a diagnostic tool. Excellent guidelines can be established for any industry, and the test is obviously a reliable quality-control tool for monitoring cleaning processes and equipment. It is also suitable for all checks where cleanliness is an indication of quality.

Remember that these tests will not measure nonionizable materials, even if the test solution contains nonpolar solvents included to make sure that no ionizable materials are masked by nonpolar dirt films.

8-23 Instrumentation for Ionic Measurements   Several commercial units are available to measure the ionic contamination on surfaces according to Sec. 8-22. They are based on a conductivity-measuring bridge, clean containers, and a prespecified solution of deionized water and alcohol of known conductivity. The biggest variation from test to test lies in the method and time of ionic contamination extraction. In its simplest form a known volume of solvent is washed over the surfaces and collected in a clean beaker. Here time of contact and uniformity of extraction are limited. More sophisticated methods involve total submersion of the parts in the test solution for extended periods of time or until no additional increase in conductivity is noted. Two industrially available instruments are described below.

Figure 8-16 depicts a compact suitcase-mounted instrument, the Ionograph.[2] The principle behind this unit is an internally mounted metering pump sending a constant volume of the test solution through the measuring chamber. The contamination level of the test solution is electrically

---

[1] MIL-P-28809, Printed-Wiring Assemblies, U.S. Government Printing Office, March 21, 1975.

[2] Trade name, Alpha Metals, Inc.

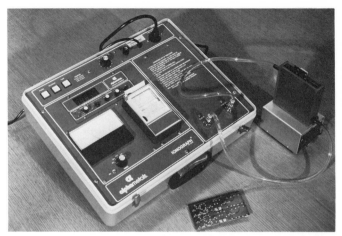

Fig. 8-16 The Ionograph. Note that the test chamber is located over the stirring mechanism to keep the solution agitated. *(Alpha Metals, Inc.)*

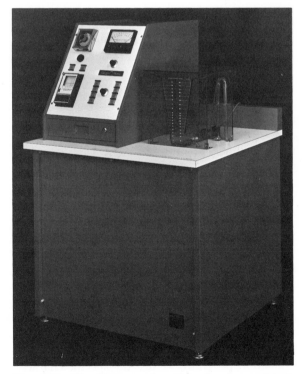

Fig. 8-17 The Omega meter. Note the shape of tank designed to accommodate various sizes of printed-circuit board. *(Kenco Alloy & Chemical Co.)*

monitored and can also be integrated on a counter. The contaminated solution is then sent through an internal ion-exchange unit and returned to the test chamber. The test is terminated when the "contaminated solution" coming out of the test chamber is considered pure.[1]

Figure 8-17 depicts the Omega meter,[2] in which an internally deionized test solution is pumped up into the test chamber to the desired level, which is a function of the surface area being tested. The solution is then agitated for a predetermined length of time to leach off the contamination from the specimen. The conductivity of the solution is measured at the end of the test cycle. After the specimen is removed, the test solution is returned through the ion exchanger into a storage reservoir.

PHYSICAL TESTING OF JOINTS

8-24 General Considerations   As described in Chap. 3, the physical strength of the alloy is mostly a function of the preparation and aging of the casting. In the tin-lead system specifically, the change of strength with time after casting can be observed. It is therefore vital to note in each case the approximate time since the solder joint was actually made in order to reproduce the same type of result in repeated tests. Another important factor that should not be overlooked is the temperature during the test period itself. Because of the low recrystallization temperature of most fusible alloys, which is close to room temperatures, the properties of a solder joint can vary sharply with small temperature variations. These alloys are far more sensitive to temperature changes than the regular base metals to which we are soldering. We have seen that the spacing between the base metals soldered influences the strength of the solder joint, and this too should be carefully recorded. Also, the duration of the soldering operation together with the thermal history of the joint have a direct bearing on the amount of intermetallic compounds formed. As discussed earlier, their influence is greatly apparent in the mechanical properties of a joint. Keeping these parameters in mind, let us consider now the individual tests and remember that the solder is a stress coupler.

8-25 Tension and Shear of Joints   When we consider the tensile and shear strength of the solder joint, the tensile strength and shear strengths of the bulk solder alloy are only a general indication of the strength we can anticipate. It might be more correct to call this by different terms such as bond holding strength or bond shear strength. Here we are testing the actual solder joint, and as long as we are not engaged in a general evaluation program, we test the specific characteristic of a tailored joint.

[1]D. Brous, Evaluation of Post-solder Flux Removal, *Weld J. Res. Suppl.*, December 1975.
[2]Trade name of Kenco Alloy & Chemical Co., Inc.

In that case, we must take the normal precautions of proper alignment and simulate the worst condition the joint will be exposed to in actual operation. We must maintain the proper speed of load application because this affects the strength greatly (a faster loading speed in a tensile test will give higher strength results for the solder joint). Finally, we should check the failed specimen and establish where it failed, in the joint or in the structural members. If the joint fails in the solder interface, we should check to see how much of this interface was wetted and how many voids in the solder are evident so that proper corrective measures can be taken.

When a bulk-solder specimen is prepared for tensile or shear measurements, the following procedure has been adopted in the general literature and can be recommended to obtain results uniform with those published by other workers.

The tensile specimen is chill-cast in a preheated mold at 100°C (212°F). The melt temperature is 50°C (122°F) over the liquidus. The usual dimensions of the specimens are ¼ by ½ in cross section and 2 in for elongation measurements, making the overall size according to the depth of insertion in the jaws. The temperatures quoted before are for tin-lead solders which have a melting range of 361 to 450°F (183 to 232°C). For higher-melting-point solders, higher preheating of the mold may be advisable to avoid nonuniform or void-filled specimens. It is important to check the specimen carefully at the point of failure to make sure no voids were present at this location to give erroneous results in converting the results to the tensile strength of the solder. If the failure always occurs in the jaw (jaw failures are not acceptable), it is recommended that a thin Mylar film be placed between the fixture and the solder specimen to minimize the danger of jaw failures.

8-26 Creep Strength of Solder Joints   The creep strength of a solder joint is an indication of the changes which will occur in a solder joint under long-term stresses, sometimes at elevated temperatures. Here the regular test methods are used, and results are recorded in the normal fashion. In fusible alloys, this type of information is very important because most of these alloys are rather weak in creep strength. Sometimes alloying additions such as antimony to the tin-lead systems will improve the creep strength of the joint to the point where it can be successfully used in most applications.

8-27 Vibration of Solder Assemblies   Since the solder alloy is usually the weakest link in the soldered assembly, the importance of vibration and its effects becomes evident. However, the low recrystallization temperature of the soldering alloys together with the large elongation that most of them have make it a desirable stress coupler. The author has performed

Fig. 8-18 Assembly including small transformer suspended
on leads by soldering. The joint did not fail the vibration test.

numerous tests on soldered assemblies using a standard vibration table
and going through an amplitude spectrum up to $10g$ without being able
to fail any solder connections. In one case, a small transformer weighing
slightly over 1 oz was soldered by its leads to the assembly (see Fig. 8-18)
and was supported only by these leads during the vibration test. The
AWG no. 18 copper wire failed by work hardening close to the solder joint
because of the mechanical stress concentrations of that location without
the solder joint itself being affected. However, no blanket insurance for
any soldered connections can be derived from this type of work. If an
assembly is in question, it should be tested thoroughly.

8-28 Environmental Testing   Here we are checking for both inherent
corrosion attack and lack of cleanliness after flux removal. Usually environmental tests are performed in humidity chambers, where the parts are
exposed for various amounts of time to temperature and humidity for
accelerated aging. Cleanliness by itself can easily be checked on the
exposed surfaces by such simple tests as a silver nitrate check for halides
or black light for rosin. However, corrosive material is often absorbed into
porous surfaces and penetrates into cracks and crevices. These materials
are not easily detected, but upon humidity exposure they slowly seep out
in what is often referred to as *bleeding.*

A word of caution. Many tests require a relative humidity of as high as
96 percent at 100°F (38°C). The slightest amount of variation in test
conditions can cause a fogging or raining condition inside the humidity
chamber, resulting in a washing operation of water-soluble contamination

from the assembly, thus improving the corrosion resistance. If water-soluble materials are present on the parts, it is highly recommended that temperature and humidity conditions inside the humidity chamber be adjusted so that no possibility of vapor condensation on the test parts exists. A relative humidity of 80 to 85 percent at 100°F produces no condensation, and the author has adopted these conditions for most environmental testing of solder assemblies with considerable success.

When we know beforehand that the environmental conditions to which the assembly will be exposed contain a specific type of dust or chemical, we must compensate for this by introducing a similar chemical material into the humidity chamber together with moisture. Several humidity chambers on the market have this additional feature.

Another aspect of environmental test is the salt spray. Here again, the use of this method depends on a knowledge of the environmental attack to which the parts will be exposed in the eventual use. However, the salt spray is an extremely hard condition for solder joints which are not protected, and experience has shown that in most cases too much galvanic corrosion around the solder joint occurs to enable one to use an unprotected solder joint in exposed marine atmospheres.

METALLURGICAL EXAMINATION

8-29 General Instructions   Most of the tests described previously are of such a nature that any general handbook can give test data easily. This is not the case with the average metallurgical textbook and this section will therefore contain more specific details and data so that it can serve as a source for the metallurgically untrained engineer to perform the described group of tests.

8-30 Mounting the Specimen   When it is desirable to section a solder connection for internal examination, special care should be exercised in cutting the joint. Because of the low recrystallization temperature and low melting point of the fusible alloys, any cutting of the joint must be done with adequate cooling. The most suitable piece of equipment is the water-cooled cutoff wheel. It is suggested that the specimen be cut several thousandths away from the actual surface to be examined and mounted in a metallurgical cap to be polished down later to the surface to be examined. This initial cutting technique eliminates the possibility of physically hurting the solder-joint area, which as yet has no backing, and of over-heating the immediate area to be examined by friction. It is sometimes advantageous to copper plate the outside surfaces of the specimen, which helps preserve the contours and makes coating measurements easy.

Once the specimen is reduced to a size suitable for a metallurgical cap, it is mounted in a backing material which provides a convenient form for handling and helps conserve the physical outlines of the solder joint during any subsequent polishing and etching operations. A whole variety of shapes for metallurgical caps are available, the most common being the 1- to 1½-in-diameter cylinders. Various materials are available such as thermoplastic and thermosetting materials, which are applied under heat and pressure; two- or one-part organic polymer systems, which set up upon curing or addition of catalyst; and materials with a low melting point such as sealing wax, which are cast molten around the specimen. The cold-polymerization materials are the most suitable for soldered connections because they do not change the microscopic structure of the solder joint. However, if hardness or any other characteristic of the potting compound is important, any other capping compounds can be used with a certain amount of care if the changes they can cause are kept in mind. The author has found the transparent epoxy materials most suitable because the area under examination is visible through the transparent side of the cap.

8-31 Polishing Instructions   The capped specimen is not yet ready for examination, and the first operation is the removal of the excess material by grinding with coolant or using abrasive paper. The hard cap material helps keep the outlines of the solder joint from being damaged and smeared. The operation should be performed without heating up the abraded surface. A normal hand polishing is performed by going from rougher grades of abrasive paper to finer grades. The direction of polishing is changed from paper to paper at 90°. A specimen is polished on the same grade until the scratches left behind from the previous polish operation are no longer visible. The final polishing is then carried out either on metallurgical wheels with fine abrasives or with the use of a Syntron.

Microscopic examination of the surfaces will quickly reveal whether there is any surface smearing or deformation due to improper polishing. Should this be the case, an etching of the surface coupled with a repolishing on the final grades of polishing materials will slowly remove such undesirable surfaces to reveal the true structure of the specimen. With some practice, the average person will find it possible to eliminate the formation of such disturbed regions on the surface of the specimen.

Another technique is the electropolishing of solder specimens. Electropolishing, which is an anodic treatment of the specimen in the appropriate electrolyte by the use of current, has the advantage of being a rapid polishing method which usually does not give a false structure on the surfaces because of the working of the surfaces during polishing and abrasion. This method is usually suitable for repetitious checks on specific parts but does not give so good a finish as the mechanical polish. For the

TABLE 8-3 Electrolytic Polishing Solutions for Tin-Base Alloys*

Composition	Cathode	Voltage	Current density, A/cm^2	Time	Temp., °C	Remarks
Perchloric acid, 194 mL; acetic anhydride, 806 mL	Tin	25–40	9–15	8–10 min	<25°	Because of explosion risk, care should be taken that the temperature remains below 25°C; Tegart and Jacquet give detailed precautions for using these solutions
Perchloric acid (sp gr 1.60)	Aluminum	50–60	40	10–15 s	<35°	Puttick claims this electrolyte is safer than Jacquet; face to be polished should be vertical and rotated at 50–100 r/min
Perchloric acid, 63 mL; acetic acid, 300 mL; distilled water, 12 mL	Tin	20–30	9–15	10 min	25	Anode should be vertical and electrolyte agitated; explosion risk also present
Ethanol, 144 mL; aluminum chloride (anhydrous), 10 g; zinc chloride (anhydrous), 45 g; water, 32 mL; butanol, 16 mL	Tin or stainless steel	25–30	30	Periods of 1 min	20	Dark layer formed on specimen removed by jet of hot water after each cycle; cycle repeated 4 or 5 times before polishing is complete
Commercial-grade fluoboric acid containing 2% sulfuric acid	Stainless steel	15–21	400–700	3–5 min	20–45	Can be used for tin-lead alloys, current density varying according to composition; to avoid etching, a specially designed cell is used

*B. L. Eyre, "The Preparation of Tin and Tin Alloys for Micro-examination," Tin Research Institute. Reprinted by permission. For sources of data, see the original publication.

benefit of those who have electropolishing equipment, Table 8-3 gives the most common electrolytic polishing solutions for tin-base alloys.

8-32 Etching Once the specimen has been polished so that all the fine details on the solder joint are visible without too many scratches, the structure of the solder is visible. It is wise to examine the specimen at this stage and determine whether enough material has been removed in the polishing operation. If not, it is recommended that the solder specimen be taken back to a rough grade of abrasives and repolished. Once the specimen is completely polished, it can be examined for microstructures.

TABLE 8-4   Etching Solutions for Tin-Base Alloys*

Reagent	Uses	Effect	Etching time
Alcoholic acid ferric chloride, 5 mL HCl, 2 g $FeCl_3$, 30 mL water, 60 mL absolute alcohol	Tin and tin alloys not containing lead; tin coatings	Reveals general structure, no effect on tin-iron or tin-copper compounds	Tin and tin alloys, 10–15 s, tin coatings, $\frac{1}{2}$ s
5% aqueous ammonium persulfate solution	Tin and tin alloys	Shows up grain boundaries very well	5 s
2% alcoholic nitric acid	Tin-antimony-copper alloys	Tin-antimony and tin-copper compounds clearly revealed	3–5 s
Nascent H_2S, 20% aqueous sodium sulfide plus a few drops conc. HCl	Tin-antimony-copper alloys; differential etch between Cu_6Sn_5 and SbSn	Cu_6Sn_5 stained brown; SbSn unaffected	A few minutes
Glycerin acetic acid, 10 mL HNO_3, 30 mL acetic acid, 50 mL glycerin	Tin-lead alloys; tin coatings on steel	Reveals general structure; tin-iron compound not attacked	10 s
2% alcoholic hydrochloric acid	Pure tin	Shows grain structure	20–30 s
Romig and Rowland's† (1 drop HNO_3, 2 drops HF, 24 mL glycerin), used at 70–80°F	Tinplate	Reveals junction between $FeSn_2$ and tin	60 s
5% alcoholic picric acid	Tinplate; steel backed bearings, etc.	Reveals $FeSn_2$-iron junction by etching steel	5–10 s
Ammonia hydrogen peroxide (ammonium hydroxide + 2 drops H_2O_2)	Tin coatings on copper; bronze-backed bearings	Reveals Cu_3Sn-copper junction by etching copper	3–5 s

*L. T. Greenfield and J. E. Davis, "Preparation of Tin and Tin Alloys for Microstructure Examination," The Tin Research Institute, London, 1951.
†O. E. Romig and D. H. Rowland, *Met. Alloys,* April 1941, p. 436.

If tin-lead is the alloy used, the structure becomes apparent even without etching. However, the use of solutions as described in Table 8-4 will bring out both the metallurgical structure of the solder alloy and some of the structure of the adjoining base metals. The reader is referred to Chap. 3 for the correct interpretation of the solder structure and the solder composition.

The specimen is now examined for uniformity and structure of the solder. Things like eutectic composition and uniformity of a phase are noted. The solder matrix as well as the interface are examined closely for the appearance of intermetallic compounds. The interface is also closely examined for gaps and other indications of poor wetting. Finally, the contact angle at the extremes of the solder joints is checked for indication of poor or insufficient wetting characteristics. Such things as gas pockets and inclusions are also carefully noted.

Index